ADVANCES IN CHEMICAL PHYSICS

VOLUME LXVI

EDITORIAL BOARD

Advances in
CHEMICAL PHYSICS

EDITED BY

I. PRIGOGINE

University of Brussels
Brussels, Belgium
and
University of Texas
Austin, Texas

AND

STUART A. RICE

Department of Chemistry
and
The James Franck Institute
The University of Chicago
Chicago, Illinois

VOLUME LXVI

AN INTERSCIENCE® PUBLICATION

JOHN WILEY & SONS

NEW YORK · CHICHESTER · BRISBANE · TORONTO · SINGAPORE

An Interscience® Publication

The Library of Congress has cataloged this serial publication as follows:

Advances in chemical physics. v.1–
New York, Interscience Publishers, 1958–

v. ill., diagrs. 24 cm.

Irregular.
Includes bibliographical references.
Editor: v. 1– I. Prigogine.

1. Chemistry, Physical and theoretical—Collected works. I. Prigogine, I. Ilya) ed.
[DNLM: W1 AD53L]

QD453.A27 541 58-9935
MARC-S

Library of Congress [8305]
Printed in the United States of America

10 9 8 7 6 5 4 3 2 1

CONTRIBUTORS TO VOLUME LXVI

R. Armstrong, Sandia National Laboratories, Livermore, California, U.S.A.

Myung S. Jhon, Department of Chemical Engineering, Carnegie-Mellon University, Pittsburgh, Pennsylvania, U.S.A.

M. Munowitz, Department of Chemistry, University of California, Berkeley, and Materials and Molecular Research Division, Lawrence Berkeley Laboratory, Berkeley, California, U.S.A. *Present address:* Amoco Research Center, Naperville, Illinois, U.S.A.

A. Pines, Department of Chemistry, University of California, Berkeley, and Materials and Molecular Research Division, Lawrence Berkeley Laboratory, Berkeley, California, U.S.A.

G. Sekhon, Department of Chemical Engineering, Carnegie-Mellon University, Pittsburgh, Pennsylvania, U.S.A.

A. González Ureña, Departamento de Química, Físcia, Facultad de Ciencias Químicas, Universidad Complutense, Madrid, Spain

INTRODUCTION

Few of us can any longer keep up with the flood of scientific literature, even in specialized subfields. Any attempt to do more and be broadly educated with respect to a large domain of science has the appearance of tilting at windmills. Yet the synthesis of ideas drawn from different subjects into new, powerful, general concepts is as valuable as ever, and the desire to remain educated persists in all scientists. This series, *Advances in Chemical Physics*, is devoted to helping the reader obtain general information about a wide variety of topics in chemical physics, which field we interpret very broadly. Our intent is to have experts present comprehensive analyses of subjects of interest and to encourage the expression of individual points of view. We hope that this approach to the presentation of an overview of a subject will both stimulate new research and serve as a personalized learning text for beginners in a field.

ILYA PRIGOGINE

STUART A. RICE

CONTENTS

ADVANCES IN CHEMICAL PHYSICS

VOLUME LXVI

PRINCIPLES AND APPLICATIONS OF MULTIPLE-QUANTUM NMR

M. MUNOWITZ* and A. PINES

Department of Chemistry
University of California, Berkeley, and
Materials and Molecular Research Division
Lawrence Berkeley Laboratory, Berkeley, California.

CONTENTS

*Present address: *Amoco Research Center, Naperville, Illinois*

I. INTRODUCTION

Common to all branches of spectroscopy is the need for detailed knowledge of the structure of the Hamiltonian, without which it is impossible to relate the arrangement of energy levels to the relevant physical parameters of a system. This need is especially pronounced in nuclear magnetic resonance (NMR), where the goal typically is first to measure the spectroscopic frequencies and relaxation times, and then to attribute these observables to the internal magnetic fields and electric field gradients from which they originate. In the pulsed, or Fourier transform, experiment the internal frequencies are revealed through the transient response of the spins following a disturbance from equilibrium. The diversity of local fields and spin interactions both adds to the complexity of the response and makes the NMR spectrum a rich source of information concerning the geometric and electronic structure, dynamic as well as static, of molecules in all phases of matter. Yet beyond this diversity NMR is additionally favored by a particular set of circumstances, among them the phase coherence of radiofrequency sources and the magnitude and time scale of nuclear spin interactions, that combine to offer extraordinary opportunities for external manipulation and control of the spin Hamiltonian: where desired, the effective operator structure of a Hamiltonian or the direction of a local field can be altered, a selected interaction attenuated or eliminated, the sense of time reversed. Moreover, the Hamiltonian governing the spin dynamics can be switched rapidly between different forms during different periods of an experiment. Taken together, the many options available lend to the pattern of excitation and response an element of choice rarely encountered in other methods of spectroscopy.

This chapter offers a review and analysis of a class of "manipulative" NMR experiments designed to probe nominally forbidden degrees of freedom in a collection of nuclear spins in a large magnetic field. The modes of interest are coherent superpositions of Zeeman states for which the expectation value of the nuclear magnetic dipole moment is zero. Frequently the Zeeman quantum numbers of the superposed states will differ by values other than ± 1, indicating a condition of *multiple-quantum* or *zero-quantum* coherence; but even when single-quantum, the condition of the system will usually reflect in some way the cooperative efforts of more than one spin. A description of these collective modes, an exposition of the different methods available for their excitation and for the detection of their response, and a review of some of the spectroscopic applications are the main points to be conveyed here.

To make some of these goals more concrete, we show in Fig. 1 a Fourier transform multiple-quantum spectrum obtained from a system of eight hydrogen nuclei in a liquid crystalline phase.[1] Signals arising from transitions of all orders are clearly visible, separated into subspectra that become progressively simpler with higher net changes in the Zeeman quantum number. The apparent simplification nevertheless belies the increasing complexity of the underlying spectroscopic events, which, loosely speaking, originate with the concerted "flips" of *groups* of coupled nuclei. In contrast

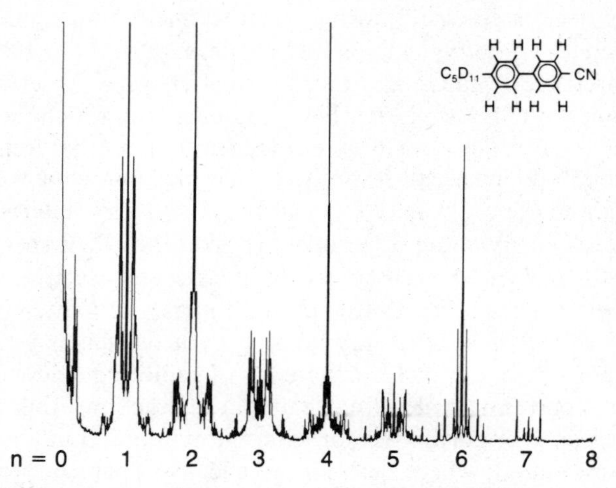

Figure 1. Multiple-quantum spectrum of an oriented system of eight spins, the hydrogen nuclei in the biphenyl portion of the nematic liquid crystal 4-cyano-4′-*n*-pentyl-d_{11}-biphenyl. Each subspectrum shown corresponds to a net change by *n* of the total magnetic quantum number characterizing the system. The integrated intensity within an order *n* is determined roughly by the number of admissible superpositions consistent with the absorption or emission of *n* quanta. The line positions reflect the response of collectively excited groups of spins to internal spin couplings. Full width shown is 500 kHz. (Reproduced from Ref. 1 with permission. © 1980 Elsevier Scientific Publishing Co.)

to the picture for conventional single-quantum resonance, where single spins are excited and allowed to respond independently, we may now visualize a set of *interdependent* spins, collectively excited as a kind of superspin, which then responds to the local fields of all the nuclei outside. Thus there is a fundamental connection between the development of multiple-quantum coherence and the formation of networks of correlated spins, a link that we try to represent symbolically in Fig. 2. Ultimately, however, a limit to collective excitation is encountered—in this example at eight quanta, with which the eight spin-$\frac{1}{2}$ particles can interact only by connecting a configuration containing all spins "up" with one containing all spins "down." Isolated from the local fields of any other nuclei, the eight correlated spins exhibit just one eight-quantum resonance; while at the other end of the spectrum, where there are far more numerous ways of realizing lesser changes in the Zeeman quantum number, some involving relatively few spins, the number of lines grows accordingly.

Though observed in continuous-wave NMR experiments as early as 1956,[2-19] time domain multiple-quantum phenomena of this kind were not investigated until the mid 1970s, for only then were the essential ingredients—

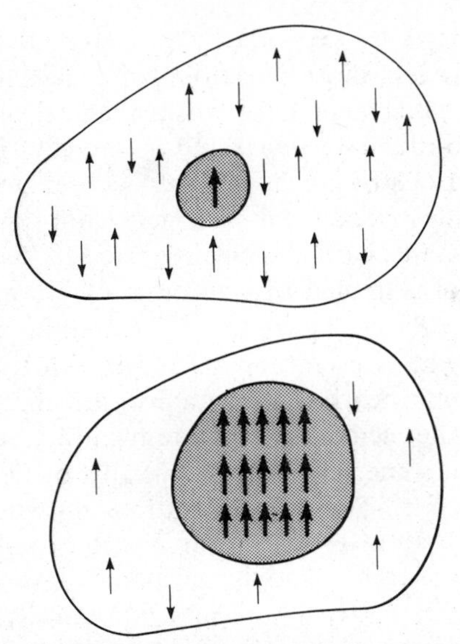

Figure 2. Idealization of a single-spin, single-quantum mode and a multiple-spin, multiple-quantum mode. In conventional NMR, a spin is excited independently and then allowed to precess in the local field of its neighbors, developing in the process a time-varying magnetic moment observable with a magnetic dipole receiver. In multiple-quantum NMR, a group of spins, excited collectively, responds to the local field without emitting dipole radiation. The multiple-quantum response is detected indirectly by means of a transfer of coherence to an allowed dipole mode.

principally the basic Fourier transform experiment[20] and the theoretical and experimental methods of average Hamiltonian theory[21,22]—firmly in place. Relevant also to the development of multiple-quantum NMR was the concurrent introduction of two-dimensional spectroscopic methods, in which the response period of the pulsed experiment is partitioned in two and the time domain analyzed by a double Fourier transformation.[23,24] Nevertheless, some of the earliest pulsed multiple-quantum experiments were performed without explicit reference to the full apparatus of two-dimensional spectroscopy.[25,26]

In preparing this chapter we have attempted to take into account the existence of other reviews, to which the reader is also directed.[27-29] The first of these summarizes most of the important methods and applications of multiple-quantum NMR through 1980, while stripping away some of the mathematical formalism.[27] A second offers a highly detailed analysis of the major theoretical concepts and experimental techniques up to the middle of 1982,[28] and still another focuses specifically on applications of multiple-quantum NMR to oriented systems.[29] Where possible, we have sought to complement these reviews by expanding upon areas that they treat only briefly and emphasizing less those areas that they describe in great detail.

The treatment presented here is thus intended to provide a comprehensive introduction to the major ideas and methods of Fourier transform multiple-quantum NMR, striking a balance between formal theory and application. The work described is taken largely from reports published by many groups throughout the world during the period 1975–1985 and blended together for the sake of a unified presentation. Our hope is to aid those outside this area who seek to acquire an understanding of it, as well as to produce a review that will be useful for specialists. For this reason we have tried to make this chapter nearly self-contained, assuming little except for some familiarity with the elementary quantum mechanics of spins, while recognizing that we run some risk, at least initially, of repeating very well-known and basic ideas.[30,31] These considerations notwithstanding, fundamentals common to all Fourier transform NMR experiments, including concepts of precession, coherence, and the mechanics of the density operator, are reviewed in Section II in order to establish a consistent language and notation. Here also are described the spin interactions that figure prominently in multiple-quantum NMR, as well as the means for averaging these interactions experimentally. Expansion of the density operator in an orthogonal basis set, the members of which represent the various degrees of freedom to be investigated, is undertaken at the end of the section. Basic mathematical tools are developed here and then used repeatedly, often without further comment.

Having introduced the notion of coherences as orthogonal modes of the density operator, we proceed to examine the properties of these modes in Section III. The dynamic evolution of coherences of different orders under resonance offsets, chemical shift interactions, and spin–spin couplings is described, together with the influence of radiofrequency pulses. The discussion here is relatively brief, intended mainly to serve as a reference for the more detailed analysis that follows, although in most instances we try to indicate how the basic properties of spin coherence ultimately will find spectroscopic application.

Section IV takes up the questions of how to excite multiple-quantum and multiple-spin coherences and how then to monitor their behavior through other, observable, modes. Various means of excitation and coherence transfer are elaborated both for narrowband frequency-selective experiments, in which only designated pairs of energy levels in a multilevel system are irradiated; and for nonselective "hard pulse" experiments, in which all the levels are irradiated simultaneously. Treated as well are methods of uniform excitation in the presence of Zeeman and spin–spin interactions and methods of order-selective excitation, either even–odd or n-quantum, whereby the spin system is forced to interact exclusively with preselected numbers of radiofrequency quanta. The spin dynamics of weakly coupled isotropic systems are compared with those of strongly coupled and anisotropic systems, including solids; and important spectroscopic considerations such as the interpretation of two-dimensional multiple-quantum/single-quantum frequency maps and the separation of the different Zeeman orders are also addressed.

Finally, in Section V the principles of multiple-quantum spectroscopy are applied in a series of examples and illustrations. The selection is admittedly incomplete, but is nonetheless intended to convey at least some of the flavor of "practical" multiple-quantum NMR. Additional formalism is kept to a minimum here in contrast with the detailed treatment it follows. The topics covered at times overlap with those of the preceding section, for we make no attempt to differentiate too sharply between a method and its application. The list of topics is ordered in no particular fashion, except for a rough grouping into applications such as multiple-quantum filtering, spin counting, and some of the methods of heteronuclear coherence transfer—which depend mostly upon the excitation pattern; and applications such as simplification of highly complex single-quantum spectra, imaging, diffusion, and relaxation studies—which depend mostly upon the response.

II. NMR FUNDAMENTALS

A. Energy Levels

The spectrum of energies of a system of nuclear spins arises from the coupling of the elementary nuclear magnetic and electric moments with various magnetic and electric fields. In the usual case of high-field NMR the spins are placed in a large, static external magnetic field that establishes a basic energy level structure via a Zeeman Hamiltonian,

$$H_{\text{ext}} = -\boldsymbol{\mu}\cdot\mathbf{B}_0, \tag{1}$$

in which $\boldsymbol{\mu}$ is the nuclear magnetic dipole moment operator and in which the external magnetic field $\mathbf{B}_0$ is understood to lie along the z direction of the laboratory frame. Since the magnetic moment is directly proportional to the spin angular momentum operator $\mathbf{I}$ through the gyromagnetic ratio γ_I, the Zeeman interaction takes the explicit form

$$H_{\text{ext}} = -\gamma_I B_0 I_z \tag{2}$$

in units of angular frequency. The eigenstates of this Hamiltonian are the familiar angular momentum states, defined by

$$\mathbf{I}^2|I, M_r\rangle = I(I+1)|I, M_r\rangle$$

and (3)

$$I_z|I, M_r\rangle = M_r|I, M_r\rangle,$$

with M_r denoting the component of the total spin angular momentum in the direction of the applied field. The resulting $2I + 1$ energy levels are equally spaced, separated by the Larmor frequency

$$\omega_0 = -\gamma_I B_0, \tag{4}$$

which for external magnetic fields on the order of 10^4–$10^5 G$ falls into the radiofrequency region for most nuclei.

The *internal* magnetic fields due to the spins themselves are considerably weaker than the Zeeman field, ranging only up to approximately $10G$ for spin-$\frac{1}{2}$ nuclei. An internal Hamiltonian describes the coupling of the spins with the available local fields through the chemical shift, direct dipole–dipole, and indirect scalar, or J, interactions. We may write

$$H_{\text{int}} = H_{\text{CS}} + H_{\text{D}} + H_J, \tag{5}$$

adding a quadrupole term, H_{Q}, to account for the coupling of the electric quadrupole moment of a nonspherical nuclear charge distribution with an electronic field gradient.

Given that the local fields are two to six orders of magnitude smaller than the external field, it is clear that the spectroscopically interesting internal contributions usually can do no more than add or subtract a small component along B_0: the internal components perpendicular to B_0 are geometrically incapable of altering the orientation of the much larger static field in the laboratory frame. In this high field, or "secular," limit the internal and external Hamiltonians commute,[32] leaving all internal interactions invariant to rotations about B_0, and leading to a common set of eigenstates defined by

$$H_{\text{ext}}|M_r\rangle = \omega_0 M_r|M_r\rangle$$

and (6)

$$H_{\text{int}}|M_r\rangle = \omega_r|M_r\rangle.$$

The internal Hamiltonian thus acts as a first-order perturbation, inducing small shifts and splittings of the Zeeman energy levels that depend upon the nature and extent of the couplings and upon the number of interacting spins. Some examples of NMR systems are shown in Fig. 3; these range from isolated nuclei with two or three energy levels to small groups of coupled spins with level structures of varying complexity to interacting systems of macroscopic size.

B. Spin-$\frac{1}{2}$ in an External Magnetic Field

1. *State Vector Description—Precession*

Simple two-level systems are the fundamental units from which are constructed all NMR experiments, whether single-quantum or multiple-quantum, time domain or frequency domain. Representative of this kind

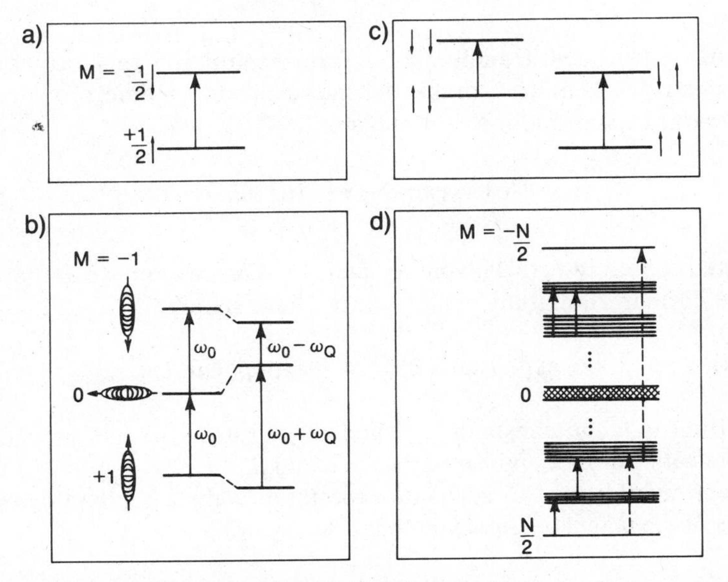

Figure 3. Energy level diagrams for typical spin systems. (*a*) $I = \frac{1}{2}$, with two states and one single-quantum transition. (*b*) $I = 1$, with a quadrupole coupling constant ω_Q. The three levels are shifted unequally by the quadrupolar interaction, thereby establishing two single-quantum transitions and one double-quantum transition. (*c*) Two weakly coupled $I = \frac{1}{2}$ nuclei with energy levels corresponding to $M = 1, 0$, and -1. There are two single-quantum transitions, one double-quantum transition, and one zero-quantum transition. (*d*) General system of N spins $I = \frac{1}{2}$ with 2^N states; selected single-quantum transitions are indicated by solid arrows and multiple-quantum transitions by dashed arrows. Each Zeeman manifold M contains $N!/[(N/2 + M)!(N/2 - M)!]$ states, their degeneracy lifted by the internal spin interactions. In a solid N is effectively infinite, and the $N + 1$ Zeeman manifolds are quasi-continuous bands.

of resonant system is an isolated spin-$\frac{1}{2}$ particle in an external magnetic field, which has accessible to it the two eigenstates $|\frac{1}{2}\rangle$ and $|-\frac{1}{2}\rangle$ corresponding to the two allowed orientations of its angular momentum. The state vector

$$|\psi\rangle = c_{1/2}|\tfrac{1}{2}\rangle + c_{-1/2}|-\tfrac{1}{2}\rangle, \tag{7}$$

in general a superposition of these basis states, obeys the time-dependent Schroedinger equation,

$$\frac{d}{dt}|\psi(t)\rangle = -iH|\psi(t)\rangle, \tag{8}$$

which may be integrated to yield

$$|\psi(t)\rangle = \exp(-iHt)|\psi(0)\rangle \tag{9}$$

for a time-independent Hamiltonian. A formal solution of this sort may also be obtained for a Hamiltonian that depends explicitly on time, provided that the operator expansion for the propagator

$$U(t) = \exp\left\{-i\int_0^t H(t')\,dt'\right\} \tag{10}$$

is ordered so that integrals involving later times always appear to the left.[33] For the Zeeman Hamiltonian, however, we have simply

$$|\psi(t)\rangle = a\exp(i\alpha)\exp(-i\omega_0 t/2)|\tfrac{1}{2}\rangle + b\exp(i\beta)\exp(i\omega_0 t/2)|-\tfrac{1}{2}\rangle, \tag{11}$$

recognizing that the constants $c_{1/2}$ and $c_{-1/2}$ can be written as complex numbers with norms a and b and phases α and β.

It is straightforward to calculate expectation values for the three components of the magnetic dipole moment

$$\begin{aligned}\langle\mu_x(t)\rangle &= \langle\psi(t)|\gamma_I\hbar I_x|\psi(t)\rangle\\ \langle\mu_y(t)\rangle &= \langle\psi(t)|\gamma_I\hbar I_y|\psi(t)\rangle\\ \langle\mu_z(t)\rangle &= \langle\psi(t)|\gamma_I\hbar I_z|\psi(t)\rangle\end{aligned} \tag{12}$$

by using the well-known properties of the raising and lowering operators, namely,

$$I_{\pm}|M\rangle = \sqrt{I(I+1) - M(M\pm 1)}\,|M\pm 1\rangle, \tag{13}$$

with the definition $I_\pm = I_x \pm iI_y$. The result is

$$\begin{aligned}\langle \mu_x(t) \rangle &= (\gamma_I \hbar ab) \cos(\alpha - \beta + \omega_0 t) \\ \langle \mu_y(t) \rangle &= (\gamma_I \hbar ab) \sin(\alpha - \beta + \omega_0 t) \\ \langle \mu_z(t) \rangle &= (\gamma_I \hbar/2)(a^2 - b^2),\end{aligned} \tag{14}$$

which indicates that the expectation value of the magnetic dipole moment for an individual spin precesses about the external field with angular frequency ω_0 and phase $\alpha - \beta$, maintaining a constant z component proportional to the difference in populations of the two spin levels.[30,31] The rotating component of the dipole moment emits dipole radiation which can, at least in principle, induce a time-varying voltage in a correctly oriented receiving coil tuned to ω_0. But the macroscopic system actually observed is always an ensemble—never just one spin. In thermodynamic equilibrium the phase difference between the $|\frac{1}{2}\rangle$ and $|-\frac{1}{2}\rangle$ states varies randomly from spin to spin, forcing the approximately 10^{20} individually precessing dipoles completely out of step with each other.[34] This randomization of the phase of precession, illustrated in Fig. 4, suggests that when the expectation values are averaged over all values of $\alpha - \beta$, only the time-independent z component survives. Such total lack of phase coherence between the $|\frac{1}{2}\rangle$ and $|-\frac{1}{2}\rangle$ eigenstates is

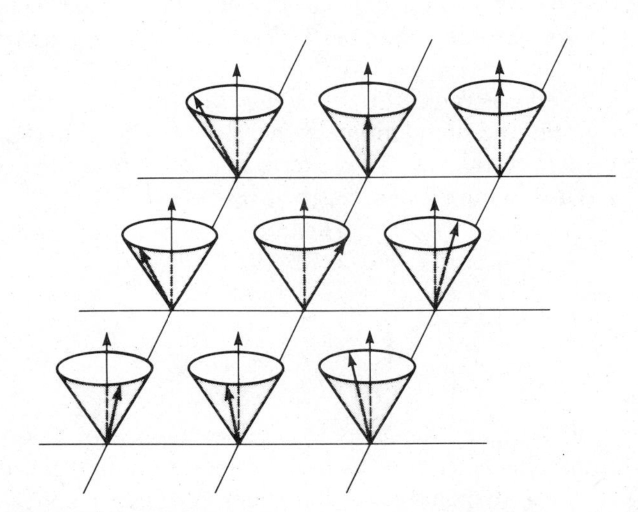

Figure 4. Random phases in an ensemble of independent oscillators in thermodynamic equilibrium. The magnetic moments precess about the external field at the same frequency but with uncorrelated phases, thus averaging to zero the macroscopic components transverse to the axis of precession.

characteristic of a collection of spins in equilibrium in high field, and suggests that the spin system is not a pure state describable by a single state vector. Instead, it is a mixture of pure states, appropriately represented by a density operator.

2. *Density Operator Description—Coherence*

The density operator for the spin system is defined as a weighted average of projection operators, specified by

$$\rho = \sum p_\psi |\psi\rangle\langle\psi|, \tag{15}$$

where p_ψ is the proportion of each pure state $|\psi\rangle$ in the mixture.[35] With the probabilities given by the Boltzmann factors

$$p(M = \pm\tfrac{1}{2}) = \frac{\exp(-M\hbar\omega_0/kT)}{Z}, \tag{16}$$

the equilibrium density operator becomes

$$\rho_{\text{eq}} = \frac{\exp(-\hbar\omega_0 I_z/kT)}{Z} \tag{17}$$

after the average over the phase differences is taken. The partition function Z is a normalization constant obtained by summing the Boltzmann factors for all possible states.

In the 2×2 representation of ρ_{eq} in the $|M_r\rangle\langle M_s|$ basis, the diagonal elements are simply the populations of the $M = \pm\frac{1}{2}$ levels, while the off-diagonal elements are the statistical averages of the cross terms $c^*_{1/2}c_{-1/2}$ and $c_{1/2}c^*_{-1/2}$. Thus in the explicit matrix form

$$[\rho_{ij}] = \begin{bmatrix} \dfrac{\exp(-\hbar\omega_0/2kT)}{Z} & 0 \\ 0 & \dfrac{\exp(+\hbar\omega_0/2kT)}{Z} \end{bmatrix} \tag{18}$$

the vanishing off-diagonal elements reflect the absence of a well-defined phase relationship between the $|\frac{1}{2}\rangle$ and $|-\frac{1}{2}\rangle$ states that constitute the ensemble. Were an off-diagonal element to differ from zero, a nonequilibrium state of phase *coherence* between the two connected levels would be indicated. Recalling the physical picture of precessing dipoles, we might then expect this *single-quantum* coherence to appear as a time-dependent magnetization transverse to the static field. In multiple-quantum NMR the concept of coherence is extended to a spin system of arbitrary size, wherein the order $n = \Delta M = M_r - M_s$ is defined as the difference in Zeeman quantum numbers between any two states in coherent superposition.

3. Time Development of the Density Operator in the Rotating Frame

The development of the system with time is governed by the Liouville-von Neumann equation

$$\frac{d\rho}{dt} = i[\rho, H], \tag{19}$$

which, together with its formal solution

$$\rho = \exp(-iHt)\rho(0)\exp(iHt), \tag{20}$$

follows directly from the Schroedinger equation.[35] Since knowledge of the density operator permits calculation of the expectation value of any operator according to the relationship

$$\langle A(t) \rangle = \mathrm{tr}[\rho(t)A], \tag{21}$$

prediction and control of the time development of ρ are always central problems for time domain NMR in general and multiple-quantum NMR in particular.

Solution of the Liouville–von Neumann equation is facilitated by viewing the problem in a frame rotating at or near the Larmor frequency in order to remove the fast precession due to the Zeeman interaction.[36] In the interaction picture defined by

$$U_z = \exp(-iH_{\mathrm{ext}}t), \tag{22}$$

the density operator is transformed to

$$\rho^R = U_Z^{-1}\rho U_Z; \tag{23}$$

a similar transformation of the Hamiltonian to

$$H^R = U_Z^{-1}HU_Z - iU^{-1}\frac{dU}{dt} \tag{24}$$

leaves the equation of motion in the same form as before, with ρ^R and H^R

substituting for ρ and H. Thus we have

$$\frac{d\rho^R}{dt} = i[\rho^R, H^R], \tag{25}$$

wherein only those components of the Hamiltonian that remain independent of time in the rotating frame are retained. For a secular internal Hamiltonian the principal consequence is that

$$H^R = H - H_{\text{ext}} = H_{\text{int}}. \tag{26}$$

Henceforth we will omit the superscripts on ρ and H, with the understanding that the laboratory frame has been replaced by a suitable rotating frame in which external interactions are absent.

C. Creation of Coherence by Radiofrequency Pulses

The simplest possible experimental manipulation of the spin system results from the application of one intense pulse of resonant or nearly resonant radiofrequency to produce single-quantum coherence. Typically offset from resonance by a small frequency $\Delta\omega = \omega - \omega_0$, the burst of electromagnetic radiation is applied at an angle ϕ relative to the x axis in a frame rotating at the carrier frequency ω, as shown in Fig. 5. The spin angular momentum couples with the radiation under the Hamiltonian

$$H_{\text{rf}} = -\gamma_I B_1 (I_x \cos\phi + I_y \sin\phi) \tag{27}$$

during which B_1, the amplitude of the magnetic component of the radiofrequency, appears stationary in the rotating frame. The effect of the resonance offset appears as an additional term,

$$H_{\text{off}} = -\Delta\omega I_z, \tag{28}$$

in the internal Hamiltonian.

Provided that $kT \gg \hbar\omega_0$, an assumption usually valid down to temperatures as low as 1 K, the initial state of the system is given to first order as

$$\rho(0) = \frac{\exp(-\beta_I I_z)}{Z} \approx \frac{1 - \beta_I I_z}{Z}, \tag{29}$$

where $\beta_I = \hbar\omega_0/kT$. The term proportional to the unit operator commutes with all operators and consequently cannot influence the time development

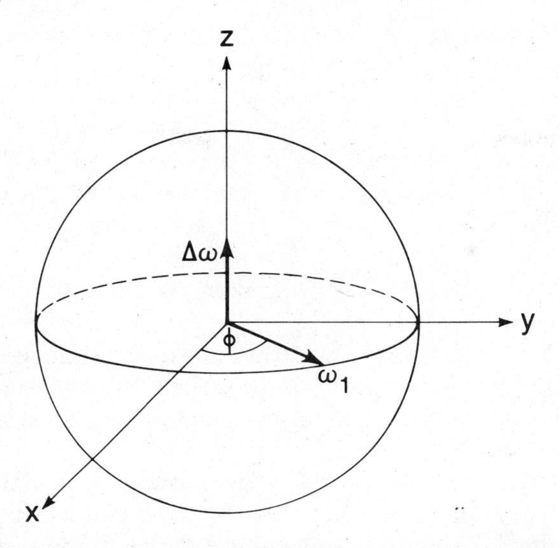

Figure 5. Rotating frame. The coordinate system for the spin degrees of freedom rotates about the z axis at the radiofrequency ω, thereby removing the time dependence of the radiofrequency Hamiltonian. The effective radiofrequency field has components of $(\omega_1 \cos \phi, \omega_1 \sin \phi, \Delta\omega)$, where the phase ϕ is measured relative to the x axis. ω_1 is proportional to the amplitude of the magnetic component of the radiofrequency, and $\Delta\omega(=\omega - \omega_0)$ is the offset from the Larmor frequency.

of ρ. For simplicity we will usually suppress this term, as well as the constant β_I/Z, and take

$$\rho(0) = I_z \tag{30}$$

as the initial condition.

Without any loss of generality we may apply the pulse along the x axis ($\phi = 0$) so that

$$H_{\text{rf}} = -\gamma_I B_1 I_x, \tag{31}$$

with the angular frequency $\omega_1 = -\gamma_I B_1$ providing a measure of the strength of the radiofrequency field. After a pulse of duration τ the system is described by

$$\rho(\tau) = \exp(-i\omega_1 \tau I_x) I_z \exp(i\omega_1 \tau I_x), \tag{32}$$

which is immediately recognizable as a rotation of the operator I_z about the

x direction through a "flip" angle $\omega_1 \tau$.[37-40] The result,

$$\rho(\tau) = I_z \cos \omega_1 \tau - I_y \sin \omega_1 \tau, \tag{33}$$

can be understood using the geometric construction shown in Fig. 6, and is a consequence of the defining commutation relations for the angular momenta, namely,

$$[I_x, I_y] = iI_z + \text{cyclic permutations.} \tag{34}$$

Any set of three independent operators exhibiting this commutation behavior will generate rotations in its own three-dimensional space, a property that will be exploited repeatedly when the discussion is extended to multiple-quantum phenomena.

When $\omega_1 \tau = \pi/2$, we speak of a $\pi/2$, or 90°, pulse; a pulse of this duration rotates ρ completely from I_z to $-I_y$, producing an equally weighted *coherent superposition* of the $|\frac{1}{2}\rangle$ and $|-\frac{1}{2}\rangle$ states. After the pulse the system evolves under the influence of the internal Hamiltonian alone, for the Zeeman interaction is removed in the rotating frame. The time development in the absence of relaxation therefore is given by

$$\rho(t) = \exp(-iH_{\text{int}}t)I_y \exp(iH_{\text{int}}t), \tag{35}$$

which in the $|M_r\rangle\langle M_s|$ basis becomes

$$\rho(t) = \sum_{r,s} \langle M_r|I_y|M_s\rangle \exp[-i(\omega_r - \omega_s)t]|M_r\rangle\langle M_s|. \tag{36}$$

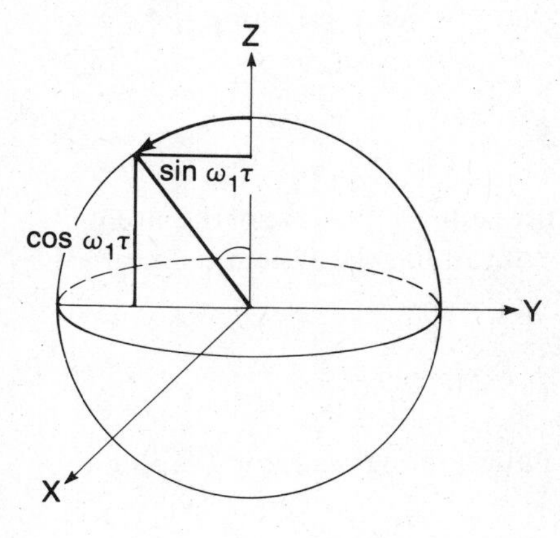

Figure 6. Nutation of the vector corresponding to the equilibrium density operator I_z about the x axis of the rotating frame during the application of a pulse of radiofrequency with phase $\phi = 0$, amplitude ω_1, and duration τ.

Since $I_y = (I_+ - I_-)/2i$, only the $|\frac{1}{2}\rangle$ and $|-\frac{1}{2}\rangle$ states are connected, and the macroscopic dipole moment thus developed oscillates at the frequency difference between the two levels. These oscillations may be detected by a heterodyne radio receiver, which in removing the carrier frequency ω mimics the rotating frame transformation. The macroscopic observable is a time-dependent magnetization proportional to I_x or I_y, written as

$$S_\alpha(t) = \mathrm{tr}[\rho(t)I_\alpha] = \sum_{\pm 1/2} |\langle M_r|I_\alpha|M_s\rangle|^2 \exp(-i\omega_{rs}t) \qquad \alpha = x, y. \quad (37)$$

When the spin system contains more than two levels, the free induction signal may oscillate at as many frequencies as there are observable single-quantum coherences.

D. Local Interactions†

The internal Hamiltonian ultimately determines both the nature of the nonequilibrim state to which a spin system may be excited, as well as the response that follows. To the extent that the internal Hamiltonian can be controlled, a wide variety of excitation and response patterns become available in a pulsed NMR experiment. In this section and the one following we consider the principal characteristics of the local interactions and the possibility of manipulating them experimentally.

The general form of any of the internal Hamiltonians,

$$H = C\mathbf{I}\cdot\mathbf{R}\cdot\mathbf{A} = C \sum_{\alpha,\beta = x,y,z} I_\alpha R_{\alpha\beta} A_\beta = C \sum R_{\alpha\beta} T_{\beta\alpha}, \quad (38)$$

is a scalar quadratic operator containing a spin angular momentum, a second-rank coupling tensor, and one other vector operator, which may or may not be a spin angular momentum. All the spin dependence is taken up in the second-rank tensor $\mathbf{T}$, a dyadic product of the two vector operators $\mathbf{I}$ and $\mathbf{A}$. The coupling tensor $\mathbf{R}$ reflects the spatial anisotropy of the interaction, and C is an appropriate collection of physical constants and other parameters.

1. Chemical Shift

The chemical shift, or shielding, interaction arises from the coupling of the spins with fields generated by the interplay of B_0 and the electronic environment surrounding a nucleus. The perturbation of the orbital motion of the electrons by the presence of the large external field induces additional electronic currents which in turn alter the net magnetic field felt at the

†See Refs. 21, 22.

nucleus. The magnitude of this effect varies with the direction of B_0 in a frame of reference defined by the anisotropy of the electronic distribution. Accordingly, the chemical shift is expressed generally as

$$H_{CS} = \gamma_I \mathbf{I} \cdot \boldsymbol{\sigma} \cdot \mathbf{B}_0, \tag{39}$$

with $-\boldsymbol{\sigma} \cdot \mathbf{B}_0$ representing the local field radiated by the electrons. The chemical shielding tensor $\boldsymbol{\sigma}$ establishes the value of the field shift for an arbitrary orientation of B_0 in the principal axis system. If the direction of B_0 is specified by polar coordinates (θ, ϕ), as in Fig. 7, the secular, or z, component of the coupling tensor becomes

$$\sigma_{zz} = \sigma + \sigma'_{xx} \sin^2 \theta \cos^2 \phi + \sigma'_{yy} \sin^2 \theta \sin^2 \phi + \sigma'_{zz} \cos^2 \theta, \tag{40}$$

where σ'_{xx}, σ'_{yy}, and σ'_{zz} are the values of the chemical shift along each of the three principal directions and σ is an isotropic component. The relevant Hamiltonian in the rotating frame,

$$H_{CS} = \sigma_{zz} \gamma_I B_0 I_z, \tag{41}$$

shows that as far as the nucleus is concerned, the chemical shift does nothing more than create a distribution of static z fields that vary depending on the local electronic environment. The linear dependence of the interaction on I_z is identical to that of the Zeeman coupling, so the chemical shielding merely *shifts* the resonant frequencies of the chemically distinct nuclei present in the sample. For convenience, it will be assumed throughout that the standard to which the resonant frequencies are referred is always chosen to ensure that all the chemical shifts in the system sum to zero.

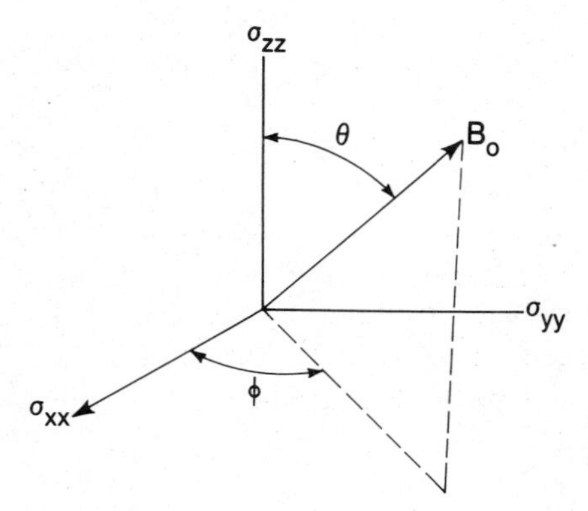

Figure 7. Principal axis system in which the anisotropic chemical shift coupling tensor is diagonal. Similar frames of reference exist for other anisotropic spin interactions.

2. *Dipole–Dipole Interaction*

The natural precessional motion of a spin generates a magnetic dipolar field,

$$\mathbf{B}_{\mathrm{D}} \sim \frac{\boldsymbol{\mu}}{r^3}, \tag{42}$$

which affects the resonance of another spin at a distance r from it. In the rotating frame this local dipolar field is resolved into a time-independent component parallel to the offset field and a time-dependent perpendicular component rotating with angular frequency ω_{D}. Since these contributions are identical in form to the externally applied B_0 and B_1 fields, a second spin can couple with them in an analogous manner.[30,31]

The interaction of a pair of spins labeled j and k is given by

$$H_{\mathrm{D}} = -\boldsymbol{\mu}_k \cdot \mathbf{B}_{\mathrm{D}j} = -D_{jk}(3I_{zj}I_{zk} - \mathbf{I}_j \cdot \mathbf{I}_k), \tag{43}$$

where the dipolar coupling constant

$$D_{jk} = \frac{\gamma_I^2 \hbar}{r_{jk}^3} \frac{3\cos^2 \theta_{jk} - 1}{2}, \tag{44}$$

depends both on the internuclear separation r_{jk} and on the angle θ_{jk} between the internuclear vector and B_0. In the compact notation of Eq. (38) the Hamiltonian takes the form

$$H_{\mathrm{D}} = -2\gamma_I^2 \hbar \mathbf{I}_j \cdot \mathsf{D} \cdot \mathbf{I}_k, \tag{45}$$

in which the axial symmetry of the interaction is manifested in a traceless, symmetric coupling tensor with principal values of $(D, -D/2, -D/2)$ and with no isotropic component.

The analogy drawn between the internal and external fields becomes clearer when the dipolar interaction is rewritten as

$$H_{\mathrm{D}} = -2D_{jk}[I_{zj}I_{zk} - \tfrac{1}{4}(I_{+j}I_{-k} + I_{-j}I_{+k})] \tag{46}$$

to incorporate the raising and lowering operators into the description: Now we see that the time-independent component of $\mathbf{B}_{\mathrm{D}j}$, represented here by the operator $-2D_{jk}I_{zj}$, increases or decreases the total offset field depending on the polarization of j. The resonance of a single spin-$\frac{1}{2}$ nucleus interacting with another spin-$\frac{1}{2}$ therefore is split into a doublet owing to the two possible values (up or down) of the static local field. The effect of the rotating component is described by the term proportional to $(I_{+j}I_{-k} + I_{-j}I_{+k})$, which connects states that differ only in the polarization of these two spins. Physically this means that if the natural frequencies are sufficiently close, then the miniature B_1 field of one can reorient the magnetic moment of the other in a resonant process. The interaction is in fact reciprocal, with the spins exchanging energy by the "flip-flop" mechanism depicted in Fig. 8. In this situation, typical in solids for nuclei with the same gyromagnetic ratio, the total Zeeman quantum number for a state $|M\rangle$ is conserved, although angular momentum components for the individual spins are not. Consequently the dipolar interaction as written above still describes a secular, or first-order, perturbation. By contrast, when the precession frequencies are appreciably different—or, put another way, when the rotating component of the local field is far from resonance—energy-conserving flip-flops cannot occur. In this limit of "weak" coupling, only the simple interaction with the static component,

$$H_D = -2D_{jk}I_{zj}I_{zk}, \tag{47}$$

remains significant. Individual Zeeman quantum numbers m_j may then be attributed to each spin, and the states may be represented by direct products of the form $\cdots |m_j\rangle|m_k\rangle \cdots$. Weak coupling is encountered when the interacting nuclei either have different gyromagnetic ratios or are separated by a difference in chemical shifts substantially greater than the dipolar coupling.

These same basic arguments apply to N interacting spins, for which the

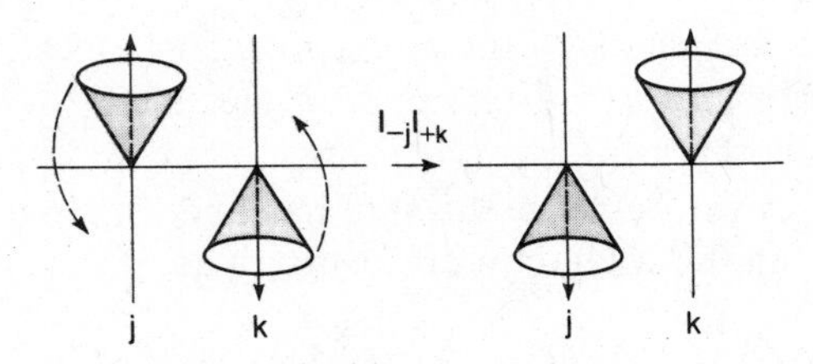

Figure 8. Classical interaction of two nuclear magnetic dipoles in an external field. Each magnetic moment experiences the static local field of the other, its magnitude determined by the length and orientation of the internuclear vector, and its sign determined by the alignment of the source spin with respect to the external field. The time-dependent component perpendicular to the axis of precession acts in the same fashion as a rotating radiofrequency field, bringing about an energy-conserving mutal reorientation, or flip-flop, when the natural frequencies are sufficiently close.

total dipolar Hamiltonian

$$H = -\sum_{\substack{j,k \\ (j<k)}} 2D_{jk}[I_{zj}I_{zk} - \tfrac{1}{4}(I_{+j}I_{-k} + I_{-j}I_{+k})] \tag{48}$$

is given as a sum over all the pairs of coupled nuclei.

3. *Scalar Coupling*

Besides the direct through-space coupling afforded by the dipole–dipole interaction, two spins are able to interact indirectly via the scalar, or J, coupling. The Hamiltonian

$$H_{\mathrm{J}} = -\mathbf{I}_j \cdot \mathbf{J} \cdot \mathbf{I}_k \tag{49}$$

is similiar in appearance to H_{D}, but the interaction, mediated by the electrons in the chemical bonds connecting the two nuclei, is usually orders of magnitude weaker. In addition, there exists an isotropic component J which makes the indirect coupling particularly important for liquid phases. We will deal with this interaction most frequently in weakly coupled, isotropic systems, in which the Hamiltonian reduces to

$$H_{\mathrm{J}} = -\sum_{j<k} J_{jk} I_{zj} I_{zk}.$$

4. *Quadrupole Coupling*

The quadrupole interaction

$$H_{\mathrm{Q}} = \frac{eQ}{6I(2I-1)} \mathbf{I} \cdot \mathbf{V} \cdot \mathbf{I} \tag{51}$$

describes the coupling of the nuclear electric quadrupole moment eQ with the electric field gradient established by the distribution of electrons near the nucleus. In a frame of reference where $\mathbf{V}$, the electric field gradient tensor, is diagonal, the quadrupole Hamiltonian becomes

$$H_{\mathrm{Q}} = \frac{eQ}{4I(2I-1)} [V_{zz}(3I_z^2 - I^2) + (V_{xx} - V_{yy})(I_x^2 - I_y^2)], \tag{52}$$

with $V_{\alpha\alpha}$ denoting the second derivative of the electric potential along the principal axis α. The coupling perturbs the Zeeman levels, shifting each energy by

$$\Delta E_m = E_m^{(1)} + E_m^{(2)} + \cdots \tag{53}$$

according to standard perturbation theory.[41] The effect depends upon the magnitude of the quadrupole interaction relative to the Zeeman interaction. When the quadrupole coupling is small, the high-field approximation remains valid, and only the first-order terms need be considered. Second-order contributions to the energy shift enter as the ratio of the quadrupole coupling

$$\omega_Q = \frac{3e^2qQ}{\hbar 2I(2I - 1)}, \qquad eq = V_{zz}, \tag{54}$$

to the Larmor frequency becomes significant. Like all spin interactions, the quadrupole coupling is anisotropic, and ω_Q exhibits the usual angular dependence in the laboratory frame.

E. Averaging of Internal Interactions

An internal Hamiltonian can be rendered time dependent if either the spatial or the spin coordinates upon which the interaction depends can be made to vary with time under the influence of external forces. When the resulting modulation of the interaction is very rapid compared to the magnitude of the static coupling, it becomes appropriate to consider only the time-independent Hamiltonian that emerges after averaging over the motion. Such a situation exists naturally in most liquids, for fast isotropic molecular motion in real space ensures that the spins see an isotropic distribution of magnetic field orientations on a suitably short time scale. Anisotropic contributions to the secular components of the various interactions are thus automatically averaged to zero in a liquid, and the system evolves under an effective internal Hamiltonian consisting solely of the isotropic chemical shifts and J couplings. In solids, by contrast, restricted molecular motion preserves the anisotropic features of the NMR spectrum in the absence of direct experimental intervention. Nevertheless, in both solids and liquids a wide variety of measures is available to alter the Hamiltonian governing the time development of the spin system. Experimental manipulation of this sort may be aimed either at the spatial dependence of the coupling, via the rapid bodily rotation of a sample; or at the spin dependence, via the rotation of the angular momentum operators by radiofrequency pulses with various phases and flip angles.

Demanded of any averaging scheme is the ability to impose the requisite time dependence upon an internal Hamiltonian when viewed from an appropriate frame of reference. Useful for this purpose is an interaction representation that removes the common motions of the spins due to the "toggling" action of the radiofrequency pulses, as well as those due to the fast Zeeman precession. In the *toggling rotating frame* defined by the transformation

$$U_T = \mathrm{U}_{\mathrm{rf}} U_Z \tag{55}$$

the axis system follows the series of discrete rotations induced by the pulses, so that dynamic evolution under the Hamiltonian

$$\tilde{H}_{\mathrm{int}} = U_T^{-1} H U_T \tag{56}$$

and its associated time-ordered propagator

$$U_{\mathrm{int}}(t) = \exp\left\{ -i \int_0^t \tilde{H}(t')\, dt' \right\} \tag{57}$$

appears to take place in the absence of any external fields or pulses.[21]

In simple cases it is sometimes possible to calculate the propagator exactly. For example, the propagator

$$U = \exp(-iH_{\mathrm{int}}\tau/2) \exp(-i\pi I_x) \exp(-iH_{\mathrm{int}}\tau/2) \tag{58}$$

for a $\tau/2$–π–$\tau/2$ "refocusing" sequence is equivalent to

$$U = \exp(-iH_{\mathrm{int}}\tau/2) \exp(-i\pi I_x) \exp(-iH_{\mathrm{int}}\tau/2) \exp(i\pi I_x) \exp(-i\pi I_x), \tag{59}$$

since $\exp(i\pi I_x) \exp(-i\pi I_x) = 1$. Taking

$$\exp(-H'_{\mathrm{int}}\tau/2) = \exp(-i\pi I_x) \exp(-iH_{\mathrm{int}}\tau/2) \exp(i\pi I_x), \tag{60}$$

where

$$H'_{\mathrm{int}} = \exp(-i\pi I_x) H_{\mathrm{int}} \exp(i\pi I_x), \tag{61}$$

we have

$$U = \exp(-iH_{\mathrm{int}}\tau/2) \exp(-H'_{\mathrm{int}}\tau/2) \exp(-i\pi I_x). \tag{62}$$

H'_{int} is related to H_{int} by a rotation through π about the x axis; consequently all the I_z operators change sign under the π pulse. For two weakly coupled spins with chemical shifts σ_j and σ_k, the Hamiltonians

$$H = \gamma_I B_0(\sigma_j I_{zj} + \sigma_k I_{zk}) - J_{jk} I_{zj} I_{zk} \tag{63a}$$

and

$$H' = -\gamma_I B_0(\sigma_j I_{zj} + \sigma_k I_{zk}) - J_{jk} I_{zj} I_{zk} \tag{63b}$$

commute with each other, leading to the overall result

$$U = \exp(iJ_{jk} I_{zj} I_{zk} \tau) \exp(-i\pi I_x). \tag{64}$$

This form of the propagator, which follows from the well-known operator expression[42]

$$\exp(A)\exp(B) = \exp(A + B + \tfrac{1}{2}[A, B] + \tfrac{1}{12}[A, [A, B]] + \tfrac{1}{12}[[A, B], B] + \cdots), \tag{65}$$

shows that the effect of the sequence may be viewed as a period of free evolution under the J coupling alone, following an initial π pulse. The effective internal Hamiltonian is therefore the *average* of the transformed Hamiltonians, so long as they all commute.

In the general case where the toggling frame Hamiltonians during the different intervals of a pulse sequence do not necessarily commute, the propagator can sometimes be evaluated according to coherent averaging theory.[21,43] The basic requirement is for the interaction to be cyclic, so that the toggling and rotating frames coincide periodically. When the condition

$$U_{\mathrm{rf}}(0) = U_{\mathrm{rf}}(t_c) = \cdots = U_{\mathrm{rf}}(Nt_c) = 1 \tag{66}$$

is satisfied over N cycles of period t_c, the propagator may be approximated by a simple exponential operator containing a time-independent effective Hamiltonian,

$$U_{\mathrm{int}}(t) = \exp(-i\bar{H}_{\mathrm{int}} t). \tag{67}$$

The effective Hamiltonian

$$\bar{H} = \bar{H}^{(0)} + \bar{H}^{(1)} + \bar{H}^{(2)} + \cdots \tag{68}$$

follows directly from the Magnus expansion,[44] and is meaningful only when the infinite series converges rapidly. The zeroth-order approximation

$$\bar{H}^{(0)} = \frac{1}{t_c} \int_0^{t_c} \tilde{H}(t)\, dt \tag{69}$$

is simply the one-cycle time average of the internal Hamiltonian in the toggling frame. The correction terms all depend upon whether or not the internal Hamiltonian commutes with itself at different times. The first two are given by

$$\bar{H}^{(1)} = \frac{-i}{2t_c} \int_0^{t_c} dt'' \int_0^{t''} dt' \, [\tilde{H}_{\rm int}(t''), \tilde{H}_{\rm int}(t')] \tag{70a}$$

and

$$\bar{H}^{(2)} = \frac{-1}{6t_c} \int_0^{t_c} dt''' \int_0^{t'''} dt'' \int_0^{t''} dt' \, \{[\tilde{H}_{\rm int}(t'''), [\tilde{H}_{\rm int}(t''), \tilde{H}_{\rm int}(t')]] + [\tilde{H}_{\rm int}(t'), [\tilde{H}_{\rm int}(t''), \tilde{H}_{\rm int}(t''')]]\}, \tag{70b}$$

from which it is clear that the cycle time should be kept short relative to the time scale of the interaction. Pulse sequences are usually designed to generate some desired average Hamiltonian while minimizing the correction terms. If observation of the system is restricted to the points where the toggling and rotating frames coincide, then the transient response appears to arise from the average Hamiltonian so created, rather than from the interaction naturally present.

The various internal Hamiltonians differ in whether they depend linearly upon the spin angular momentum operator, as do the chemical shift and offset terms; or bilinearly, as do the dipolar, scalar, and quadrupolar coupling terms. This basic difference permits the two classes of interactions to be distinguished in spin space, despite the overall common operator structure established in Eq. (38). For example, the four-pulse WHH-4 sequence sketched in Fig. 9 along with two other similar sequences effectively eliminates the homonuclear dipole–dipole coupling, which is bilinear in I_z, but only attenuates the linear chemical shift, resonance offset, and heteronuclear dipolar interactions.[43,45] If in the rotating frame the four $\pi/2$ pulses are applied along the x, y, $-x$, and $-y$ directions with the timing as specified, then in the toggling frame the angular momentum operators appear to move through the sequence as I_z, I_y, I_x, I_x, I_y, I_z. The average dipolar

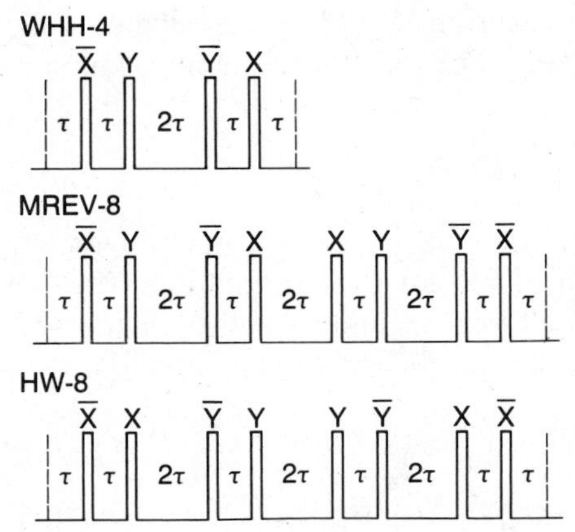

Figure 9. Multiple-pulse line narrowing sequences designed to average homonuclear dipole–dipole interactions to zero while preserving chemical shifts and heteronuclear couplings. The pulse trains illustrated differ in their susceptibility to errors and imperfections, the degree to which they eliminate correction terms in the Magnus expansion, and their effect on the linear interactions.

Hamiltonian over the cycle of length 6τ therefore is

$$\bar{H}_{\mathrm{D}} = H_{xx} + H_{yy} + H_{zz} = 0,$$

where

$$H_{\alpha\alpha} = -\sum_{j<k} D_{jk}(3I_{\alpha j}I_{\alpha k} - \mathbf{I}_j \cdot \mathbf{I}_k). \tag{71}$$

However, since the average value of I_z,

$$\bar{I}_z = \frac{1}{3}(I_x + I_y + I_z) = \frac{\sqrt{3}}{3} I', \tag{72}$$

is not zero, the linear internal Hamiltonians are merely scaled by the factor $\sqrt{3}/3$ to yield

$$H_{z,\mathrm{int}} = \frac{\sqrt{3}}{3} C\mathbf{I}' \cdot \mathbf{R} \cdot \mathbf{A}. \tag{73}$$

The average direction of the scaled operator $\mathbf{I}'$ is given by a unit vector,

$$\mathbf{e} = \frac{\sqrt{3}}{3}(\mathbf{x} + \mathbf{y} + \mathbf{z}), \tag{74}$$

pointing along the (1, 1, 1) axis at an angle $\theta' = \cos^{-1}(\sqrt{3}/3)$ relative to B_0.

Subjected to a WHH-4 sequence directed at the I spins, a *heteronuclear* coupling described by the operator $I_z S_z$ behaves like any interaction linear in the spin angular momentum of the irradiated species. In this context the S spin merely acts as the source of a static local field, indistinguishable from other internal and external sources except by the magnitude of its contribution. The heteronuclear interaction can be averaged to zero with a variety of decoupling methods, the simplest of which consists of a continuous irradiation at the Larmor frequency of one of the spins in a direction transverse to the external field.[46-48] Under such treatment the z component of the irradiated spin nutates about the direction of the radiofrequency field, presenting to the other spin a vanishing time-averaged local field if the nutation rate ω_1 exceeds the magnitude of the coupling.

F. Expansion of the Density Operator in an Orthogonal Basis

Given the ability to direct the time development of the spin system through control of the internal Hamiltonian, we will frequently need to evaluate the operator expression

$$\rho(t) = \rho(t_0 + \Delta t) = \exp(-iH\Delta t)\rho(t_0)\exp(iH\Delta t) \tag{75}$$

for different initial conditions and for different Hamiltonians. This problem often is simplified considerably when the density operator is represented by an expansion such as

$$\rho(t) = \sum_i b_i(t) B_i, \tag{76}$$

in which the operators B_i form a complete orthogonal basis set.[49] If the set is chosen carefully, the commutation properties of the basis operators may be exploited for the solution of the Liouville–von Neumann equation. It then becomes possible to project out of the expansion a desired component in accordance with the orthogonality condition

$$\mathrm{tr}(B_i^\dagger B_j) = \begin{cases} 0, & i \neq j \\ C, & i = j. \end{cases} \tag{77}$$

In this section we consider three basis sets that figure prominently in the treatment of multiple-quantum NMR: the outer product eigenbasis, the fictitious spin-$\frac{1}{2}$ (or single-transition) operators,[50-53] and the single-spin cartesian product operators.[54-57] (A related basis employing spherical tensor operators[58-60] is also useful, but not treated explicitly here.) Having acquired most of the necessary tools, we can then begin to discuss the properties of multiple-quantum coherence in detail, employing the equivalent operator descriptions as needed.

1. Outer Product Eigenbasis

The dyadic operators $\rho^n = |M_r\rangle\langle M_s|$ constitute a complete orthonormal set from which arises the expansion

$$\rho(t) = \sum_{r,s} \langle M_r|\rho(t)|M_s\rangle|M_r\rangle\langle M_s| \tag{78}$$

after two applications of the unit operator

$$\mathbf{1} = \sum_r |M_r\rangle\langle M_r| \tag{79}$$

to $\rho(t)$. This basis, already introduced earlier in the discussion, is particularly useful for describing evolution under a secular internal Hamiltonian, during which

$$\rho(t) = \sum \langle M_r|\rho(t_0)|M_s\rangle \exp[-i(\omega_r - \omega_s)\Delta t]|M_r\rangle\langle M_s|. \tag{80}$$

Each off-diagonal element, representing a coherence of order $n = M_r - M_s$, evolves independently as it oscillates at the frequency spanning the two connected levels. The diagonal terms ($r = s$) give the populations of the levels, which in the absence of spin-lattice relaxation remain constant during free evolution under a secular Hamiltonian.

2. Fictitious Spin-$\frac{1}{2}$ Operators

The fictitious spin-$\frac{1}{2}$ operators derive from the three Pauli operators

$$\sigma_x = \tfrac{1}{2}(|\tfrac{1}{2}\rangle\langle -\tfrac{1}{2}| + |-\tfrac{1}{2}\rangle\langle \tfrac{1}{2}|) \tag{81a}$$

$$\sigma_y = -\tfrac{1}{2}i(|\tfrac{1}{2}\rangle\langle -\tfrac{1}{2}| - |-\tfrac{1}{2}\rangle\langle \tfrac{1}{2}|) \tag{81b}$$

$$\sigma_z = \tfrac{1}{2}(|\tfrac{1}{2}\rangle\langle \tfrac{1}{2}| - |-\tfrac{1}{2}\rangle\langle -\tfrac{1}{2}|) \tag{81c}$$

used to describe a two-level system. The corresponding matrix representations

$$\sigma_x = \begin{pmatrix} 0 & 1 \\ 1 & 0 \end{pmatrix}, \quad \sigma_y = \begin{pmatrix} 0 & -i \\ i & 0 \end{pmatrix}, \quad \sigma_z = \begin{pmatrix} 1 & 0 \\ 0 & -1 \end{pmatrix} \tag{82}$$

form a linearly independent basis for any traceless 2×2 matrix.[49] With the appropriate rearrangements of the nonzero elements and with the insertion of zeros where needed, a complete basis set of extended Pauli matrices can obviously be generated for a matrix of any dimension. In a similar fashion, each of the operators above can be generalized to span two levels, $|M_r\rangle$ and $|M_s\rangle$, in a multilevel system.[50,52,53] Since these operators

$$I_x^{r-s} = \tfrac{1}{2}(|M_r\rangle\langle M_s| + |M_s\rangle\langle M_r|) \equiv \rho_x^{[n]} \tag{83a}$$

$$I_y^{r-s} = -\tfrac{1}{2}i(|M_r\rangle\langle M_s| - |M_s\rangle\langle M_r|) \equiv \rho_y^{[n]} \tag{83b}$$

$$I_z^{r-s} = \tfrac{1}{2}(|M_r\rangle\langle M_r| - |M_s\rangle\langle M_s|) \equiv \rho_z^{[n]} \tag{83c}$$

retain the angular momentum commutation property

$$[I_x^{r-s}, I_y^{r-s}] = iI_z^{r-s} + \text{cyclic permutations,} \tag{84}$$

they transform as the three components of a pseudomagnetization vector in a subspace pertaining only to levels r and s. The transverse components correspond to n-quantum coherence, and the longitudinal component corresponds to the population difference.

It should be noted that the various z operators are not linearly independent since $I_z^{r-s} + I_z^{s-t} = I_z^{r-t}$. Two more useful commutators involving these operators are

$$[I_z^{r-s}, I_z^{s-t}] = 0 \tag{85a}$$

and

$$[I_\alpha^{r-s}, I_z^{r-t} + I_z^{s-t}] = 0, \qquad \alpha = x, y. \tag{85b}$$

Additional cyclic commutation relations of the type $[P, Q] = i\kappa R$, as for example,

$$[I_x^{r-s}, I_x^{s-t}] = \tfrac{1}{2}iI_y^{r-t}, \qquad \kappa = \tfrac{1}{2}, \tag{86a}$$

$$[I_y^{r-s}, I_y^{s-t}] = -\tfrac{1}{2}iI_y^{r-t}, \qquad \kappa = -\tfrac{1}{2}, \tag{86b}$$

$$[I_x^{r-s}, I_y^{s-t}] = -\tfrac{1}{2}iI_x^{r-t}, \qquad \kappa = -\tfrac{1}{2}, \tag{86c}$$

link different three-dimensional subspaces and permit the straightforward evaluation of the general expression

$$\exp(-i\theta P)Q\exp(i\theta P) = Q\cos\kappa\theta + R\sin\kappa\theta. \tag{87}$$

3. *Single-Spin Product Operators*

The 2^{2N} single-spin product operators

$$B_i = 2^{q-1} \prod_j (I_{\alpha j})^{a_j}, \qquad a_j = 0 \text{ or } 1; \alpha = x, y, z, \tag{88}$$

span the operator space of a system of N coupled spin-$\frac{1}{2}$ nuclei, each term generally describing some combination of coherences and populations for a group of q spins taken from the total collection of N.[54,55] A given basis operator contains one angular momentum factor from each of the q spins in the product for which $a_j = 1$, there being a total of

$$\zeta = \frac{N!}{q!(N-q)!} 3^q \tag{89}$$

independent product operators for $1 \leqslant q \leqslant N$.

For the case of two coupled spins, the basis set contains 15 operators, exclusive of the unit operator. These are the six single-spin terms I_{xj}, I_{yj}, I_{zj}, I_{xk}, I_{yk}, I_{zk}, and the nine double-spin terms $2I_{xj}I_{xk}$, $2I_{yj}I_{yk}$, $2I_{zj}I_{zk}$, $2I_{xj}I_{yk}$, $2I_{xj}I_{zk}$, $2I_{yj}I_{xk}$, $2I_{yj}I_{zk}$, $2I_{zj}I_{xk}$, $2I_{zj}I_{yk}$. It is easy to verify that these operators are orthogonal, and that they are normalized to $C = 2^{N-2}$.

The product basis is convenient primarily because many of the operators that figure prominently both in the internal Hamiltonians and in the density operator commute as angular momenta. For example, we have for the two-spin case

$$[I_{yj}, 2I_{zj}I_{zk}] = i2I_{xj}I_{zk}, \tag{90}$$

which shows that a density operator proportional to I_{yj} will evolve into a mode proportional to $I_{xj}I_{zk}$ in the presence of weak bilinear couplings. Furthermore, since product operators involving different spins always commute, the general problem of N spins can be formulated as a succession of independent product terms.

The single-spin operators involving either I_x or I_y,

$$I_{xj} = \frac{1}{2}(I_{+j} + I_{-j}) \equiv \rho_x^{[1]} \tag{91a}$$

$$I_{yj} = \frac{1}{2i}(I_{+j} - I_{-j}) \equiv \rho_y^{[1]}, \tag{91b}$$

are single-quantum coherences that connect states for which $n = \pm 1$, without distinguishing the sign. Together with $I_{zj}(\equiv I_{0j})$, the single-spin raising and lowering operators I_{+j} and I_{-j} transform under rotations as the components of a first-rank spherical tensor operator, and are alternatively expressed as the dyads $|m_j = \frac{1}{2}\rangle\langle m_j = -\frac{1}{2}|$, $|m_j = -\frac{1}{2}\rangle\langle m_j = \frac{1}{2}|$, $|m_j = \frac{1}{2}\rangle\langle m_j = \frac{1}{2}|$, and $|m_j = -\frac{1}{2}\rangle\langle m_j = -\frac{1}{2}|$.

Cartesian quadrature components of pure double-quantum and zero-quantum coherence are defined in an analogous fashion as

$$\rho_x^{[2]} = \frac{1}{2}(I_{+j}I_{+k} + I_{-j}I_{-k}) \tag{92a}$$

$$\rho_y^{[2]} = \frac{1}{2i}(I_{+j}I_{+k} - I_{-j}I_{-k}) \tag{92b}$$

and

$$\rho_x^{[0]} = \frac{1}{2}(I_{+j}I_{-k} + I_{-j}I_{+k}) \tag{93a}$$

$$\rho_y^{[0]} = \frac{1}{2i}(I_{+j}I_{-k} - I_{-j}I_{+k}), \tag{93b}$$

with

$$\rho_x^{[n]} = \frac{1}{2}(\rho^n + \rho^{-n}) \tag{94a}$$

$$\rho_y^{[n]} = \frac{1}{2i}(\rho^n - \rho^{-n}) \tag{94b}$$

the obvious extension to higher order coherences in larger systems. In addition, there exist n-quantum combination terms such as

$$\rho_x^{[n]} = \frac{1}{2}(\cdots I_{+j}I_{-k}I_{+l}I_{+m}\cdots + \cdots I_{-j}I_{+k}I_{-l}I_{-m}\cdots) \tag{95a}$$

$$\rho_y^{[n]} = \frac{1}{2i}(\cdots I_{+j}I_{-k}I_{+l}I_{+m}\cdots - \cdots I_{-j}I_{+k}I_{-l}I_{-m}\cdots),$$

$$|n| = |\sum \Delta m_j|, \tag{95b}$$

in which opposing flip-flops cause the coherence order to be less than the number of spins participating. Diagonal z operators may also enter a coherence term, as in the $I_{xj}I_{zk}$ operator mentioned above.

Since the only degrees of freedom directly observable by an NMR magnetic dipole receiver are the transverse components of the macroscopic magnetization,

$$\langle \mu_x(t) \rangle = \langle \gamma_I \hbar I_x \rangle, \qquad I_x = \sum_{j=1}^{N} I_{xj}, \tag{96a}$$

and

$$\langle \mu_y(t) \rangle = \langle \gamma_I \hbar I_y \rangle, \qquad I_y = \sum_{j=1}^{N} I_{yj}, \tag{96b}$$

the single-spin/single-quantum coherences are the only detectable components of the density operator. Yet it is clear that these are just a small subset of the degrees of freedom available to the spins; the more numerous multiple-spin/multiple-quantum modes, though nominally forbidden, may nonetheless be excited by radiofrequency and observed through their effect on the allowed modes. We therefore turn our attention now to the special properties of multiple-quantum coherence.

III. PROPERTIES OF SPIN COHERENCE

A. Multiple-Quantum Coherence—Physical Picture

Though the existence of coherence between a pair of states $|M_r\rangle$ and $|M_s\rangle$ is evidence that the spins have at some earlier time interacted with at least $\pm(M_r - M_s)$ quanta of electromagnetic radiation, the events leading to the production of spin coherence are not necessarily transitions in the traditional spectroscopic sense of promotions and demotions between energy levels. In a picture where the radiofrequency field is treated classically, the excitation appears instead to bring about a dynamic evolution of the system through which is established a new phase relationship between the two superposed states. Each mode of coherence available to the spins is entirely independent, and accordingly is represented by an individual component of the density operator orthogonal to all the rest. The complete system therefore breaks up into subsystems of different order, each of which evolves in its own fashion.

The concept of a fictitious spin-$\frac{1}{2}$ allows any superposition of two states to be treated as a vector in an abstract three-dimensional space. A coherence in this space develops from a generalized rotation of the appropriate density operator component away from its z axis in a manner analogous to that depicted in Fig. 6 for the more familiar single-quantum rotation. Since rotation by π inverts the populations of the two levels, the presence of coherence is sometimes said to indicate that a transition is "in progress."

We can formulate a simple physical picture of coherence in a weakly coupled system, where it is meaningful to assign Zeeman quantum numbers to the individual spins. With the total Zeeman quantum number given simply by the algebraic sum of the separate components, any weakly coupled state may be visualized as a loose collection of spins with different polarizations. In Fig. 10, for example, two six-spin states with $M_r = 2$ and $M_s = -3$ are represented symbolically by up and down arrows within circles. These arrangements of the spins are only two of 2^6 possible, but nevertheless they can be isolated as one two-level fictitious spin-$\frac{1}{2}$ system out of an overall pattern of energy levels similar, perhaps, to that shown in Fig. 3*d*. All the basic concepts of precession and phase coherence applicable to a real spin-$\frac{1}{2}$ may then be used to understand the behavior of the 5-quantum coherence that can be realized between the two states. These arguments extend to strongly coupled systems as well, the only difference being that a strongly coupled state must generally be described as a sum of direct product terms with fixed Zeeman quantum numbers.

B. Enumeration of Multiple-Quantum Coherences*

In the general case of N coupled spin-$\frac{1}{2}$ nuclei, the 2^N energy levels are partitioned into $N + 1$ Zeeman manifolds, each containing

$$\binom{N}{p} = \frac{N!}{(N/2 + M_r)!(N/2 - M_r)!} \tag{97}$$

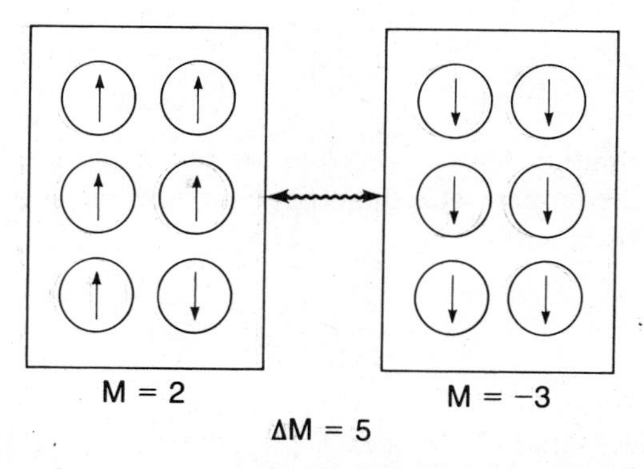

Figure 10. Direct product states with magnetic quantum numbers of 2 and -3 in a system of six weakly coupled spins. A coherent 5-quantum superposition of the two states is suggested by the wavy arrow. There are a total of twelve ± 5-quantum coherences possible, corresponding to the different ways of placing the odd spin in the $M = \pm 2$ state.

†See Refs. 61, 62.

distinct states. The integer p ($= N/2 + M_r$) denotes the excess of spins up over down needed to yield a particular value of M_r, and the combinatorial factor specifies the number of different ways these p spins can be selected from the group of N.

Degenerate in the Zeeman energy, the levels within a manifold are shifted and split by the internal interactions, thereby making possible a large number of spectroscopic transitions or coherences. The number of pairs of levels over which coherence can develop is simply a sum over the products of the dimensions of the appropriate manifolds, equal to

$$\sum_{p=0}^{N-n} \binom{N}{p} \binom{N}{n+p}, \qquad n \neq 0$$

and

$$\frac{1}{2} \sum_{p=1}^{N-1} \binom{N}{p} \left[\binom{N}{p} - 1 \right], \qquad n = 0.$$

Population terms, which involve the same state, are excluded in the counting of zero-quantum coherences. When simplified, these expressions reduce to

$$\binom{2N}{N-n} = \frac{(2N)!}{(N-n)!(N+n)!}, \qquad n \neq 0,$$

and

$$\frac{1}{2} \left[\binom{2N}{N} - 2^N \right], \qquad n = 0.$$

Stirling's approximation may be invoked for sufficently large N and small n, in which case the number of n-quantum coherences is well approximated by the Gaussian distribution

$$I(n, N) = \frac{4^N}{\sqrt{N\pi}} \exp\left(-\frac{n^2}{N} \right). \tag{98}$$

C. Response to Resonance Offsets and Chemical Shifts

The response of an n-quantum coherence to a uniform resonance offset is determined by the equation of motion

$$\frac{d\rho^n}{dt} = i[\rho^n, H_{\text{off}}], \tag{99}$$

where

$$H_{\text{off}} = -\Delta\omega I_z \tag{28}$$

and, say, $\rho^n = |M_r\rangle\langle M_s|$. Evaluation of the commutator in any basis leads to the first-order differential equation

$$\frac{d\rho^n}{dt} = (in\Delta\omega)\rho^n \tag{100a}$$

whose solution is

$$\rho^n(t) = \rho^n(0)\exp(in\Delta\omega t). \tag{100b}$$

The quadrature components

$$\rho_x^{[n]}(t) = \rho_x^{[n]}(0)\cos(n\Delta\omega t) + \rho_y^{[n]}(0)\sin(n\Delta\omega t) \tag{101a}$$

and

$$\rho_y^{[n]}(t) = \rho_y^{[n]}(0)\cos(n\Delta\omega t) - \rho_x^{[n]}(0)\sin(n\Delta\omega t) \tag{101b}$$

oscillate harmonically at n times the actual frequency offset (Fig. 11*a*), thereby enhancing the effect of a static magnetic field inhomogeneity when $|n| > 1$ and eliminating it when $n = 0$.[24,26,51,63] On the one hand, zero-quantum spectra, which remain sharp even in poor external magnetic fields, offer a means of overcoming line broadening due to field inhomogeneity;[64] while on the other hand, multiple-quantum spectra offer a means of amplifying the effect of a field gradient when one is introduced intentionally as, for example, in imaging experiments and studies of molecular diffusion (see Section V.E).

When the offset derives from the chemical shift interaction

$$H_{\text{CS}} = \sum \Omega_j I_{zj}, \qquad \Omega_j = \sigma_{zz,j}\gamma_I B_0, \tag{102}$$

the frequently shift generally varies for each of the spins. Under these conditions an n-quantum coherence evolves according to the Liouville–von Neumann equation as

$$\begin{aligned}\rho_x^{[n]}(t) &= \exp(-i\sum_j \Omega_j I_{zj}t)\rho_x^{[n]}(0)\exp(i\sum_j \Omega_j I_{zj}t)\\ &= \rho_x^{[n]}(0)\cos(\Omega_{\text{eff}}t) + \rho_y^{[n]}(0)\sin(\Omega_{\text{eff}}t),\end{aligned} \tag{103}$$

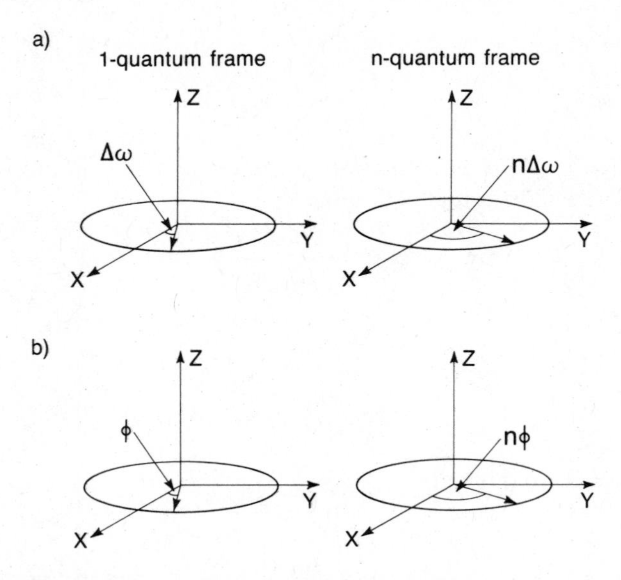

Figure 11. n-fold amplification of the effect of the resonance offset and radiofrequency phase in multiple-quantum spectroscopy. Shifts of $\Delta\omega$ and ϕ in the irradiation sequence appear as $n\Delta\omega$ and $n\phi$ to an n-quantum transition.

rotating about the z direction at a frequency Ω_{eff}. In the absence of strong couplings the effective precession frequency, given by

$$\Omega_{\mathrm{eff}} = \sum_j \Delta m_j \Omega_j, \tag{104}$$

is a sum over all the chemical shifts, weighted by $\Delta m_j = 1, -1$, or 0, depending on the change in quantum number for each spin. Thus for a pair of spins a zero-quantum coherence oscillates at the difference of the two chemical shifts, and a double-quantum coherence oscillates at their sum. The form of the effective precession frequency results from the products of independent trigonometric factors such as $\sin(\Omega_j t)$ and $\cos(\Omega_j t)$ that are implicit in Eq. (103).[62]

D. Response to Weak Couplings

Ubiquitous, spin–spin couplings are used to shape both the excitation of multiple-quantum coherence as well as its subsequent response. In the presence of weak indirect (or dipolar) couplings,

$$H_J = -\sum_{j<k} \tfrac{1}{2} J_{jk} 2 I_{zj} I_{zk}, \tag{105}$$

where the coupling constants J_{jk} (or $2D_{jk}$) are understood to be expressed in units of radians per second, an n-quantum coherence evolves as

$$\rho_x^{[n]}(t) = \rho_x^{[n]}(0)\cos(J_{\text{eff}}t/2) + 2I_{zk}\rho_y^{[n]}(0)\sin(J_{\text{eff}}t/2). \tag{106}$$

This basic form for the time development follows immediately from the cyclic commutation relations for operators such as I_{xj}, $2I_{zj}I_{zk}$, and $2I_{yj}I_{zk}$.

Owing to relationships such as

$$[I_{xj}I_{xk}, I_{zj}I_{zk}] = 0,$$

the effective coupling constant

$$J_{\text{eff}} = -\sum_j \Delta m_j J_{jk} \tag{107}$$

specifically excludes any interactions between nuclei directly involved in the coherence, that is, pairs of nuclei in which each member contributes ± 1 quantum to the total.[62] Couplings between "active" spins ($I_{\pm j}$) and "passive" spins (I_{zk}) alone influence the dynamical evolution of a previously existing coherence, determining not only the frequency of oscillation but introducing new modes as well. Hence the interaction of an active spin j within a coherence $\rho_x^{[n]}$ with an external, or passive, spin k leads to the development of a new term $2I_{zk}\rho_y^{[n]}$. The effect of a weak bilinear coupling, in general, is to expand the number of spins participating either directly or indirectly in a coherence via the regular exchange of amplitude between old and new terms as time progresses. This effect was alluded to in Fig. 2, which was used in the Introduction to suggest how a group of spins in a multiple-quantum state behaves in some sense as a superparticle, responding only to the local fields of those spins outside the group.

The growth of a complex network of correlated spins will be a recurring theme in the dynamics of multiple-quantum excitation. The simplest example of the phenomenon, already treated briefly in Section II.F.3, is just the formation of $I_{xj}I_{zk}$ from a single-quantum initial condition in a two-spin system. In turn, the two-spin/single-quantum coherence $I_{xj}I_{zk}$ is able to evolve back into *observable* single-quantum coherence via the transformation

$$\begin{aligned}\exp[(-iJ_{jk}t)I_{zj}I_{zk}](2I_{xj}I_{zk})\exp[(iJ_{jk}t)I_{zj}I_{zk}]\\ = 2I_{xj}I_{zk}\cos(J_{jk}t/2) + I_{yj}\sin(J_{jk}t/2).\end{aligned} \tag{108}$$

The observable term I_{yj} oscillates at $\pm J_{jk}/2$, its two frequency components differing in phase by π, so that Fourier transformation of the waveform yields a pair of lines equal in intensity but opposite in sign.

E. Effects of Radiofrequency Pulses and Phase Shifts

Application of a radiofrequency x pulse,

$$H_{\mathrm{rf}} = \omega_1 \tau \sum_{j=1}^{N} I_{xj}, \tag{109}$$

to the spin system causes the various density operator components to nutate through an angle $\omega_1 \tau$ about the x axis of the rotating frame. Denoting by $\rho(\tau^-)$ and $\rho(\tau^+)$ the state of the system before and after the pulse, we have the factored form

$$\rho(\tau^+) = \cdots \exp(-i\omega_1 \tau I_{xj}) \exp(-i\omega_1 \tau I_{xk}) \rho(\tau^-) \exp(i\omega_1 \tau I_{xj}) \exp(i\omega_1 \tau I_{xk}) \cdots, \tag{110}$$

since operators for different spins always commute. Each single-spin operator is individually rotated by the pulse according to the rules

$$\exp(-i\omega_1 \tau I_{xj}) I_{zj} \exp(i\omega_1 \tau I_{xj}) = \cos(\omega_1 \tau) I_{zj} - \sin(\omega_1 \tau) I_{yj} \tag{111a}$$

$$\exp(-i\omega_1 \tau I_{xj}) I_{yj} \exp(i\omega_1 \tau I_{xj}) = \cos(\omega_1 \tau) I_{yj} + \sin(\omega_1 \tau) I_{zj} \tag{111b}$$

$$\exp(-i\omega_1 \tau I_{xj}) I_{xj} \exp(i\omega_1 \tau I_{xj}) = I_{xj}, \tag{111c}$$

with a similar set of relations holding for y pulses. The effect of a pulse usually is to transform a component of the density operator either partially or entirely into another mode. For example, the antiphase term $I_{xj} I_{zk}$ is rotated into a linear combination of itself and the zero- and double-quantum term $I_{xj} I_{yk}$ by an x pulse with flip angle $\omega_1 \tau$, yielding

$$\begin{aligned} &\exp(-i\omega_1 \tau I_{xj}) \exp(-i\omega_1 \tau I_{xk}) I_{xj} I_{zk} \exp(i\omega_1 \tau I_{xj}) \exp(i\omega_1 \tau I_{xk}) \\ &\quad = \cos(\omega_1 \tau) I_{xj} I_{zk} - \sin(\omega_1 \tau) I_{xj} I_{yk}. \end{aligned} \tag{112}$$

The transfer of coherence is complete when $\omega_1 \tau = \pi/2$.

A pulse of arbitrary phase is related to an x pulse through a rotation of ϕ about the z axis. The effect of a phase shift on a density operator component, given by

$$\rho(\tau^+) = \exp\Big(-i\phi \sum_j I_{zj}\Big) \rho(\tau^-) \exp\Big(i\phi \sum_j I_{zj}\Big), \tag{113}$$

derives from the rotations

$$\exp(-i\phi I_{zj})I_{\pm j}\exp(i\phi I_{zj}) = \exp(\mp i\phi)I_{\pm j}; \tag{114}$$

and from these relationships arises the important result

$$\rho_x^{[n]}(\tau^+; \phi) = \rho_x^{[n]}(\tau^-)\cos(n\phi) + \rho_y^{[n]}(\tau^-)\sin(n\phi), \tag{115}$$

which shows that when the phase of the rf is shifted by ϕ, an n-quantum coherence sees an apparent phase shift of $n\phi$ (Fig. 11*b*).[63] This property is exploited in numerous spectroscopic techniques, some of which include methods to separate signals from different orders and to excite and detect signals from different orders selectively.

F. Total Spin Coherence

The total spin coherence,

$$|-N/2\rangle\langle N/2| = \prod_{j=1}^{N} I_{+j}$$

or

$$|N/2\rangle\langle -N/2| = \prod_{j=1}^{N} I_{-j},$$

connects the two outermost levels of a system of N coupled spin-$\frac{1}{2}$ nuclei. Neither of these levels is split by the internal Hamiltonian, for regardless of the coupling, there exists only one configuration of the spins (all up or all down) that can realize the extreme Zeeman energies. The coherent superposition of the $|N/2\rangle$ and $|-N/2\rangle$ eigenstates, arising from an interaction with a full complement of N quanta, is of course the highest order possible in the coupled system, and necessarily demands the active participation of all N spins. It frequently finds application in experiments designed either to ascertain the number of spins constituting a system or to isolate the response of a selected subsystem.

The total spin coherence belongs to the broader class of coherence between "spin-inversion" states, eigenstates that differ in the polarization of each of the N spins.[28,65] Here all the spins "flip," some up and some down, so the total change in the Zeeman quantum number need not be N. Coherences in this class commute with any bilinear Hamiltonian, and therefore evolve independently of the spin–spin couplings. Their development is generally determined only by the resonance offset interaction and those chemical shift differences that commute with the bilinear couplings. The total spin coherence is distinguished by its commutation with all the internal Hamiltonians except the full sum of the linear terms which, given the convention of zero total chemical shift, leaves just the resonance offset term to determine its development. Referring again back to Fig. 2, we can understand this behavior by noting that with N spins out of N tied up in the coherence, there are no passive spins available to which they can respond.

IV. EXCITATION AND DETECTION OF MULTIPLE-QUANTUM COHERENCE

A. General Form of the Multiple-Quantum Experiment

Any experimental attempt to monitor the dynamic evolution of a spin system in which there exists a condition of multiple-quantum coherence must inevitably be based upon the detection of single-quantum transverse magnetization. However complex the structure of a multilevel system may be, these relatively few magnetic dipole modes still remain the only coherences directly observable with conventional radiofrequency technology. Operating within this constraint, multiple-quantum experiments typically employ a method of indirect detection using two-dimensional spectroscopy and coherence transfer first to excite coherences of desired orders, and then to record their response to either naturally occurring or externally manipulated local fields.

The basic scheme of a two-dimensional multiple-quantum experiment is illustrated in Fig. 12.[24] Partitioned into four distinct periods of time, the pulse sequence creates a nonequilibrium condition of multiple-quantum coherence during a *preparation* period; allows the system to respond during a subsequent *evolution* period; and then transfers during a *mixing* period some or all of the existing coherence to observable single-quantum modes for

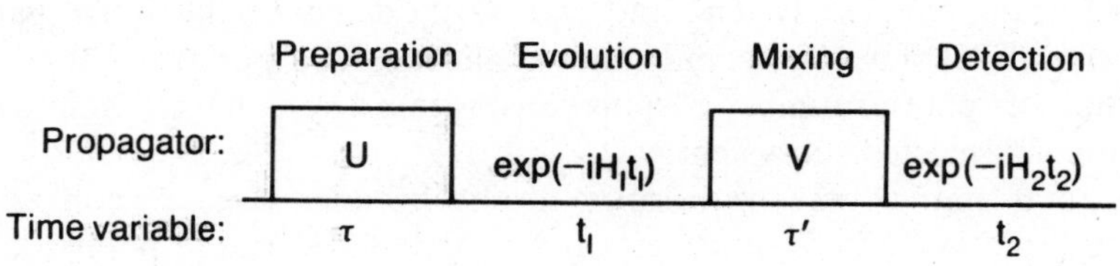

	Preparation	Evolution	Mixing	Detection
Propagator:	U	exp(−iH$_1$t$_1$)	V	exp(−iH$_2$t$_2$)
Time variable:	τ	t$_1$	τ'	t$_2$

Figure 12. General form of a two-dimensional multiple-quantum experiment. Multiple-quantum coherences are created by the preparation period propagator $U(\tau)$ and respond to local fields during the evolution period t_1. The mixing period propagator $V(\tau')$ converts multiple-quantum coherence to observable single-quantum coherence for detection during t_2.

observation during a final *detection* period. The signal obtained in this manner is necessarily determined by the detailed history of the spins up to the point of detection, and therefore provides a window through which the otherwise "invisible" degrees of freedom can be viewed.

1. *Excitation and Response*

The preparation period propagator

$$U(\tau) = \exp(-iH\tau) \tag{116a}$$

arises from a combination of pulses and delays appropriate for the energy level spacings and internal couplings of a particular multilevel system. Under its influence the system is brought to the condition

$$\rho(\tau) = U(\tau)\rho(0)U^{-1}(\tau), \tag{116b}$$

which unlike the state obtained following one $\pi/2$ pulse, generally reflects far more of the available degrees of freedom than just the single-spin/single-quantum modes. The preparation sequence is usually designed to be either narrowband or broadband over a range of frequencies, depending upon whether it is necessary or desirable to excite coherences only between specific pairs of levels. The excitation, even when broadband with respect to frequency, may still be selective or nonselective with respect to the orders of coherence created.

A nonequilibrium distribution of multiple-quantum coherence having been established, the system is allowed to develop, freely or otherwise, for an evolution period of length t_1. During this time the different modes of coherence oscillate at the eigenfrequencies determined by the effective Hamiltonian H_1, with all but the single-quantum coherences I_x and I_y unrecognized by the electronic detection apparatus. The hidden response, made manifest through the spectral distribution of the multiple-quantum coherence, is revealed after the time development is frozen at t_1 and the system taken back to single-quantum modes. Following the transfer of coherence during the mixing period, the density operator becomes

$$\rho(\tau, t_1, \tau') = V(\tau') \exp(-iH_1 t_1)\rho(\tau) \exp(iH_1 t_1)V^{-1}(\tau'), \tag{117a}$$

in which

$$V(\tau') = \exp(-iH'\tau') \tag{117b}$$

denotes the mixing period propagator.

2. *Time Reversal and Conjugate Mixing*

The signal observable immediately after mixing,

$$S_\alpha(\tau, t_1, \tau') = \mathrm{tr}[\rho(\tau, t_1, \tau')I_\alpha], \qquad \alpha = x, y, \tag{118}$$

is a transverse component of the total spin angular momentum, which in practice is often combined with another component in quadrature to yield a complex signal $S = S_x + iS_y$. With the initial condition $\rho(0)$ taken to be proportional to I_z, however, it is sometimes convenient to regard I_z as an observable operator as well. No inconsistency is created by so doing, for I_z, though not measureable directly, is nonetheless related to I_x or I_y by a simple $\pi/2$ pulse. Defining a coherence amplitude

$$Z_{rssr} = \langle M_r|U\rho(0)U^{-1}|M_s\rangle\langle M_s|V^{-1}I_z V|M_r\rangle, \tag{119}$$

we then express the trace as

$$S_z(\tau, t_1, \tau') = \sum_{r,s} Z_{rssr} \exp[-i\omega_{rs}^{(1)}t_1] \tag{120}$$

to reveal the oscillations $\omega_{rs}^{(1)} = \omega_r - \omega_s$ at the energy level differences of the evolution period Hamiltonian. The complete multiple-quantum frequency response is mapped out point by point through repetition of the experiment for a series of regularly incremented values of t_1. Fourier transformation with respect to t_1 of the interferogram that results provides the multiple-quantum frequency spectrum, detected indirectly via the modulation of the single-quantum signal present during the detection period.

In general the amplitude and the phase of each frequency component are determined by the combined effects of preparation and mixing through the complex factor Z and may be different for every multiple-quantum oscillation frequency.[66] But if the inverse propagator $V^{-1}(= V^\dagger$ for the unitary operator $V)$ can be made either equal to U or different from it only by a phase factor ϕ, then with

$$V^\dagger = \exp(-i\phi I_z)U \exp(i\phi I_z), \tag{121}$$

the signal

$$S_z(\tau, t_1) = \sum_n \sum_{r,s} |\langle M_r|I_z|M_s\rangle|^2 \exp(in\phi) \exp[-i\omega_{rs}^{(1)}t_1] \tag{122}$$

reduces to a Fourier series with real coefficients. Under these circumstances all spectral lines within a given order have the same phase, and lines belonging to adjacent orders differ in phase by $\pm\phi$. In addition, the signal detected at $t_2 = 0$ is now the largest possible, theoretically equal to that which follows a simple $\pi/2$ pulse.[28,67]

One way of satisfying the requirement that V and U be Hermitian conjugates is to arrange for the preparation and mixing Hamiltonians to be equal in magnitude but opposite in sign. This condition results in an apparent time reversal for the propagator, and is generally brought about using coherent averaging methods of the kind discussed in Section II.E.[68-71] Even where actual time reversal is impossible, a conjugate mixing period of a sort still can be realized simply by shifting the phase of each pulse in the preparation period by π and reversing the order of application.[28] The signal energy averaged over the entire detection period is maximized under such a treatment, but the phases of the different components may vary. Nevertheless, the spectrum of the *spin-inversion* coherences is obtained in absorption mode.

3. *Two-Dimensional Fourier Transformation*

Either one point or many may be sampled during t_2 to obtain the modulated signal. When a complete single-quantum signal is recorded, the additional period under the Hamiltonian H_2 leaves the spins described by

$$\rho(\tau, t_1, \tau', t_2) = \exp(-iH_2 t_2)\rho(\tau, t_1, \tau') \exp(iH_2 t_2), \tag{123}$$

from which originates the two-dimensional time domain signal

$$S(t_1, t_2) = \sum_{rstu} Z_{rstu} \exp[-i\omega_{rs}^{(1)} t_1] \exp[-i\omega_{tu}^{(2)} t_2]. \tag{124}$$

Fourier transformation with respect to both time variables produces a two-dimensional frequency map showing the connection between the multiple-quantum frequencies present during evolution and the single-quantum frequencies present following mixing. An example of a two-dimensional multiple-quantum spectrum, obtained from the six ^{1}H nuclei in oriented benzene, is given in Fig. 13 together with an expanded view of the four-quantum region in Fig. 14.[72] Peaks appearing at the intersection of two frequencies indicate a transfer of coherence between multiple-quantum and single-quantum components oscillating at $\omega_{rs}^{(1)}$ and $\omega_{tu}^{(2)}$. The correlation is especially clear in Fig. 14, where each peak marks a four-quantum $\rightarrow$ one-quantum pathway. Projection of the spectrum in either direction, accomplished by summing all cross sections parallel to the desired axis,[73] yields either the complete multiple-quantum spectrum $S[\omega^{(1)}]$, or the single-quantum spectrum, $S[\omega^{(2)}]$, "filtered" through the various orders of multiple-quantum coherence.

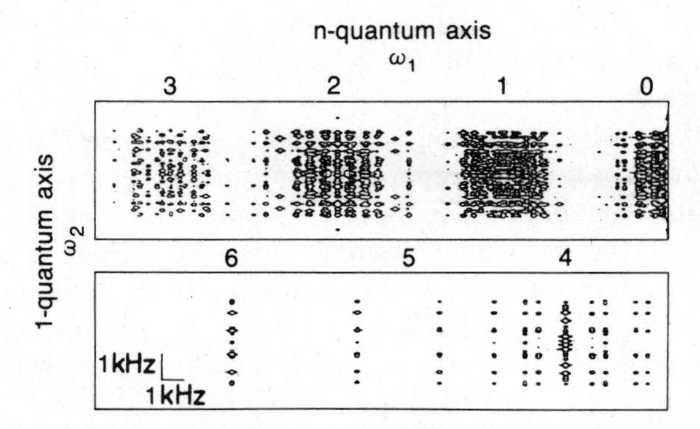

Figure 13. Contour plot of the two-dimensional multiple-quantum spectrum of oriented benzene, obtained with preparation and mixing times of 9 ms. Dipole forbidden n-quantum coherences oscillating at frequencies ω_1 $[\equiv \omega^{(1)}]$ during the evolution period are observed as allowed single-quantum coherences oscillating at frequencies ω_2 $[\equiv \omega^{(2)}]$ during the detection period. The scale pertains to the linear frequency $\omega/2\pi$. (Reproduced from Ref. 72 with permission. © 1984 Elsevier Scientific Publishing Co.)

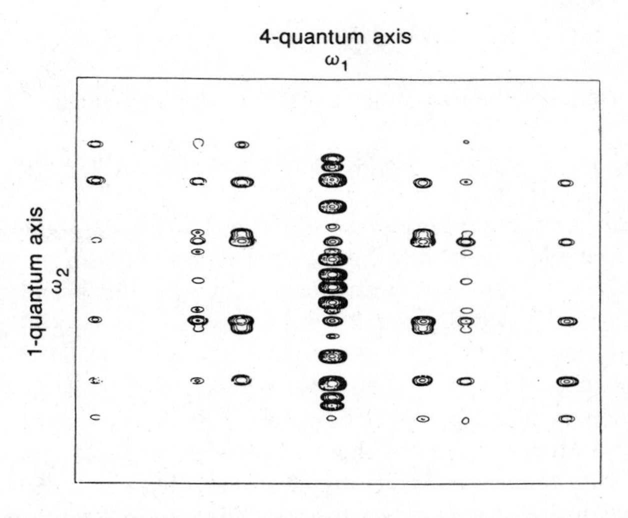

Figure 14. Expanded view of the 4-quantum region of Fig. 13. The projection onto the ω_2 axis yields the spectrum of those single-quantum coherences present during t_2 derived from 4-quantum coherences present during t_1. The projection onto the ω_1 axis yields the spectrum of the 4-quantum coherences as it would have appeared during t_1 had direct detection been possible. (Reproduced from Ref. 72 with permission. © 1984 Elsevier Scientific Publishing Co.)

B. Narrowband Excitation

The internal interactions of nuclei with $I > \frac{1}{2}$ frequently are dominated by the electric quadrupole coupling, which can range up to and even beyond 10 MHz in some systems of interest. In many of these systems the energy levels are spaced too widely to be covered by a single pulse, owing simply to practical limitations on radiofrequency power. When the nutation frequency, proportional to the inverse of the duration of a $\pi/2$ pulse, is significantly less than the full spectrum of internal frequencies, it becomes necessary to employ narrowband excitation techniques to create coherences between selected pairs of levels.

In this section we treat three major examples of frequency-selective multiple-quantum excitation, representative of a larger class of phenomena observed not only in NMR but in EPR and optical spectroscopy as well.[74-77] We therefore consider, without prejudice, a generic system of unequally spaced levels $r, s, t, u, \ldots$, as depicted in Fig. 15*a*. It is assumed that the frequency differences are such that irradiation of two of the levels does not affect any of the others. The first case treated, the excitation of coherence by a single pulse at half the frequency spanning a pair of levels, may be viewed loosely as an interaction of the spin system with two radiofrequency quanta simultaneously. In the second example, coherence established by a selective pulse between one pair of levels (say, r and s) is subsequently relayed to another pair (say, r and t) by means of a second selective pulse. The third example, which serves to introduce a discussion of *broadband* excitation, demonstrates how coherence between levels s and t can be transferred to r and u via the simultaneous irradiation of r–s and t–u. Each of these coherence transfer processes is illustrated schematically in Fig. 15.

1. *Selective Double-Quantum Pulse*

a. Formalism. The first-order rotating frame Hamiltonian for a quadrupolar nucleus in the presence of a weak radiofrequency field is given by

$$H = H_{\mathrm{rf}} + H_{\mathrm{int}} = -\Delta\omega I_z + \tfrac{1}{3}\omega_{\mathrm{Q}}[3I_z^2 - I(I+1)] + \omega_1 I_x, \quad (125)$$

where

$$\omega_{\mathrm{Q}} = \tfrac{1}{4}e^2qQ[\tfrac{1}{2}(3\cos^2\theta - 1) + \eta \sin^2\theta \cos 2\phi]$$

and

$$\eta = \frac{V_{xx} - V_{yy}}{V_{zz}}.$$

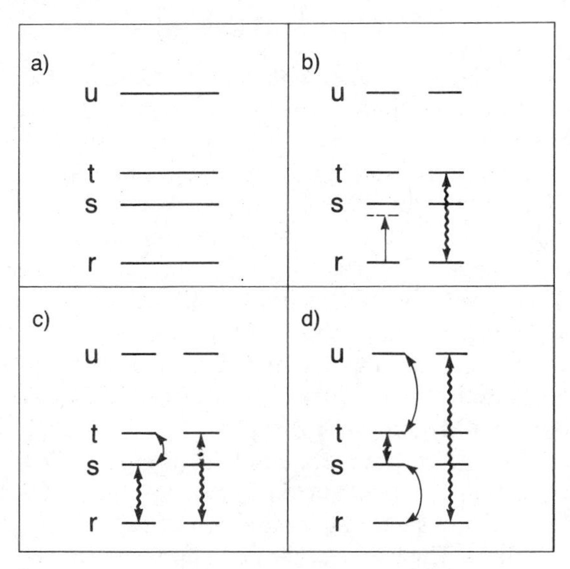

Figure 15. Frequency-selective coherence transfer processes in a multilevel system. (*a*) Generalized set of well-separated, unequally spaced energy levels . . . , *r*, *s*, *t*, *u*, ... (*b*) Excitation of dipole-forbidden coherence ($n \neq 1$) by a single weak pulse at half the frequency difference between two levels. The straight arrow represents the pulse; the wavy arrow symbolizes the coherence that results. (*c*) Transfer of coherence originally existing between eigenstates $|r\rangle$ and $|s\rangle$ to eigenstates $|s\rangle$ and $|t\rangle$ by means of a selective pulse at ω_{st}. By superposing states $|s\rangle$ and $|t\rangle$, the pulse induces a harmonic exchange of coherence between the *r–s* and *r–t* modes with period 4π. (*d*) Transfer of coherence existing between eigenstates $|s\rangle$ and $|t\rangle$ to eigenstates $|r\rangle$ and $|u\rangle$ by simultaneous irradiation at the frequencies ω_{tu} and ω_{rs}. Modulation techniques are used to generate sidebands at the desired frequencies. As the number of levels to be covered increases and as the spacing between them decreases, the selective pulses may be replaced by strong nonselective pulses with Fourier components sufficient to excite the entire spectrum. (Adapted from Ref. 27. Original illustration © 1981 Pergamon Press, Ltd.)

For $I = 1$ the three Zeeman levels are shifted by $\omega_Q/3$, $-2\omega_Q/3$, and $\omega_Q/3$, the magnitude of the shift determined by the orientation of the electric field gradient tensor in the laboratory frame. The offset term becomes negligible when a pulse is applied near the unshifted center frequency, as in Fig. 15*b*, and the remaining interaction then can be recast in the fictitious spin-$\frac{1}{2}$ formalism as

$$H = \tfrac{2}{3}\omega_Q(I_z^{1-2} - I_z^{2-3}) + \sqrt{2}\omega_1(I_x^{1-2} + I_x^{2-3}), \qquad (126)$$

where the superscripts 1, 2, and 3 refer to the Zeeman basis states $|1\rangle$, $|0\rangle$, and $|-1\rangle$, respectively.

The motion of the spin-1 density operator under this Hamiltonian is most easily viewed in a tilted reference frame defined by

$$H_T = U^{-1}HU$$

and

$$\rho_T = U^{-1}\rho U$$

wherein the Hamiltonian is diagonal. The ratio of the radiofrequency amplitude to the quadrupole coupling enters into the above transformation, which is represented in matrix form by

$$U = \begin{bmatrix} \frac{\sqrt{2}}{2}\cos(\theta/2) & \frac{-\sqrt{2}}{2}\sin(\theta/2) & \frac{-\sqrt{2}}{2} \\ \sin(\theta/2) & \cos(\theta/2) & 0 \\ \frac{\sqrt{2}}{2}\cos(\theta/2) & \frac{-\sqrt{2}}{2}\sin(\theta/2) & \frac{-\sqrt{2}}{2} \end{bmatrix}, \qquad (127)$$

through the angle $\theta = \tan^{-1}(2\omega_1/\omega_Q)$. Evaluation of the Liouville–von Neumann equation in the tilted frame, followed by a transformation back to the rotating frame, yields the result

$$\rho(\tau) = 2[I_z^{1-3}\cos(\omega_1^2\tau/\omega_Q) - I_y^{1-3}\sin(\omega_1^2\tau/\omega_Q)] \qquad (128)$$

for weak irradiation ($\omega_1 \ll \omega_Q$). Recalling the commutation properties of the single-transition operators, we see that this time dependence describes a nutation of the density operator about I_x^{1-3} at a frequency of ω_1^2/ω_Q. When $\omega_1^2\tau/\omega_Q = \pi/2$, the equilibrium population difference I_z^{1-3} is completely converted into double-quantum coherence, in analogy to the effect of a single-quantum $\pi/2$ pulse.[51-53]

Since to first order in perturbation theory the $M = 1$ and $M = -1$ levels in the spin-1 are shifted equally by $\omega_Q/3$, the double-quantum coherence evolves independently of the quadrupole Hamiltonian. Consequently coherence created by the preparation pulse subsequently develops only under the influence of the smaller internal interactions, such as the chemical shift and dipole–dipole couplings.[78] This evolution may be halted at any time and the progress of the system monitored by application of a mixing pulse. For example, either a $\pi/2$ or a π pulse applied at one of the single-quantum frequencies will convert the double-quantum coherence into observable magnetization. Where possible, the monitoring pulse might even be a "hard" pulse, with $\omega_1 \gg \omega_Q$, able to cover more than one frequency.[79]

b. Examples: Resonance Offsets, Double-Quantum Nutation, and Double-Quantum Decoupling. Shown in Fig. 16*a* is the double-quantum interferogram $S(t_1)$ exhibited by the carboxyl deuterons in a crystal of perdeuterated oxalic acid dihydrate.[78] Obtained point by point 2 kHz off resonance, the indirectly detected signal oscillates at 4 kHz, twice the offset frequency, in accord with the basic properties of multiple-quantum coherence outlined in Section III.C. The double-quantum response largely ignores the ≈ 31 kHz quadrupolar splitting, thus revealing upon Fourier transformation a spectrum of chemical shifts uncomplicated by the quadrupole couplings. Relatively narrow deuterium chemical shift anisotropy *powder* patterns, normally obscured by the large quadrupole interaction, become visible in this fashion.[79]

Selective double-quantum pulses may also be applied to nuclei with spins larger than $I = 1$ if three of the levels behave effectively independently of the rest. The principles remain the same, although the effective nutation frequencies depend upon the value of I. An example of double-quantum behavior among three levels in an oriented single crystal of $^{27}Al_2O_3$, where $I = \frac{5}{2}$, is illustrated in Fig. 16*b*.[26] In this particular experiment a selective $\pi/2$ pulse is applied at half the frequency difference between the $M = \frac{1}{2}$ and

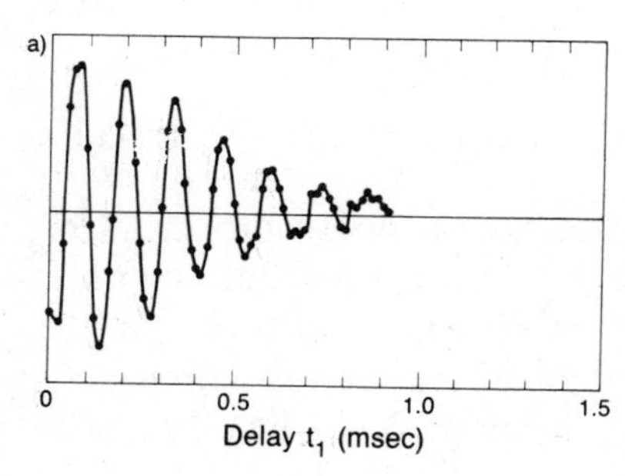

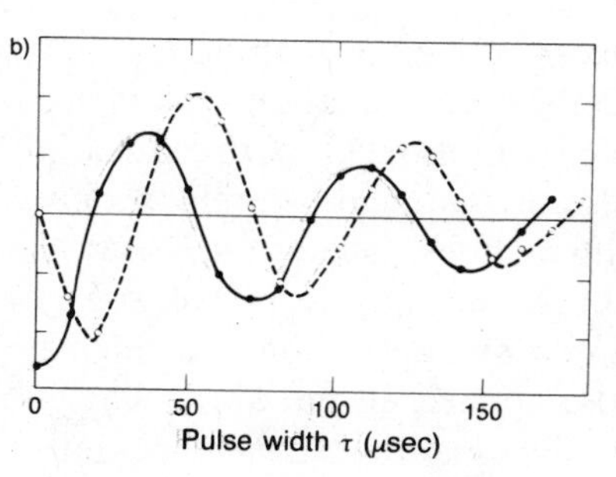

Figure 16. Preparation and evolution of double-quantum coherence in experiments employing a single weak excitation pulse. (*a*) Double-quantum free induction decay exhibited by the carboxyl deuterons in a crystal of perdeuterated oxalic acid dihydrate. A double-quantum $\pi/2$ pulse was applied as in Fig. 15*b*, and the response monitored by observing the single-quantum signal following a second pulse at a later time t_1. The 2 kHz offset from resonance is perceived as 4 kHz by the double-quantum coherence. (*b*) Transient nutation in a three-level subsystem of ^{27}Al during the application of a double-quantum pulse of duration τ. The open and filled circles show the harmonic oscillation of the double-quantum coherence and the population difference, respectively. (Reproduced from Refs. 26 and 78 with permission (*a*) © 1976 American Physical Society; (*b*) from *J. Phys. Soc. Jpn.*, **39** (1975) 1139, *Fig.* 2.)

$M = -\frac{3}{2}$ levels, after which the state of the system is immediately monitored by a single-quantum π pulse. The mixing pulse inverts the population across r and s or s and t of Fig. 15*b*, thereby creating observable single-quantum coherence. Repetition of the experiment for different preparation pulse lengths reveals a double-quantum transient nutation similar to that predicted in Eq. (128) to be visualized. Double-quantum analogues of familiar NMR phenomena such as spin echoes, rotary echoes, spin locking, rotary saturation, and population inversion have been observed both in this system and in ^{23}Na ($I = \frac{3}{2}$) after preparation by a selective pulse.[26,80-82]

The concept of a rotation in a double-quantum space finds further application in the decoupling of the heteronuclear dipolar interaction between hydrogen and deuterium.[83-87] Conventional decoupling, for which the radiofrequency amplitude must greatly exceed the quadrupole coupling constant, is difficult to achieve here because ω_Q for ^2H is usually on the order of 100 kHz. But if the free induction decay of the hydrogen, or S, magnetization is observed in the presence of a weak continuous-wave decoupling field at the deuterium, or I, Larmor frequency, then the Hamiltonian in a frame rotating simultaneously at ω_{0I} and ω_{0S},

$$H = \omega_1 I_x + \frac{\omega_Q}{3}[3I_z^2 - I(I+1)] - 2D_{IS}I_zS_z, \tag{129}$$

reduces to

$$H = (\omega_1^2/\omega_Q)I_x^{1-3} - 4D_{IS}I_z^{1-3}S_z + \tfrac{2}{3}\omega_Q(I_z^{1-2} - I_z^{2-3}) \tag{130}$$

in the limit that $\omega_1 \ll \omega_Q$. Both double-quantum operators commute with the last term above; the irradiation thus causes a continuous double-quantum nutation at an effective angular velocity of

$$\omega_e = \sqrt{(2D_{IS})^2 + (\omega_1^2/\omega_Q)^2}. \tag{131}$$

Rotated about the x axis in the double-quantum space, the dipolar interaction to lowest order averages to zero when ω_e exceeds D_{IS}, a condition far weaker than the conventional requirement that ω_1 cover the full width of the quadrupolar-broadened I spectrum.

The same result emerges from a second-order perturbation treatment, which shows that the probability for the double-quantum transition, $W(1 \to -1)$, is approximately equal to ω_1^4/ω_Q^2. This transition rate must at least be comparable to the dipolar coupling for the averaging to be effective. With the quadrupole interaction bypassed under double-quantum rotation, however, now it is only necessary to irradiate with sufficient power to ensure that $W(1 \rightarrow -1) \sim D_{IS}^2$. Hence a radiofrequency amplitude

$$\omega_1 \sim \sqrt{\omega_Q D_{IS}}$$

rather than $\omega_1 \sim \omega_Q$ suffices.[83–85]

A deuterium-decoupled chemical shift anisotropy powder pattern of the residual ^1H nuclei in deuterated ice, shown in Fig. 17, illustrates the ability of the technique to improve the resolution of spectra obtained from dilute ^1H systems in the solid state.

2. *Coherence Transfer by Multistep Processes*

Coherence across one single-quantum transition in a multilevel system results from the application of a frequency-selective pulse of the appropriate duration. If, for example, a spin-1 is irradiated weakly at one of the satellite frequencies ($\Delta\omega = \omega_Q$), the Hamiltonian of Eq. (125) reduces to

$$H \approx \sqrt{2}\omega_1 I_x^{1-2} - \tfrac{4}{3}\omega_Q(I_z^{1-3} + I_z^{2-3}). \tag{132}$$

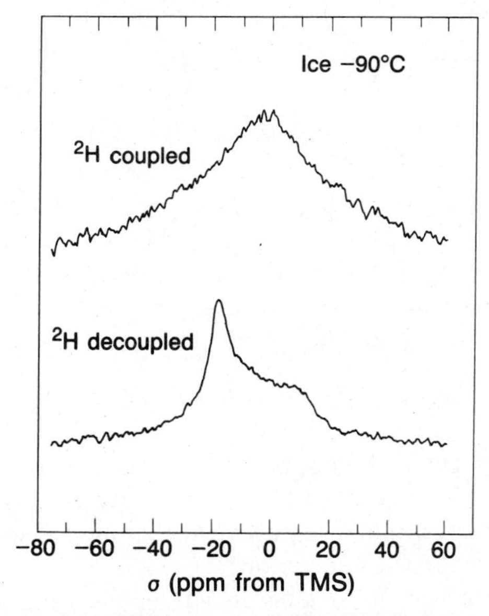

Figure 17. Fourier transform NMR spectra of the residual protons in 99% deuterated ice. The spectrum at the top, significantly broadened by the hydrogen–deuterium direct dipole–dipole interaction, exhibits no fine structure. The deuterium-decoupled spectrum below, acquired in the presence of double-quantum irradiation at the deuterium Larmor frequency, shows the characteristic features of an anisotropic chemical shift powder pattern. (Reproduced from Ref. 85 with permission. © 1976 American Physical Society.)

Noting that

$$[I_x^{1-2}, (I_z^{1-3} + I_z^{2-3})] = 0$$

and

$$\rho(0) = 2I_z^{1-3} = I_z^{1-2} + (I_z^{1-3} + I_z^{2-3}),$$

we see that under this Hamiltonian the density operator evolves strictly in the three-dimensional single-quantum subspace of $(I_x^{1-2}, I_y^{1-2}, I_z^{1-2})$, nutating about I_x^{1-2} at a frequency $\sqrt{2}\omega_1$.[51-53] In general the effective single-quantum nutation frequencies of a spin with angular momentum **I** are scaled by the matrix element $2\langle M_r|I_x|M_s\rangle$.

Coherence already existing between levels r and s can be transferred to r and t by a selective pulse applied at the frequency ω_{st}, as in Fig. 15*c*.[25] The pulse superposes states s and t, bringing r and t into coherent superposition at the same time. The process is viewed formally as a rotation of the component I_x^{r-s} about I_x^{s-t} in a subspace $(I_x^{r-s}, I_x^{s-t}, I_y^{r-t})$, defined by

$$[I_x^{r-s}, I_x^{s-t}] = \tfrac{1}{2}iI_y^{r-t}, \tag{86a}$$

and in which a rotation by θ yields the combination

$$\rho = \exp(-i\theta I_x^{s-t})I_x^{r-s}\exp(i\theta I_x^{s-t}) = I_x^{r-s}\cos(\theta/2) + I_y^{r-t}\sin(\theta/2). \tag{133}$$

A complete transfer to r–t coherence is effected by a selective pulse with a flip angle of π.

If undamped, the two components in Eq. (133) oscillate with period $\theta = 4\pi$, displaying rotational symmetry characteristic of a particle with half-integral spin.[88-94] Spinor behavior of this sort is observed for any pair of levels in a multilevel system, regardless both of the difference in Zeeman quantum numbers and of the total spin angular momentum, provided that the transition is excited selectively, as is suggested in Fig. 18. Acting as a spin-$\frac{1}{2}$ object, an individual coherence even in a boson system such as ^{2}H exhibits the rotational properties of a fermion, changing sign under rotation by 2π while remaining invariant under rotation by 4π.

With the sign of the state vector for a fictitious spin-$\frac{1}{2}$ system thus observable, it is possible to design *interferometric* experiments that would otherwise be impossible: left undisturbed, the quantum mechanical phase of one state in a multilevel system serves as a reference for the measurement of the relative phase between it and one of the other states, thereby allowing a transition to be detected indirectly. The two-step (r–s, s–t) coherence transfer

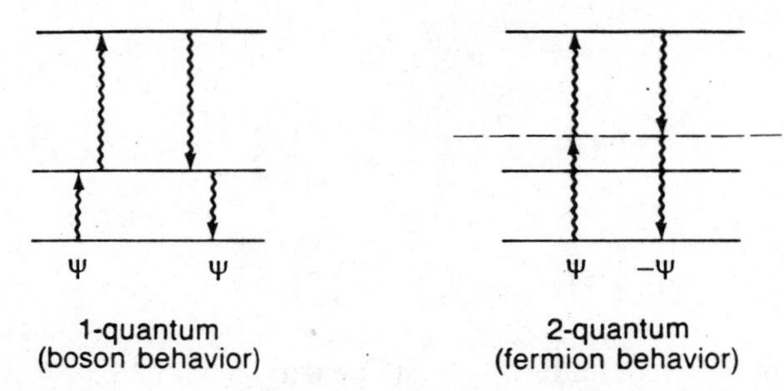

Figure 18. Transformation characteristics of the state vectors for three-level and fictitious spin-$\frac{1}{2}$ systems. Simultaneous irradiation of all transition frequencies in a spin-1 system (symbolized by the single-quantum coherences in the diagram at the left) allows the particle to behave as a boson, with its state vector invariant to a rotation of 2π. The isolation of one pair of levels as a fictitious spin-$\frac{1}{2}$, however, creates a pseudo-fermion, with a state vector that changes sign under a rotation of 2π as the system goes from the ground state to an excited state and back again.

offers a specific example, in which transitions induced by the selective s–t pulse are observed through their effect on coherence already established between $|r\rangle$ and $|s\rangle$. This principle has been used to advantage in the measurement of single-quantum satellites in very broad quadrupolar spectra and in studies of multiple-quantum relaxation.[95–98]

The expected oscillatory exchange of single- and double-quantum coherence following selective $\pi/2$ and π pulses is illustrated experimentally in Fig. 19, which contains more data for the pseudo-three-level system in $^{27}Al_2O_3$.[25] Each filled circle represents the initial amplitude of the free induction signal observed at ω_{rs} after a pulse of length τ is applied at ω_{st}.

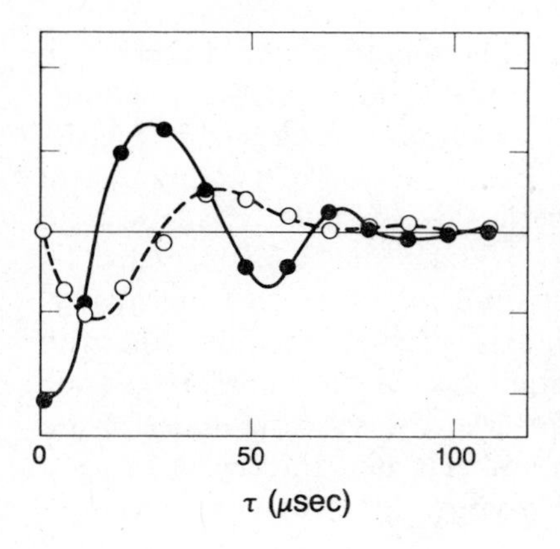

Figure 19. Harmonic exchange of single-quantum coherence (filled circles) and double-quantum coherence (open circles) in $^{27}Al_2O_3$, observed according to the method of Fig. 15*c*. Signals are plotted in arbitrary units as a function of the duration of the coherence transfer pulse at the frequency ω_{st}. The period of the oscillations 4π shows that the two-level systems r–s and r–t transform as spinors. (Reproduced from *J. Phys. Soc. Jpn.*, **39** (1975) 835, Fig. 4.)

Reflecting the residual I_x^{r-s} component, this signal varies cosinusoidally with the duration of the pulse. Lagging in phase by $\pi/2$ is the sinusoidally oscillating double-quantum coherence I_y^{r-t}, which is monitored at ω_{st} after a selective π pulse between levels r and s. These signals, plotted as open circles in Fig. 19, are not observed following a pulse applied directly at the dipole-forbidden r–t frequency.

Many other examples of multistep coherent phenomena have been reported.[99–107] Among these are experiments designed to create coherence exclusively between two levels by means of a sequence of nonselective hard pulses,[100,105,106] to convert single-quantum coherence to triple-quantum coherence,[107] and to probe systems of coupled spins.[99,101,104] Moreover, multistep coherence transfer processes analogous to those described here for NMR occur at optical frequencies as well. For example, in four-wave-mixing spectroscopy, coherences among the rovibronic energy levels of a molecular crystal are created and probed with laser light applied at one, two, or three distinct frequencies. One interesting class of experiments involves the study of four such levels, with $|r\rangle$ and $|s\rangle$ taken to be the ground state and a vibrationally excited state in the S_0 electronic manifold of a dilute guest molecule, and $|t\rangle$ and $|u\rangle$ taken to be the corresponding states in the electronically excited S_1 manifold. It is possible to prepare vibrational coherence, say $|r\rangle\langle s|$, in the four-level system by simultaneously exciting the material with two pulses of light with a frequency difference equal to ω_{rs}. This coherence, the result of anti-Stokes–Raman scattering, can be transferred to r–u by means of a pulse applied at ω_{su}, consistent with the picture of fictitious spin-$\frac{1}{2}$ behavior given above. Introduction of a variable delay between the "pump" and the "probe" pulses allows the relaxation of the vibrational coherence to be studied.[77]

3. *Multistep Coherence Transfer with Amplitude-Modulated Pulses*

Multistep processes involving single-quantum coherence transfer pulses may be employed to connect selected levels in systems of any size, but the number of steps necessary increases with the order of coherence to be observed. What might be a cumbersome procedure is simplified considerably, however, if more than one frequency is irradiated simultaneously during the coherence transfer phase.[108,109] One way of realizing this objective experimentally, pictured schematically in Fig. 15*d*, is to use an amplitude-modulated transfer pulse to generate sidebands at the desired frequencies. Modulation methods of this sort have been employed successfully to excite triple-quantum coherence in the spin-$\frac{3}{2}$ ^{23}Na system: after excitation of the central $(-\frac{1}{2}, \frac{1}{2})$ transition, a pulse modulated at a frequency matched to the satellite spacing stirs the $(-\frac{3}{2}, -\frac{1}{2})$ and $(\frac{1}{2}, \frac{3}{2})$ transitions simultaneously but independently to effect the transfer into triple-quantum coherence.[109]

C. Broadband Excitation

Frequency-selective excitation schemes become difficult or impossible to to implement in a variety of circumstances, particularly when the energy level structure is either unknown or too complex to permit efficient step-by-step coherence transfer. Narrowband excitation may also prove to be inconvenient when the levels are spaced so closely that it is impractical to treat any pair in isolation. These conditions commonly exist among nuclei coupled by dipolar or scalar interactions and also among quadrupolar nuclei, where a spread of resonant frequencies might arise from an anisotropic distribution of molecular orientations. Such systems are usually treated experimentally with an approach based upon the idea of simultaneous multifrequency irradiation that incorporates two or more intense radio-frequency pulses of short duration, each with a Fourier spectrum sufficient to cover the frequency range demanded by the internal Hamiltonian. These broadband excitation methods are built around the natural bilinear internal couplings, frequently manipulated to create a nonsecular effective Hamiltonian during the preparation period. The sudden switching to a nonsecular Hamiltonian renders the Zeeman states nonstationary, allowing the system to develop coherence over a time scale comparable to the inverse of the coupling.

1. Nonselective Pulse Methods

The simple pulse sequence shown in Fig. 20*a*, historically the first example of nonselective pulsed multiple-quantum excitation,[24] is capable of creating coherences of orders zero through N in a system of N coupled spins-$\frac{1}{2}$.[79,110] To understand the workings of this very basic experiment, once again we first consider a pair of weakly coupled spin-$\frac{1}{2}$ nuclei with chemical shifts Ω_j and Ω_k and coupling constant J_{jk}, deferring until Section IV.E an analysis of multiple-quantum excitation in strongly coupled systems. For simplicity we use the sequence

$$(\pi/2)_x\text{–}\tau\text{–}(\pi/2)_{-x}\text{–}t_1\text{–}(\pi/2)_x\text{–}t_2,$$

retaining the option to alter both flip angles and phases in later variations.

Zero-, single-, and double-quantum coherences develop during the preparation period, which consists of an interval of free evolution sandwiched between two $\pi/2$ pulses. During this time the equilibrium density operator,

$$\rho(0^-) = I_{zj} + I_{zk}, \tag{134}$$

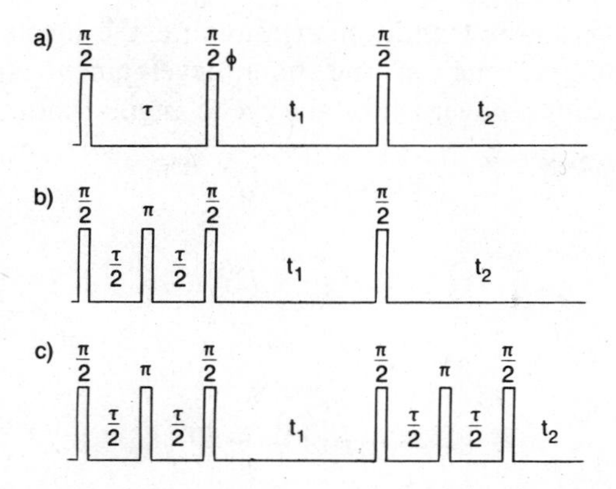

Figure 20. Nonselective pulse sequences for multiple-quantum NMR. (*a*) Two intense $\pi/2$ pulses, separated by an interval τ, are used to prepare the multiple-quantum coherence. The free period must be approximately equal to the inverse of the spin–spin coupling for the excitation to be effective. Free evolution during t_1 is terminated by a $\pi/2$ mixing pulse, after which a conventional single-quantum signal is detected. Provision is made for a phase difference ϕ between the two pulses of the preparation period, or between these pulses and the mixing pulse, or both. (*b*) As (*a*), but with a π echo pulse to eliminate resonance offset and chemical shift interactions during the preparation period. A phase difference of 0 or π between the first two $\pi/2$ pulses leads to selective excitation of even-order coherences. (*c*) Similar, but with a symmetric mixing period of duration τ to match the preparation period. The last pulse, drawn to highlight the symmetry, creates a component proportional to I_z. In practice this pulse is usually omitted, and I_x or I_y is observed instead.

is transformed first by the initial pulse to

$$\rho(0^+) = -(I_{yj} + I_{yk}) \tag{135}$$

and then to a sum of eight terms,

$$\begin{aligned}\rho(\tau^-) = & -\cos(\Omega_j\tau)[\cos(J_{jk}\tau/2)I_{yj} + \sin(J_{jk}\tau/2)2I_{xj}I_{zk}] \\ & -\cos(\Omega_k\tau)[\cos(J_{jk}\tau/2)I_{yk} + \sin(J_{jk}\tau/2)2I_{xk}I_{zj}] \\ & +\sin(\Omega_j\tau)[\cos(J_{jk}\tau/2)I_{xj} - \sin(J_{jk}\tau/2)2I_{yj}I_{zk}] \\ & +\sin(\Omega_k\tau)[\cos(J_{jk}\tau/2)I_{xk} - \sin(J_{jk}\tau/2)2I_{yk}I_{zj}]\end{aligned} \tag{136}$$

after the free period τ under the Hamiltonian

$$H = \Omega_j I_{zj} + \Omega_k I_{zk} - J_{jk} I_{zj} I_{zk}. \tag{137}$$

Since all operators in the Hamiltonian commute, the effects of the chemical shift and coupling terms on the time development appear as three independent rotations, governed by the cyclic angular momentum relations

$$[I_{xj}, I_{yj}] = iI_{zj},$$

$$[I_{yj}, 2I_{zj}I_{zk}] = i2I_{xj}I_{zk},$$

and

$$[I_{xj}, 2I_{zj}I_{zk}] = -i2I_{yj}I_{zk}.$$

The second pulse leaves the system described by

$$\begin{aligned}\rho(\tau^+) = {} & \cos(\Omega_j\tau)\cos(J_{jk}\tau/2)I_{zj} + \cos(\Omega_k\tau)\cos(J_{jk}\tau/2)I_{zk} \\ & + \sin(\Omega_j\tau)\cos(J_{jk}\tau/2)I_{xj} + \sin(\Omega_k\tau)\cos(J_{jk}\tau/2)I_{xk} \\ & - \sin(\Omega_j\tau)\sin(J_{jk}\tau/2)2I_{zj}I_{yk} - \sin(\Omega_k\tau)\sin(J_{jk}\tau/2)2I_{zk}I_{yj} \\ & - \cos(\Omega_j\tau)\sin(J_{jk}\tau/2)2I_{xj}I_{yk} - \cos(\Omega_k\tau)\sin(J_{jk}\tau/2)2I_{xk}I_{yj}, \quad (138)\end{aligned}$$

a combination of single-quantum transverse magnetization terms proportional to I_{xj}, population terms proportional to I_{zj}, antiphase single-quantum terms proportional to $I_{zj}I_{yj}$, and zero- and double-quantum terms proportional to $I_{xj}I_{yj}$. These modes oscillate during the evolution period t_1 according to the principles discussed in Sections III.C and III.D, after which the third pulse converts some of them into observable transverse magnetization.

2. *Offset-Independent Excitation*

Although the hard pulses used in the basic experiment will excite to some extent any spin resonating at a frequency within their spectral bandwidth, it is clear from Eq. (138) that the coherence amplitudes are determined by the magnitudes of both the chemical shifts and the scalar couplings. No single value of τ can simultaneously maximize the trigonometric functions $\cos(\Omega_j\tau)$, $\cos(\Omega_k\tau)$, and $\sin(J_{jk}\tau/2)$, but a simple modification of the pulse sequence, shown in Fig. 20*b*, ameliorates the problem by eliminating the influence of the chemical shift and offset interactions upon the excitation dynamics: insertion of a π pulse midway into the preparation period as shown refocuses the linear terms in the Hamiltonian to generate a propagator

$$U(\tau) = \exp[i(\pi/2)I_x] \exp(iI_{zj}I_{zk}J_{jk}\tau) \exp[-i(\pi/2)I_x]$$
$$= \exp(iI_{yj}I_{yk}J_{jk}\tau) \tag{139}$$

under which the state

$$\rho(\tau) = \cos(J_{jk}\tau/2)(I_{zj} + I_{zk}) + \sin(J_{jk}\tau/2)(2I_{xj}I_{yk} + 2I_{yj}I_{xk}) \tag{140}$$

develops independently of all chemical shifts and offsets.[79,111] The term proportional to $\sin(J_{jk}\tau/2)$ is the pure double-quantum coherence operator $\rho_y^{[2]}$, which oscillates at the frequency $(\Omega_j + \Omega_k)$ during the subsequent evolution period. A complete transfer of coherence for the spin pair is effected when τ is chosen to equal π/J_{jk}. Any diminution of the coherence magnitude owing to transverse relaxation can be measured in a series of double-quantum spin-echo experiments, where a refocusing pulse inserted into the evolution period removes the oscillation due to the chemical shift and resonance offset interactions.[26,59,65]

Just before the mixing pulse, the double-quantum component now has evolved to the point

$$\rho(\tau, t_1) = 2\sin(J_{jk}\tau/2)[\rho_y^{[2]} \cos(\Omega_j + \Omega_k)t_1 - \rho_x^{[2]} \sin(\Omega_j + \Omega_k)t_1]$$
$$= \sin(J_{jk}\tau/2)[\cos[(\Omega_j + \Omega_k)t_1](2I_{xj}I_{yk} + 2I_{yj}I_{xk})$$
$$- \sin[(\Omega_j + \Omega_k)t_1](2I_{xj}I_{xk} - 2I_{yj}I_{yk})], \tag{141}$$

whereupon it is transformed by the final pulse to

$$\rho(\tau, t_1, \tau') = \sin(J_{jk}\tau/2)[\cos[(\Omega_j + \Omega_k)t_1](2I_{xj}I_{zk} + 2I_{zj}I_{xk})$$
$$- \sin[(\Omega_j + \Omega_k)t_1](2I_{xj}I_{xk} - 2I_{zj}I_{zk}]. \tag{142}$$

Only the I_xI_z terms, proportional to $\sin(J_{jk}\tau/2)\cos(\Omega_j + \Omega_k)t_1$, are able to develop into observable magnetization I_y under a bilinear Hamiltonian during the detection period. Accordingly, the signal detected as a function of t_1 is a pair of "up, down" antiphase lines, modulated at the sum frequency of the two chemical shifts. The complete sequence of events from preparation through detection under this sequence is summarized in the diagram shown in Fig. 21, which is representative of a general method for the analysis of coherence transfer pathways.[112,113]

Extension of the foregoing analysis to larger systems is straightforward, though tedious algebraically, with the principal physical consequences of expanding the network of weakly coupled spins deriving from the properties

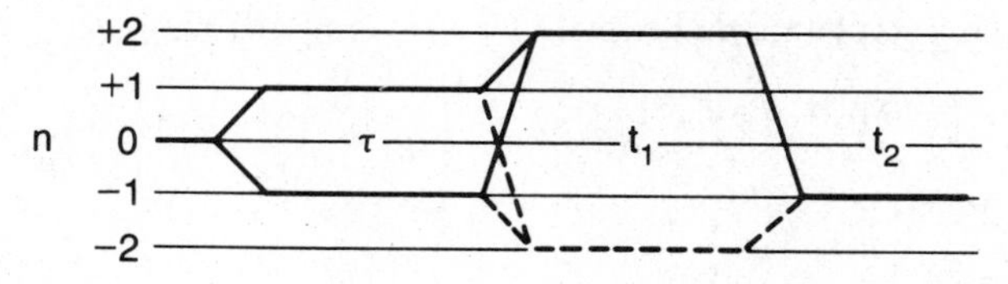

Figure 21. Coherence transfer pathways for the double-quantum experiment of Fig. 20*b*. The sequence begins with the density operator proportional to a zero-quantum mode I_z corresponding to the population difference at equilibrium. Single-spin coherence of order $n = 1$ is created by the first $\pi/2$ pulse, and then develops into two-spin single-quantum modes over the interval τ. The second $\pi/2$ pulse converts these into double-quantum coherence, $n = 2$, which persists during the free period t_1. The mixing pulse regenerates $n = 1$ components for detection during t_2. (Reproduced from Ref. 113 with permission. © 1984 Academic Press, Inc.)

of coherence discussed in Section III.D. A new spin begins to influence the excitation dynamics as soon as its coupling to a central spin from which the coherence originates becomes nonnegligible relative to the inverse of the preparation time. Having been prepared as a coherent group, the spins actively coupled during the preparation phase then interact with the spins outside, splitting the multiple-quantum resonance into multiplets according to Eq. (106).

3. *Radiofrequency Pulses and Coherence Transfer Selection Rules*

The ultimate development of the spin system during a multiple-quantum experiment is directed largely by the radiofrequency pulses, which, owing to their ability to modify the free Hamiltonian and to rotate density operator components almost instantaneously, are instrumental in transferring coherence between different modes. Consequently a change in the flip angle or phase of even one pulse in a sequence often leads to an entirely different outcome.

Consider for instance the effect of introducing a relative phase difference ϕ between the first two $\pi/2$ pulses in the excitation scheme of Fig. 20*b*. With this modification, the propagator

$$U(\tau) = \exp[i(\pi/2)I_x] \exp(-iH_{zz}\tau) \exp[-i(\pi/2)I_\phi] \tag{143}$$

differs from the form expressed in Eq. (139) only by the rotation of one of the pulses,

$$exp[-i(\pi/2)I_\phi] = \exp(-i\phi I_z) \exp[-i(\pi/2)I_x] \exp(i\phi I_z), \tag{144}$$

by ϕ about the z axis. Nevertheless, application of the identity operator

$$\mathbf{1} = \exp[-i(\pi/2)I_x]\exp[i(\pi/2)I_x]$$

to the left of the exponentials above shows that the effective propagator is now

$$U(\tau) = \exp(-iH_{yy}\tau)\exp(-i\phi I_y)\exp(i\phi I_z), \tag{145}$$

under which the density operator, initially I_z, develops as

$$\rho(\tau) = (\cos\phi)\exp(-iH_{yy}\tau)I_z\exp(iH_{yy}\tau) + (\sin\phi)\exp(-iH_{yy}\tau)I_x\exp(iH_{yy}\tau). \tag{146}$$

The phase shift effectively rotates the initial condition from thermal equilibrium to

$$\rho(0) = I_z\cos\phi + I_x\sin\phi, \tag{147}$$

through which is opened up a completely new channel for the formation of *odd-order* coherences. [79] These develop from the "initial" I_x term according to the cyclic relations

$$[I_{xj}, I_{yj}I_{yk}] = iI_{zj}I_{yk}, \qquad \text{etc.},$$

while the even-order coherences continue independently to develop from the I_z term according to the relations

$$[I_{yj}I_{yk}, I_{zj}] = iI_{xj}I_{yk}, \qquad \text{etc.}$$

When the phases of the pulses either are identical or differ by π, as in the treatment in Section IV.C.2, the excitation is purely even selective. Odd-order selectivity is achieved when one of the $\pi/2$ pulses differs in phase from the other $\pi/2$ pulse and from the refocusing pulse by $\pi/2$. In addition to these routes, other pathways, perhaps through coherences only of a given sign[113] or order (see Section IV.F), become available through the dynamic control afforded by radiofrequency phase shifting.

In general the rotation of the density operator by a pulse with flip angle θ and phase ϕ is given by the unitary transformation

$$\rho(t^+) = U(\theta, \phi)\rho(t^-)U^{-1}(\theta, \phi), \tag{148}$$

where

$$U(\theta, \phi) = \prod_j \exp[-i\theta(I_{xj}\cos\phi + I_{yj}\sin\phi)] \equiv \prod_j \exp(-i\theta I_{\phi j}). \quad (149)$$

We have already seen several examples illustrating the action of the single-spin operators in coherence transfer processes between single-quantum and multiple-quantum modes. Now expansion of the rotated density operator in the Zeeman eigenbasis to yield

$$\rho(t^+) = \sum_{rstu} U_{tr}\rho_{rs}(t^-)U_{su}^{-1}|M_t\rangle\langle M_u| \quad (150)$$

highlights the spectroscopic consequences accompanying, for example, the redistribution of coherence induced by a single pulse such as a final mixing pulse. Every single-quantum eigenfrequency observed during the detection period is associated with a component $\rho_{tu}(t^+)$ generated from a set of components $\rho_{rs}(t^-)$ oscillating at the multiple-quantum eigenfrequencies $\omega_{rs}^{(1)}$ during the evolution period. The transfer of multiple-quantum coherence to a particular mode of single-quantum coherence proceeds as

$$\rho_{tu}(t^+) = \sum_{rs} Z_{rstu}\rho_{rs}(t^-), \quad (151)$$

with the amplitudes and phases of the complex transfer coefficients,

$$Z_{rstu} = U_{tr}U_{su}^{-1}, \quad (152)$$

determining the pathway.[114]

Evaluation of these matrix elements in the direct product basis $[|M_r\rangle = |j\rangle|k\rangle|l\rangle|m\rangle\cdots]$ provides the "selection rules" for one-pulse coherence transfer in weakly coupled systems. With the substitution $\alpha = \theta/2$ and with I_ϕ as defined in Eq. (149), the exponential operator for the pulse is first expanded as

$$U(\theta, \phi) = \exp[-i\alpha(2I_\phi)] = \prod_j [1 - i\alpha(2I_{\phi j}) - \tfrac{1}{2}\alpha^2(2I_{\phi j})^2 - \cdots] \quad (153)$$

and then simplified to

$$U(\theta, \phi) = \prod_j [\cos(\theta/2)\mathbf{1} - i\sin(\theta/2)(2I_{\phi j})] \quad (154)$$

after repeated application of the identity $I_{\phi j}^2 = \mathbf{1}/4$. The term proportional to the unit operator in each factor is diagonal in the direct product basis, connecting only those states where every individual Zeeman quantum number is unchanged. The term proportional to $I_{\phi j}$ connects only those states where m_j changes by ± 1. We therefore have three classes of matrix elements,

$$\langle \tfrac{1}{2}|\mathbf{1}|\tfrac{1}{2}\rangle = 1$$

$$\langle \tfrac{1}{2}|2I_{\phi j}|-\tfrac{1}{2}\rangle = \exp(-i\phi)$$

$$\langle -\tfrac{1}{2}|2I_{\phi j}|\tfrac{1}{2}\rangle = \exp(i\phi),$$

contributing to a total product

$$U_{tr} = [-i\sin(\theta/2)]^{\Delta_{tr}}[\cos(\theta/2)]^{N-\Delta_{tr}}\exp[-i\phi(M_t - M_r)] \qquad (155)$$

in which Δ_{tr} denotes the number of spins that are inverted in going from $|M_t\rangle$ to $|M_r\rangle$.[115,116] The full coherence transfer coefficient,

$$Z_{rstu} = U_{tr}^{-1}U_{su} = i^{\Delta_{us}-\Delta_{tr}}[\sin(\theta/2)]^{\Delta_{tr}+\Delta_{us}}[\cos(\theta/2)]^{2N-\Delta_{tr}-\Delta_{us}}$$
$$\times \exp[-i\phi(M_t - M_r - M_s + M_u)], \qquad (156)$$

is obtained by noting that $U_{su}^{-1} = U_{us}^{*}$ for the unitary operator $U(\theta, \phi)$.

The observed phase of the coherence is determined by the phase of the complex coefficient, the amplitude by its absolute value. Consequently the flip angle of the pulse contributes only to the coherence amplitude through the trigonometric factors $\sin(\theta/2)$ and $\cos(\theta/2)$, while the radio-frequency phase enters into both the coherence amplitude and phase through the complex exponential. An additional phase shift of 0, $\pi/2$, π, or $3\pi/2$ is also induced during the transfer of coherence, depending on the value of $\Delta_{us} - \Delta_{tr}$.

To illustrate the mechanics of coherence transfer we consider the two weakly coupled six-spin states,

$$|M_r = 2\rangle = |+ + + + + -\rangle$$

and

$$|M_s = -3\rangle = |- - - - - -\rangle, \qquad + = \tfrac{1}{2}, \qquad - = -\tfrac{1}{2},$$

introduced as examples in Section III.A, and attempt to transfer the five-quantum coherence to an allowed single-quantum coherence between two other states, say,

$$|M_t = 0\rangle = |+ - + - + -\rangle$$

and

$$|M_u = 1\rangle = |+ - + - + +\rangle.$$

Here $\Delta_{us} = 4$ since the two states differ in the polarization of the first, third, fifth, and sixth spins; correspondingly, $\Delta_{tr} = 2$ for the other pair. The overall phase factor $i^{\Delta_{us}-\Delta_{tr}}$ is -1, indicating a shift by π relative to the pulse phase of ϕ. Therefore if an x pulse ($\phi = 0$) effects the transfer, then the transverse magnetization will appear along the $-x$ axis in the rotating frame.

The general result is that $\Delta_{us} - \Delta_{tr}$ is even during the conversion of odd-order coherence and odd during the conversion of even-order coherence. Hence single-quantum modes originating from odd orders appear either parallel or antiparallel to the direction of the mixing pulse, but those originating from even orders appear perpendicular to it.

Although the coherence transfer amplitude is nonzero for any flip angle greater than zero and less than π, degenerate single-quantum components with intensities equal in magnitude but opposite in sign will nevertheless interfere destructively during their detection. For cancellation to occur, however, there must be two pathways, $\rho_{rs} \rightarrow \rho_{tu}$ and $\rho_{rs} \rightarrow \rho_{t'u'}$, with

$$\omega_{tu} = \omega_{t'u'}$$

and

$$Z_{rstu} = -Z_{rst'u'}.$$

The latter circumstances exist when

$$\Delta_{us} - \Delta_{tr} = \Delta_{u's} - \Delta_{t'r} \pm 2(1 + 2k), \qquad k = 0, 1, 2, \ldots, \tag{157}$$

and if θ is not equal to $\pi/2$, when

$$\Delta_{us} + \Delta_{tr} = \Delta_{u's} + \Delta_{t'r}. \tag{158}$$

For a specific illustration we return to our model system of six spins and consider another pair of states, such as

$$|M_{t'} = -1\rangle = |- - + - + -\rangle$$

and

$$|M_{u'} = 0\rangle = |- - + - + +\rangle,$$

which differs from the pair $|M_t\rangle$ and $|M_u\rangle$ only in the polarization of the first spin, a spin actively involved in the five-quantum coherence between $|M_r\rangle$ and $|M_s\rangle$. Both superpositions, ρ_{tu} and $\rho_{t'u'}$, describe coherences belonging to the same single-quantum multiplet, wherein the sixth spin responds to the *static* local field produced by the other five. Now the transfer of coherence from ρ_{rs} to $\rho_{t'u'}$ proceeds with $\Delta_{u's} = 3$ and $\Delta_{t'r} = 3$, satisfying the two preceding conditions for $k = 0$. Accordingly, cancellation of the corresponding spectral lines in $\omega^{(2)}$ is expected if the two single-quantum frequencies are identical, owing perhaps to molecular symmetry or simply to a failure to resolve the individual components. In general, for a multiple-quantum coherence involving a group of q spins to be transferred successfully to the single-quantum coherences of one particular spin A, the weak couplings between spin A and all q spins must be resolved. Spin A may exist either outside the group of q spins or inside it, in which case there are $q - 1$ relevant couplings. If two single-quantum coherences differ just in the polarization of a *passive* spin, then $\Delta_{us} - \Delta_{tr}$ and $\Delta_{u's} - \Delta_{t'r}$ are always equal, thereby precluding any possibility of destructive interference between the signals. When the difference is limited to the polarization of an active spin, however, the signals *will* interfere if the coupling between A and the spin in question is not resolved. Moreover, since isotropic J couplings between nuclei with the same chemical shifts are automatically strong, and therefore ineffective dynamically owing to the relationship

$$[\sum_{j<k} J_{jk}\mathbf{I}_j\cdot \mathbf{I}_k, I_\alpha] = 0, \qquad \alpha = x, y, z \qquad I_\alpha = \text{total spin}, \tag{159}$$

it is impossible to transfer multiple-quantum coherence involving two or more equivalent spins to the single-quantum coherences of any of the spins within the group.[114]

4. *Uniform Excitation over a Range of Coupling Constants*

a. τ Averaging. Wherever there exists a distribution of coupling constants, the question of multiple-quantum excitation efficiency is complicated further by the dependence of each double-quantum term upon the factor $\sin(J_{jk}\tau/2)$. This situation frequently arises both in two-spin systems, where the coupling constant may vary depending upon the chemical environment or molecular orientation; and in multispin systems, where a coherence originating with an active spin ($j = 1$) and involving a total of n weakly coupled spins acquires the amplitude factor

$$f^n(\tau) = \prod_{j=2}^{n} \sin(J_{1j}\tau/2) \prod_{k=n+1}^{N} \cos(J_{1k}\tau/2) \tag{160}$$

after τ seconds of offset-independent excitation.[117] The original spin may be passive as well, in which case one of the sine factors is changed to a cosine. Either way, if the amplitude is an even function of τ, then the variation in excitation efficiency can be smoothed experimentally by averaging directly over τ. In this approach the length of the preparation period is incremented, randomly or systematically, and the resulting spectra are added together to approximate the effect of a uniform excitation over the entire range of coupling constants.[59] The method is equally applicable to strongly coupled systems, in which the coherence amplitude, given perhaps by an expression more complicated than that above, need not vanish when averaged over τ. In general, however, τ averaging requires the use of magnitude spectra,

$$S = \sqrt{S_x^2 + S_y^2},$$

for the phase of each frequency component usually depends on τ, and addition of phase-sensitive spectra typically brings with it a loss of signal owing to destructive interference between resonances with opposite phases.

Incorporation into the pulse sequence of a mixing period symmetrically disposed with respect to the preparation period greatly extends the utility of τ averaging, for weakly coupled systems in particular.[28,66,117] An example of such a scheme is depicted in Fig. 20*c*. Here, without the final $\pi/2$ pulse (which is shown only to emphasize the symmetry of the design), the sinusoidally modulated double-quantum operator for a spin pair, $(2I_{xj}I_{yk} + 2I_{yj}I_{xk})$ $\sin(J_{jk}\tau/2)$, develops further into

$$\rho = (2I_{xj}I_{zk} + 2I_{zj}I_{xk}) \sin(J_{jk}\tau/2) \cos(J_{jk}\tau/2) + (I_{yj} + I_{yk}) \sin^2(J_{jk}\tau/2) \quad (161)$$

during the mixing period, acquiring an additional factor of $\sin(J_{jk}\tau/2)$ and thus generating an *absorption-mode* observable term proportional to $\sin^2(J_{jk}\tau/2)$ that can be averaged over τ in a series of experiments. The method also works for larger weakly coupled systems, for which the matched preparation and mixing periods similarly produce a set of absorption-mode spectra. When N spins are involved, however, only those antiphase magnetization components that rephase on the same spin during the mixing period accumulate the amplitude factor $[f^n(\tau)]^2$ needed for a nonvanishing average.[117]

b. Uniform Double-Quantum Excitation with Composite Preparation Sequences. An entirely different approach to the problem of uniform excitation uses the concept of a composite pulse, an irradiation sequence formed by combining a number of pulses with various flip angles and phases in order to bring about some condition otherwise inaccessible to the spins.[118] Originally developed for single-quantum spectroscopy, these pulses are usually designed to function over a chosen range of resonance offsets, couplings, and radiofrequency field strengths. As an illustrative example we may consider the problem of the spatial variation of the radiofrequency amplitude ω_1 in a transmitting coil. This inhomogeneity frequently degrades the ability of a single π pulse to invert all the spins in a sample uniformly, for the actual flip angle of the pulse at any point within the coil depends upon the local nutation frequency, and equals π only in some relatively small region. Nevertheless, application of a string of *three* pulses $(\pi/2)_x$–π_y–$(\pi/2)_x$ to a system of uncoupled spin-$\frac{1}{2}$ nuclei brings about a nearly uniform population inversion over a far wider range of flip angles.[119] The effect of this composite π pulse is illustrated in Fig. 22 with a vector diagram showing magnetization trajectories for spins experiencing flip angles less than the nominal values associated with optimal ω_1. The vector picture is rigorously correct here, since the density operator describing an isolated spin-$\frac{1}{2}$ particle is specified completely by the three components I_x, I_y, and I_z. The diagram shows graphically how the three rotations, each individually less than $\pi/2$ or π, bring I_z very close to $-I_z$, with minor variation in the terminal points for the eight paths considered. Although this type of graphic interpretation rapidly becomes unwieldy when highly versatile pulses are demanded, the basic idea of employing composite rotations to shape the density operator remains the same, however complex the scheme may become.

When applied to a spin-1 system with bilinear internal couplings, the refocused preparation sequences of Fig. 20*b* and *c* confine the density operator to rotations within a three-dimensional double-quantum subspace

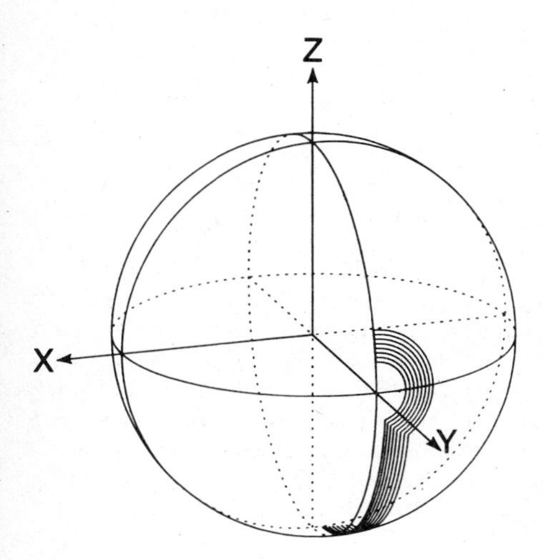

Figure 22. Effect on a nuclear magnetization vector of the composite pulse $(\pi/2)_x$–π_y–$(\pi/2)_x$, with flip angles between 0.8 and 0.9 of the values stated. Though the first rotation falls short of the y axis, much of the difference is made up by the second rotation of approximately π about the y axis. The third pulse, with its flip angle less than $\pi/2$, then terminates the sequence, bringing the vector nearly to the south pole. (Reproduced from Ref. 119 with permission. © 1979 Academic Press, Inc.)

$(I_x^{1-3}, I_y^{1-3}, I_z^{1-3})$. In such a fictitious spin-$\frac{1}{2}$ system, formed perhaps by a quadrupolar nucleus with $I = 1$ or a pair of spin-$\frac{1}{2}$ nuclei, the extent of the rotation induced by the combined effect of the pulses and the free periods depends upon the magnitude of the coupling. According to the formalism, pulses shifted from the x direction by ϕ create a propagator

$$U(\tau; \Theta, \phi) = \exp(-i\phi I_z)\exp(-iH_{xx}\tau)\exp(i\phi I_z) \tag{162}$$

that rotates the equilibrium density operator about an axis at azimuth 2ϕ through a double-quantum flip angle Θ, which is proportional to the product of the coupling constant with τ. In this picture the variation in double-quantum excitation efficiency arising from a distribution of coupling constants is analogous to the variation in single-quantum excitation efficiency arising from a distribution of flip angles. Consequently the same sequence of flip angles and phases $[(\Theta_1, \phi_1)(\Theta_2, \phi_2)\cdots(\Theta_m, \phi_m)]$ used to construct a composite pulse that compensates for radiofrequency inhomogeneity can also be used to construct an efficient double-quantum excitation scheme.[120]

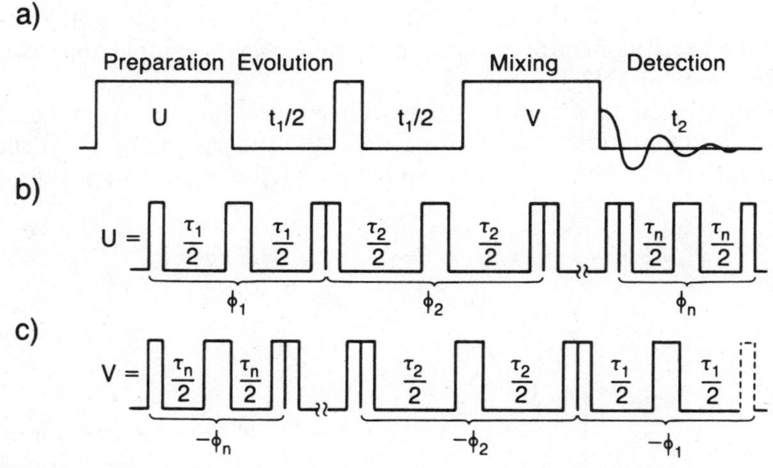

Figure 23. Composite sequences for uniform excitation of double-quantum coherence over a range of spin coupling constants. (*a*) General scheme of the two-dimensional multiple-quantum experiment, as in Fig. 12. The optional π pulse at $t_1/2$ is used to eliminate resonance offset and chemical shift interactions from the evolution period. (*b*) A composite preparation sequence built from three-pulse units of the kind shown in Fig. 20*b*. The overall radiofrequency phases ϕ_i and durations τ_i for the subcycles are derived from existing composite pulses designed to invert a magnetization vector uniformly despite variations in the flip angles due to radiofrequency inhomogeneity. (*c*) Matching mixing sequence, used to generate single-quantum signals with maximum amplitudes and uniform phases. (Reproduced from Ref. 120 with permission. © 1985 Academic Press, Inc.)

The design of a matched composite preparation and mixing sequence is illustrated in Fig. 23. In this experiment the excitation periods are formed from a series of elementary $\pi/2-\tau/2-\pi-\tau/2-\pi/2$ sequences, each with pulse phases and delays chosen to produce the overall propagator

$$U = U(\tau_1; \Theta_1, \phi_1/2)U(\tau_2; \Theta_2, \phi_2/2) \cdots U(\tau_m; \Theta_m, \phi_m/2). \qquad (163)$$

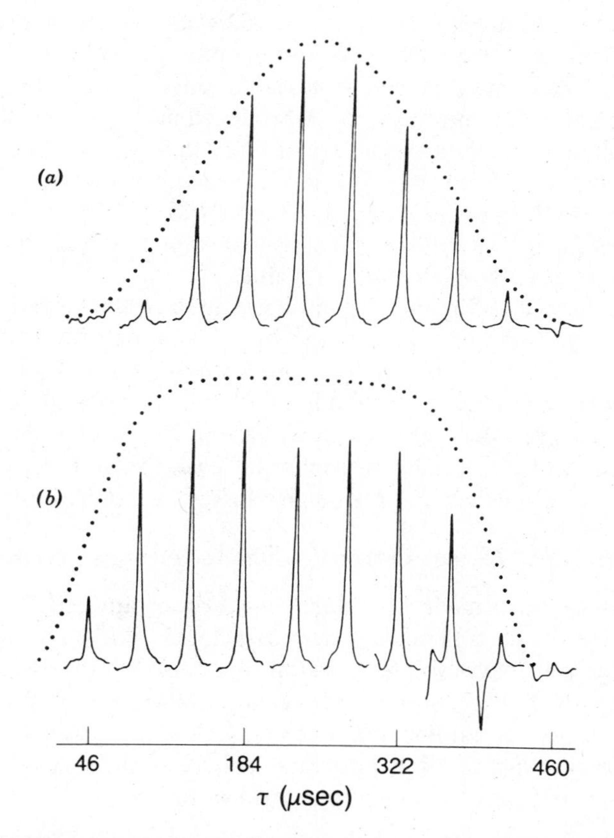

Figure 24. Double-quantum ^1H NMR resonances of CH_2Cl_2 oriented in a liquid crystal solvent. The signals are plotted as a function of the excitation time for experiments employing (*a*) simple preparation and simple mixing and (*b*) composite preparation and composite mixing. The dotted lines show the theoretical amplitudes, normalized to those obtained with simple preparation and simple mixing. In this system, where there is just one dipole coupling constant, the composite sequence promotes uniform excitation over a range of excitation times. In a system with more than one coupling constant, the same sequence promotes uniform excitation over a broad range of couplings for a fixed excitation time. (Reproduced from Ref. 120 with permission. © 1985 Academic Press Inc.)

The double-quantum flip angles are calibrated with reference to some nominal coupling constant about which the excitation profile is to be centered.

A direct comparison of the performance of simple and composite double-quantum excitation is presented in Fig. 24, in which the intensity of the double-quantum signal from the two strongly coupled ^1H nuclei in CH_2Cl_2 is plotted as a function of τ for both methods. With its dipolar splitting reduced to 2.18 kHz while diffusing anisotropically in a liquid crystal solvent, this pair of spins requires an excitation time of 230 μs for optimal production of double-quantum coherence under a simple preparation sequence. Deviations from this time are responsible for the $\sin^2 \Theta$ variation in intensity evident in the double-quantum spectra reproduced in Fig. 24*a*. Much of this variation is removed from the spectrum of Fig. 24*b* by the use of composite sequences composed of four $\pi/2$–$\tau/2$–π–$\tau/2$–$\pi/2$ subcycles with flip angles and phases of $[(3\pi/2, 0)(2\pi, 0.939\pi)(\pi, 0.183\pi)(\pi, 0.989\pi)]$. Now the intensities of the double-quantum signals are nearly constant over excitation times ranging from 50 to 150% of the optimal value.

The particular set of flip angles and phases used in this experiment was selected for its ability to engineer a composite $\pi/2$ rotation without distorting the phase of the observed signal, but in other ways is not unique. Reduced composite sequences, which replace adjoining pulses belonging to different subcycles by a single joined pulse with a phase of 0, $\pi/2$, π, or $3\pi/2$, can be constructed as well. Besides being somewhat easier to implement experimentally, reduced sequences minimize signal losses from pulse imperfections.

D. Separation of Zeeman Orders in Multiple-Quantum Spectroscopy

To this point we have made no explicit mention of any special methods needed to identify the coherence orders associated with the different frequency components in a multiple-quantum spectrum. Yet the assignment of the orders, a matter of tremendous practical importance in multiple-quantum spectroscopy, is neither automatic nor obvious when the system is viewed in the rotating frame. Since for each n-quantum coherence the rotating reference frame is effectively accelerated by n in the transformation

$$\rho^{n,R} = \exp(iI_z\omega_0 t)\rho^n \exp(-iI_z\omega_0 t) = \exp(in\omega_0 t)\rho^n, \qquad (164)$$

the large laboratory frame spacing of $n\omega_0$ shown in Fig. 3 is lost. The observed multiple-quantum frequencies are determined by the eigenvalues of the internal Hamiltonian, with lines belonging to different orders separated only by multiples of the resonance offset. Thus the one-quantum signals are centered about the frequency $\Delta\omega$, the two-quantum signals are centered about the frequency $2\Delta\omega$, and so forth. If the offset is too small compared to the spread of internal frequencies, however, then the subspectra will overlap.

Methods designed to separate the coherence orders in a multiple-quantum interferogram generally rely upon the characteristic n-fold response of multiple-quantum coherence to resonance offsets and phase shifts. Most of the approaches seek either to introduce a sufficiently large offset term into the evolution period or to cancel unwanted orders through the coaddition of spectra obtained with phase-shifted excitation sequences. These techniques, which address mainly the question of detection, are to be compared with the methods of selective *excitation* taken up later in Sections IV.F and V.A.3.

1. Time-Proportional Phase Incrementation

Reliance upon an actual resonance offset to generate an order-specific response

$$\rho(\tau, t_1) = \sum_n \rho^n(\tau, t_1) \exp(-in\Delta\omega t_1) \tag{165}$$

unnecessarily limits the degree to which the internal Hamiltonian can be manipulated during the evolution period. Since any type of Zeeman term, whether it originates externally or internally, depends linearly on the I_{zj} operators, variations in the resonant frequency due both to chemical shifts and to any spatial inhomogeneity of the magnetic field must by necessity be retained during the evolution period along with the overall resonance offset. Setting the receiver reference frequency off resonance to separate the coherence orders thus precludes the use of π pulses to refocus any linear spin interactions during t_1. Consequently, broadening of the multiple-quantum resonances due to magnetic field inhomogeneity, an effect that increases linearly with the order of coherence, cannot be removed by a spin echo. The inability to refocus the linear terms poses similar problems for multiple-quantum relaxation studies.

These difficulties are eliminated by time-proportional phase incrementation (TPPI), a method of excitation that generates the desired order-dependent response without ever introducing a "real" resonance offset.[59,65] Instead, an overall phase shift ϕ applied to the pulses in the preparation period is used to obtain the condition

$$\rho(\tau, t_1) = \sum_n \rho^n(\tau, t_1) \exp(-in\phi), \tag{166}$$

which for $\phi = \Delta\omega t_1$ is formally identical to the desired density operator of Eq. (165). The necessary proportionality between ϕ and t_1 is achieved by advancing the phase of the preparation sequence by $\Delta\phi = (\Delta\omega)(\Delta t_1)$ each time the length of the evolution period is increased by Δt_1. Fourier transformation of the signal $S_\alpha = \mathrm{tr}(I_\alpha \rho)$ with respect to t_1 distributes the orders over a bandwidth in $\omega^{(1)}$ equal to $1/\Delta t_1$, separating adjacent subspectra by the apparent offset frequency $\Delta\omega$. The number of orders detected, $\pm n_{\mathrm{max}}$, is determined by the size of the phase increment according to the relationship $\Delta\phi = 2\pi/2n_{\mathrm{max}}$ Both Δt_1 and $\Delta\phi$ must be chosen so that all signals from different coherence orders fit into the available bandwidth without aliasing and without overlapping.

The effect of phase advancement is evident in the introductory spectrum of eight spins shown in Fig. 1, which has been separated cleanly into its eight component subspectra by time-proportional phase incrementation. The ability to control the effective offset through the radiofrequency phases alone makes the method especially useful for large systems, where the higher orders of coherence are accommodated simply with smaller phase shifts.

Related closely to time-proportional phase incrementation, is the method of parameter-proportional phase incrementation, in which the value of an excitation parameter, for instance τ, is increased systematically along with the radiofrequency phase, while the evolution time t_1 is held constant.[121] In this way the excitation function is revealed empirically, and an optimal preparation time may be selected for each order. The procedure provides a useful alternative to the averaging methods discussed above.

2. *Phase Cycling Techniques*

Maintenance of a constant phase difference ϕ between the preparation and mixing pulses for all values of t_1 associates the factor $\exp(-in\phi)$ with each frequency component in the multiple-quantum interferogram. Upon Fourier transformation an n-fold phase shift is propagated throughout the frequency domain, rotating the real, or absorptive, part of the complex multiple-quantum spectrum into a mixture of quadrature components given by

$$S^n[\omega^{(1)}] = \cos(n\phi)S_x^n[\omega^{(1)}] + i\sin(n\phi)S_y^n[\omega^{(1)}]. \tag{167}$$

Each line in the spectrum thus acquires a dispersive character, the extent of which is determined by the order of coherence.

Order-dependent shifts in spectral phase can be exploited to cancel selected signals, as illustrated in Fig. 25 for $n = 1, 2, 3, 4$ and $\phi = 0, \pi/2, \pi, 3\pi/2$. In this example all but the four-quantum components are eliminated when the four separately acquired spectra are added together. Many different phase cycles may be designed to accomplish this kind of order-selective detection, with the complexity of the cycle necessarily increasing with the number of orders to be suppressed.[63]

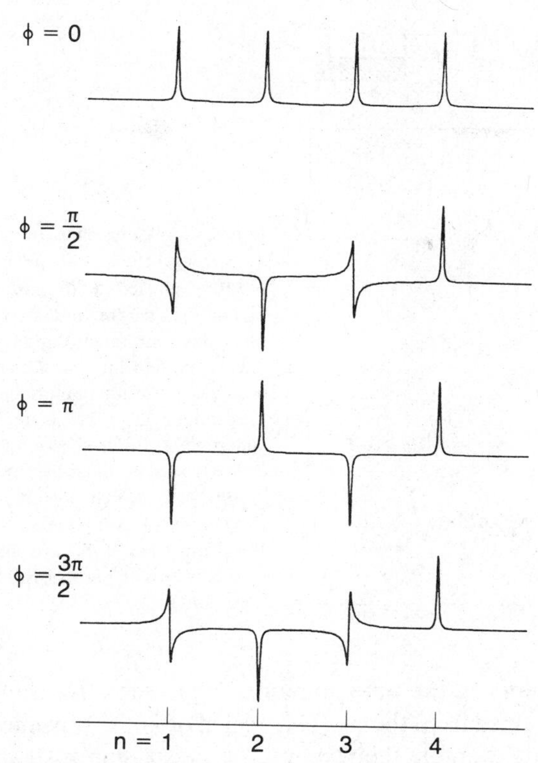

Figure 25. *n*-fold dependence of the phase of an *n*-quantum quadrature component on the radiofrequency phase difference between pulses of the preparation and mixing periods. Various cycles can be designed to detect signals from one or more orders selectively.

3. *Coherence Transfer Echo Filtering*

Selective detection is also possible with a modified spin-echo experiment that exploits the differences in dephasing rates existing among the various orders of coherence.[122–125] In the version of the coherence transfer echo filtering experiment illustrated schematically in Fig. 26*a*, the mixing period is preceded by a fixed interval T during which all transverse components are allowed to dephase in an external magnetic field. The spatial variation in resonant frequency, $\omega(\mathbf{r})$, which may either exist naturally or be imposed with an external gradient, labels each component with the dephasing factor $\exp[in\omega(\mathbf{r})T]$. Upon conversion to single-quantum coherence by the pulse, all the labeled components begin to rephase at the common rate of $\omega(\mathbf{r})$, individually canceling their dephasing factors when the refocusing period is equal to nT. As a result, signals originating from different orders are isolated

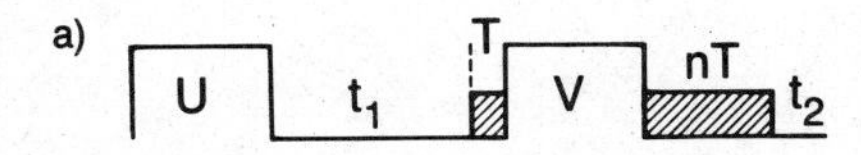

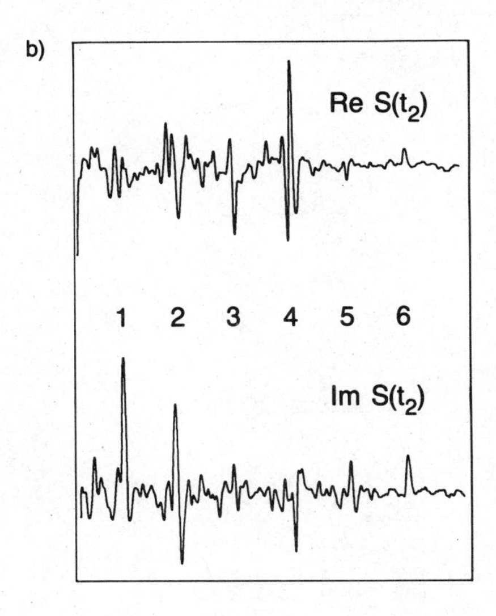

Figure 26. Coherence transfer echo filtering. (*a*) General form of the experiment, incorporating pulsed field gradients of durations T and nT before and after mixing. Each n-quantum component dephases at the rate $n\Delta\omega$ during the first interval T and then rephases as a single-quantum component at the rate $\Delta\omega$ to produce an echo at the time nT. (*b*) ^1H coherence transfer echoes of orders 1 through 6 observed in oriented benzene. The echoes may be sampled selectively to separate the subspectra. Both components of the quadrature spectrum ($S = S_x + iS_y$) are shown. (Reproduced from Ref. 28 with permission. © 1983 Academic Press, Inc.)

as separate echoes in the time domain. A representative multiple-quantum echo train, obtained from the six ^1H nuclei in oriented benzene, is provided in Fig. 26*b*. In this example the response of the six-spin system is partitioned into six individual segments which then may be removed and Fourier transformed separately to yield a spectrum for each order of coherence.

E. Excitation Dynamics in Strongly Coupled Systems

Strongly coupled spins, interacting under the Hamiltonian

$$H_D = -\sum D_{jk}(3I_{zj}I_{zk} - \mathbf{I}_j \cdot \mathbf{I}_k)$$

or

$$H_J = -\sum J_{jk}\mathbf{I}_j \cdot \mathbf{I}_k,$$

are able to exchange energy via the conservative flip-flop processes allowed by the zero-quantum operators $I_{+j}I_{-k}$ and $I_{-j}I_{+k}$. As a result, Zeeman quantum numbers for the individual spins are no longer conserved, although the total component of angular momentum along the z axis, $M = \sum m_j$, remains well defined. Under these conditions each eigenstate $|M_r\rangle$ is a linear combination of direct-product basis states in which every term of the form

$\cdots |m_j\rangle|m_k\rangle|m_l\rangle|m_m\rangle\cdots$ describes a spin configuration where the algebraic sum of the individual z components totals M_r.

Many of the dynamic properties of multiple-quantum coherence remain the same whether the spin–spin coupling is strong or weak. In particular, since the response of the spins to phase shifts and resonance offsets is unaffected by bilinear couplings, methods used to separate the orders of coherence work equally well in the presence of strong couplings. Moreover, when the preparation period is built around two $\pi/2$ pulses, the phase difference between the pulses still determines the mixture of odd and even orders excited, provided that the linear spin interactions are effectively eliminated. A technical difference arises here because a single π pulse does not always refocus the linear terms completely, since the internal Hamiltonians before and after the pulse,

$$H = H_z + H_{zz}$$

and

$$H' = -H_z + H_{zz},$$

do not always commute.[126] Trains of π pulses, or other coherent averaging methods, must then be employed instead.[28]

Although the introduction of strong couplings does not alter in any fundamental way the dynamics of multiple-quantum excitation, it does complicate evaluation of the time development of the system. An expression such as

$$\rho(\tau) = \exp(-iH_{xx}\tau)I_z \exp(iH_{xx}\tau),$$

which for weak couplings describes a succession of simple bilinear rotations, may require a computer for its solution when $H_{xx} = -\sum D_{jk}(3I_{xj}I_{xk} - \mathbf{I}_j \cdot \mathbf{I}_k)$. Yet sometimes the dynamics are simplified to the point of triviality, as we have noted earlier for isotropic systems, where if there are no differences in chemical shifts, then the relationship

$$[H_J, I_\alpha] = 0 \qquad \alpha = x, y, z$$

prevents multiple-quantum coherence from developing at all.

1. *Early Time Development*

Except for very simple cases, calculation of both the dynamics of multiple-quantum excitation in a strongly coupled system and the spectral response that follows must be undertaken numerically.[66,127] To date, computations have been performed on systems containing between four and eight spin-$\frac{1}{2}$ nuclei, limited mainly by constraints on the consumption of computer time. Even without an exact calculation, however, considerable insight into the excitation dynamics at early times may be gained by expanding the prepared density operator in a power series according to the Baker–Cambell–Hausdorf formula.[42,128] For a $(\pi/2)_y$–τ–$(\pi/2)_{-y}$ preparation sequence the expansion takes the form

$$\rho(\tau) = I_z + i\tau[I_z, H_x] - \tfrac{1}{2}\tau^2[[I_z, H_x], H_x] + \cdots, \tag{168}$$

where H_x, the internal Hamiltonian after rotation by the two $\pi/2$ pulses, is assumed to contain both linear and bilinear terms. Explicit evaluation of the commutators reveals that the different orders of coherence grow in at different times, with higher orders always requiring longer excitation times than lower orders.[128]

The development of multiple-quantum coherence in *isotropic* systems is governed by the Hamiltonian

$$H_x = -\Delta\omega I_x + \sum \Omega_j I_{xj} - \sum J_{jk}\mathbf{I}_j \cdot \mathbf{I}_k, \tag{169}$$

a sum of resonance offset, chemical shift, and scalar interactions. The general result under these interactions is that *n*-spin/single-quantum coherences grow in with an initial time dependence of τ^n, but that *n-quantum* coherences grown in more slowly as τ^{2n-1}. At very short times, when

$$\rho(\tau) \sim I_z - i\tau(\Delta\omega I_y - \sum \Omega_j I_{yj}), \tag{170}$$

only single-quantum coherence can exist, for the necessary multiple-spin operators have had insufficient time to emerge. The first two-spin operators grow in as τ^2 through the commutator

$$\begin{aligned}[][[I_z, H_x], H_x] = \Delta\omega^2 I_z - (2\Delta\omega)\sum \Omega_j I_{zj} + \sum \Omega_j^2 I_{zj} \\ + \sum J_{jk}(\Omega_j - \Omega_k)(I_{zj}I_{xk} - I_{xj}I_{zk}),\end{aligned} \tag{171}$$

but these too are just single-quantum. Only with the term associated with τ^3, whose nested commutator pairs the two-spin terms above with H_J, does the first double-quantum coherence appear. Proportional to

$$\tfrac{1}{12}\tau^3 \sum J_{jk}(\Omega_j - \Omega_k)^2(I_{j+}I_{k+} - I_{j-}I_{k-}),$$

this early double-quantum coherence requires excitation times sufficiently long to fulfill the condition $[\tau^3 J_{jk}(\Omega_j - \Omega_k)^2] \sim 1$ in order to acquire a significant amplitude.

The inverse relationship between the duration of excitation and the magnitude of the coupling reflects the propagation time that must elapse before any two spins become fully aware of each other. A network of interactions thus develops around each initially independent spin, expanding with the passage of time as partners with smaller and smaller coupling constants are added. The network may radiate outward from a shared central spin as depicted in Fig. 27*a*, or develop as a chain of couplings along the lines of Fig. 27*b*. By contrast, the radial pattern is the only option available in weakly coupled systems, where any two Hamiltonian terms with one spin in common always commute. Under these circumstances the relationship

$$[H_{jk}, H_{kl}] = 0$$

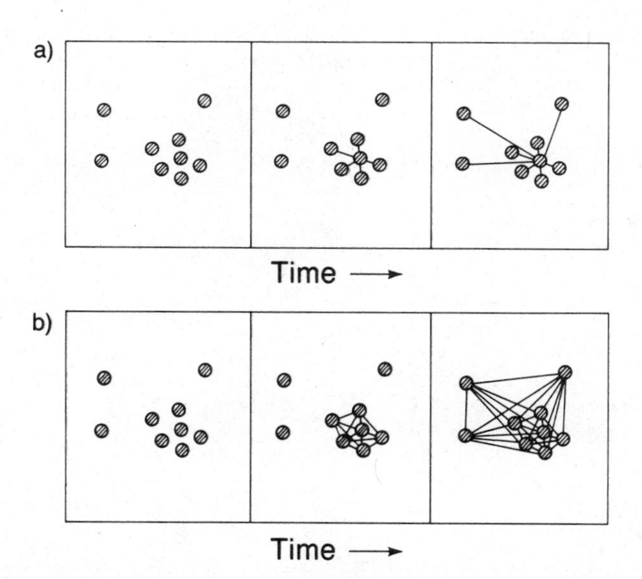

Figure 27. Symbolic representations of the growth of a coupling network with time. Initially the spins act independently, having had insufficient time to communicate via dipole–dipole or scalar interactions. Pairwise couplings gradually become effective as time passes, either developing from a central spin and expanding outward (*a*), or moving from spin to spin (*b*), or both. Roughly speaking, small couplings require longer propagation times than large ones. The highly efficient sequential pathways shown in (*b*), which are available only to strongly coupled spins, allow two spins to be correlated indirectly through a common coupling partner. (Adapted from Ref. 131.)

causes the propagator

$$U(\tau) = \exp(-iH_x\tau) = \prod \exp[-i(\Omega_j I_{xj} - J_{jk}\mathbf{I}_j \cdot \mathbf{I}_k)\tau] \qquad (172)$$

to factor into a product of *independent* pair coupling operators, each of which is able to influence only its own coupled spins as the density operator changes with time.

Excitation dynamics are qualitatively similar in anisotropic phases, but the dipolar coupling

$$H_{D,xx} = -\sum D_{jk}(3I_{xj}I_{xk} - \mathbf{I}_j \cdot \mathbf{I}_k),$$

usually dominant, excites coherences more efficiently than does the J coupling. Here the first n-spin term is proportional to τ^{n-1}; when n is even, the term contains *n-quantum* operators as well. Odd orders appear as τ^n.

TABLE I
Dependence on τ of Multiple-Quantum Coherences Produced by the Sequence $(\pi/2)_\phi$–τ–$(\pi/2)_y$.[a]

		Initial τ power dependence		
	n	$\phi = -\pi/2$[b]	$\phi = 0$[b]	$\phi = -\pi/4$[c]
Anisotropic systems	0	2	1	1[d]
	1	1	0	0
	2	1	2	1
	3	3	2	2
	4	3	4	3
	n (even)	$n-1$	n	$n-1$
	n (odd)	n	$n-1$	$n-1$
Isotropic systems	0	2	1	1
	1	1	0	0
	2	3	4	3
	3	5	6	5
	4	7	8	7
	n (even)	$2n-1$	$2n$	$2n-1$
	n (odd)	$2n-1$	$2n$	$2n-1$

[a] From Ref. 66.

[b] For systems without chemical-shift differences, $\phi = -\pi/2$ is even-selective and $\phi = 0$ is odd-selective when $\Delta\omega = 0$.

[c] $\phi = -\pi/4$ is equivalent to a superposition of $\phi = -\pi/2$ and $\phi = 0$ results.

[d] In systems without chemical-shift differences, only even-selective terms can prepare zero-quantum coherence. For these systems, the appearance of zero-quantum coherence is therefore proportional to τ^2.

The development of multiple-spin correlations in anisotropic phases is discussed further in Section IV.E.3 when we treat infinitely extended dipolar solids. For now we simply summarize in Table I the initial time dependence of the first few coherence orders in both isotropic and anisotropic systems, including for comparison the exponents appropriate for odd-quantum ($\phi = \pi/2$) and nonselective ($\phi = \pi/4$) sequences.

Oriented benzene, a system that has been thoroughly studied both by single-quantum and multiple-quantum NMR, provides a convenient illustration of the manner in which multiple-quantum coherence develops in a small spin system. When dissolved in a nematic liquid crystal solvent, the benzene molecule is free to diffuse linearly but not to tumble isotropically. As a result, all *inter*molecular dipolar couplings are averaged to zero, effectively isolating the six ^1H nuclei on one molecule from all other spins. Restricted anisotropic reorientation of the molecules partially averages the intramolecular couplings, attenuating but not eliminating them.

Average coherence magnitudes computed numerically as a function of excitation time for $n = 2$, 3, and 6 are graphed in Fig. 28; an expanded

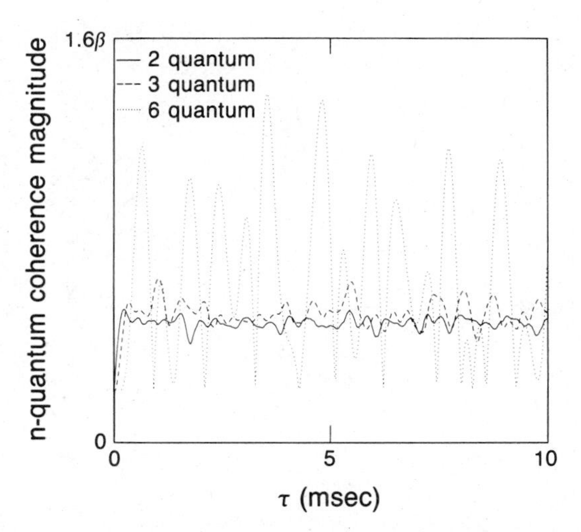

Figure 28. Average prepared coherence magnitudes computed as a function of preparation time for oriented benzene. The calculation assumes that even and odd orders of coherence are excited by two $\pi/2$ pulses and that the resonance offset is zero. The lower orders of coherence ($n = 2$ and 3) grow in steeply over an interval approximately equal to the inverse of the average dipolar coupling, after which their magnitudes remain relatively flat. The dynamics of the total spin coherence ($n = 6$) are determined by the fewest matrix elements, and as a result, the $n = 6$ component oscillates to a greater extent during the preparation period than does either the $n = 2$ or the $n = 3$ component. Magnitudes are measured in units of $\beta = \hbar\omega_0/k_B T$. (Reproduced from Ref. 66 with permission. © 1984 Academic Press, Inc.)

view of the *early* time development of all the orders, in which each curve is normalized relative to its own maximum amplitude, is presented in Fig. 29.[66] These calculations assume that the coherences have been excited by a nonselective even–odd sequence, and that the scaled dipolar coupling between ortho hydrogens is 817.1 Hz. The curves show that the initial development of each order proceeds according to the expected power law, with the higher orders requiring longer excitation times to attain their limiting values. After approximately 570 μs, when $2\pi\tau \times$ (average dipolar coupling in Hz) $= \pi/2$, all orders of coherence are present, and there is little further variation in coherence magnitudes for all but the highest orders.

2. *Multiple-Quantum Intensities after Long Excitation Periods—Statistical Model*

Shown in Fig. 30*a* is an experimental multiple-quantum spectrum of oriented benzene, obtained with the sequence

$$\pi/2 - \tau - (\pi/2)_{\phi=\pi/4} - t_1 - \pi/2 - t_2,$$

and averaged over four values of $\tau (= 4, 6, 8, \text{ and } 10 \text{ ms})$. The magnetization was sampled once during the detection period at $t_2 = \tau$ in order to

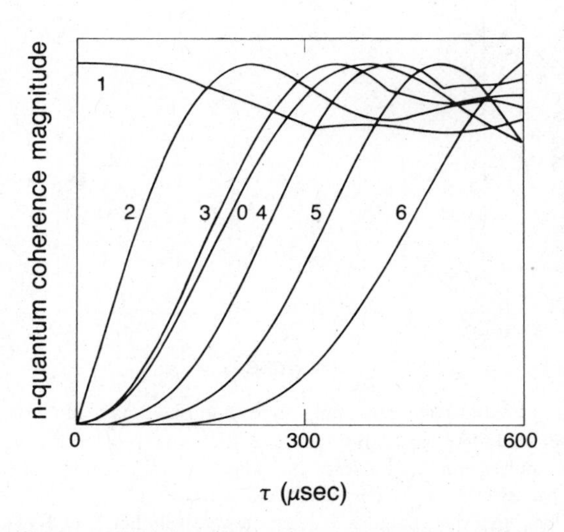

Figure 29. Plots of the average prepared magnitude per allowed coherence in oriented benzene for short preparation times, showing the increasing induction times needed to excite higher orders. The development follows the power laws given in Table I. Each curve is normalized to its maximum amplitude. (Reproduced from Ref. 66 with permission. © 1984 Academic Press, Inc.)

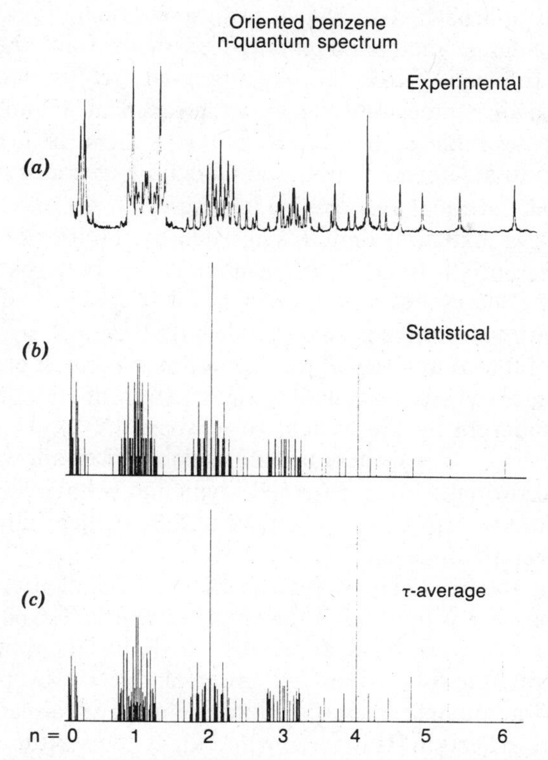

Figure 30. Experimental and simulated *n*-quantum spectra of oriented benzene. (*a*) Experimental spectrum obtained with a nonselective even–odd pulse sequence and separated by time-proportional phase incrementation. The subspectra become sparser as the order of coherence increases, reflecting the progressive decrease in the number of admissible superpositions. (*b*) Simulated spectrum based on a statistical-limit calculation in which each allowed coherence is weighted equally. The integrated intensities of the subspectra fall off according to a Gaussian distribution. (*c*) Simulation based on exact density operator dynamics during the preparation and mixing periods, integrated over the excitation time. (Reproduced from Refs. 66 and 130 with permission. (*a*)© 1980 American Institute of Physics (*b*) and (*c*)© 1984 Academic Press, Inc.)

symmetrize the preparation and mixing periods, and the resulting interferogram was Fourier transformed to yield a complex (quadrature) spectrum, which is displayed here in absolute value mode free of any variation in the phases of the lines.

The intensities of the signals in the multiple-quantum spectrum are determined by the coherence magnitudes attained during the preparation and mixing periods, and their positions are determined by the eigenvalues of the Hamiltonian during the evolution period. Solution of the eigenvalue problem is usually approached iteratively, using a set of couplings based upon a presumed molecular geometry and refining the structure until the calculated frequencies are sufficiently close to the observed frequencies. Although calculation of the line intensities can be undertaken in a similar fashion, it is nonetheless desirable to be able to estimate them, at least roughly, without resorting to a detailed computation. Such an option is provided by the *statistical model* of equal average line intensities.[66]

The concept of a statistical limit is suggested by the long-term stability of low-order coherence, a trend clearly evident in the patterns of development exhibited by the modes $n = 2$ and $n = 3$ in Fig. 28. The absence of variation in the average coherence magnitude after the initial growth period is symptomatic of the complexity of the excitation dynamics: except for the total spin coherence, which regardless of the strength of coupling can only be realized by superposing the extreme eigenstates $|N/2\rangle = |+ + \cdots +\rangle$ and $|-N/2\rangle = |- - \cdots -\rangle$, there are many coherences allowed within a given order. Accordingly, the observed dynamic behavior reflects the existence of numerous frequencies of oscillation, which ultimately are damped out to a common level.

The statistical model assumes, for the sake of simplicity, that in the limit of long preparation times all allowed coherences are excited with the same magnitude but with random phases. With this simplification the integrated spectral intensity within an order is directly proportional to the number of admissible superpositions. We have already counted coherences earlier in Section III.B, finding that for systems without symmetry there are $\binom{2N}{N-n}$ possibilities when n is unequal to zero and $[\binom{2N}{N} - 2^N]/2$ possibilities otherwise. Hence when N is large (in practice $\gtrsim 6$) and n is small, the intensities of the orders in a magnitude spectrum are described by a Gaussian distribution, falling off as

$$I(n) = \frac{4^N}{\sqrt{N\pi}} \exp\left(-\frac{n^2}{N}\right).$$

In systems where the eigenstates are symmetry-adapted linear combinations, the allowed coherences are restricted to states in the same representation, each of which develops as a separate system. For reference, the energy level of benzene according to its irreducible representations is given in Fig. 31.

The statistical-limit spectrum of oriented benzene is shown in Fig. 30*b*. Agreement with the experimental spectrum directly above is reasonably good, in view of all that the statistical model neglects. Discrepancies are most pronounced in the tail of the distribution, where there are the fewest

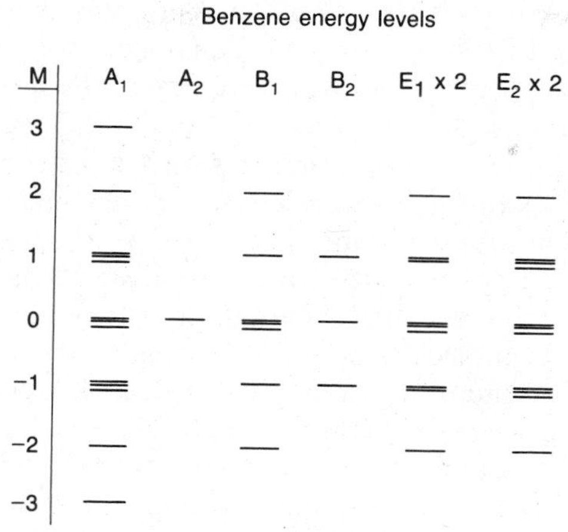

Benzene energy levels

Figure 31. Energy level diagram for oriented benzene, assuming D_{6h} symmetry. ^1H spin states are classified according to how they transform under the molecular symmetry operations. The E_1 and E_2 irreducible representations are two-dimensional: each energy level is doubly degenerate. In all systems the $|\pm N/2\rangle$ states, formed by taking all spins up or all spins down, belong to the totally symmetric representations A_1. (Reproduced from Ref. 133 with permission.)

admissible coherences and where the assumption of a flat coherence magnitude is therefore least likely to be valid. The intensities of the higher orders are in fact consistently underestimated in calculations of this sort, which may be regarded as worst-case scenarios.[66] Nevertheless, the practical difficulties of creating coherences of the highest orders are still severe enough to warrant the use of order-selective excitation (*see* Section IV.F).

To be compared with the statistical-limit spectrum is the simulated spectrum shown in Fig. 30*c*, which results from a numerical treatment of the exact density operator dynamics. This kind of calculation approximates the effect of uniform excitation by averaging the τ dependence contained in the expression for coherence magnitude, and is usually required if the intensities are to be accounted for accurately.

3. *Multiple-Quantum Dynamics in Strongly Coupled Solids*

Throughout the entire discussion up to this point, our assumption of a small, well-defined group of spins has been justified by the nature and extent of the bilinear couplings in the systems considered. In isotropic liquids, for example, relatively short-range J interactions typically limit the coupling network to only those nuclei within a few chemical bonds of a central nucleus.

Intramolecular groupings of two, three, or four spins are most common under these conditions, which prevail in homonuclear and heteronuclear systems involving a variety of isotopes in solution. In the specific case of ^{1}H, a nucleus of general chemical interest, isotropic couplings range up to approximately 40 Hz, and preparation times on the order of tens of milliseconds are therefore required for efficient multiple-quantum excitation. Constraints on the effective size of the system and on the optimum excitation time exist in anisotropic fluids as well, where partial averaging of the through-space dipolar coupling prevents the formation of long-range correlations. We have seen that for molecules oriented in nematic liquid crystals, translational freedom is retained but reorientation via isotropic tumbling is restricted, so that what remains of the *intra*molecular dipolar coupling is the principal interaction among the spins. Though these direct couplings may be up to three orders of magnitude greater than the through-bond scalar couplings, multiple-quantum dynamics are still confined within molecular boundaries and still can involve only a limited number of spins.

Restrictions on the number of spins among which multiple-quantum coherence can be created are lessened considerably in dipolar *solids*, wherein the overall coupling network includes all the spins in the sample. Although multiple-spin events in solids develop along lines qualitatively similar to those described above for small strongly coupled systems, the sheer numbers of spins involved lead to dynamic effects worth noting. To grasp the essential features peculiar to solids we first consider the origin of the familiar free induction decay, a phenomenon that, though it may be strictly single-quantum, nonetheless mirrors the development of multiple-quantum coherence under a nonsecular bilinear Hamiltonian. Both processes depend upon the formation of spin correlations of high order through a sequence of events that can be visualized explicitly with multiple-quantum spectroscopy.

a. Multiple-Spin Processes and the Free Induction Decay. In any strongly coupled, rigid solid the free induction signal observed after a $(\pi/2)_y$ pulse,

$$S_x = \mathrm{tr}[\rho(\tau)I_x] = \sum |\langle M_r|I_x|M_s\rangle|^2 \exp(i\omega_{rs}\tau),$$

reflects the response of initially independent single spins to the collective local field of all the other spins. The internal dipolar coupling

$$H_{zz} = -\sum D_{jk}(3I_{zj}I_{zk} - \mathbf{I}_j \cdot \mathbf{I}_k),$$

which determines the frequencies of response, includes contributions from even the smallest pair interactions. With N typically falling within a few orders of magnitude around 10^{19}, however, there can now be no question of computing the eigenstates and eigenvalues of the full dipolar Hamiltonian either numerically or analytically. The complexity of the system is such that it is possible merely to assert that the eigenvalues exist and that their spectrum of frequencies must be quasi-continuous. Thus the free induction decays with a time constant T_2, the transverse relaxation or dephasing time, simply as the result of destructive interference among the numerous modes of oscillation.[129]

Associated with the loss of observable magnetization during free evolution is the progressive appearance of multiple-spin correlations of increasingly high order. This phenomenon is readily understood when the density operator

$$\rho(\tau) = \exp(-iH_{zz}\tau)I_x \exp(iH_{zz}\tau) \tag{173a}$$

is expanded in a power series according to Eq. (168) to yield the product form

$$\rho(\tau) = I_x + i\tau[I_x, H_{zz}] - \tfrac{1}{2}\tau^2[[I_x, H_{zz}], H_{zz}] + \cdots . \tag{173b}$$

As we have noted earlier, the nested commutators produce multiple-spin terms such as $I_{zq}I_{z(q-1)}\cdots I_{z2}I_{x1}$, which describe single-quantum coherence among q interacting spins and which are associated with various powers of $D_{jk}\tau$. Single-quantum combination terms, incorporating one or more pairs of zero-quantum flip-flops, also arise under the strong coupling. Hence the system, represented by a density operator initially composed entirely of independent single-spin/single-quantum modes, undergoes a unitary conversion into multiple-spin/single-quantum degrees of freedom that are not directly associated with the macroscopic magnetic dipole moment. The decay of the free induction with time therefore is just a symptom of a growing interdependence among the spins, a relationship fostered under the influence of the bilinear couplings. In contrast to the situation in smaller systems, however, where spin–spin couplings may sometimes induce an oscillatory exchange of amplitude between observable and nonobservable modes, the loss of observable magnetization in a solid is nearly monotonic because any oscillations are quickly damped out.

What distinguishes a dipolar solid from any other collection of coupled spins is the absence of a clear-cut, practical value for N, the number of *effectively* coupled nuclei. Superficially the basic dynamic patterns remain the same: the influence of a coupling between two spins on the development of the system depends on the time elapsed, with the value of $D_{jk}\tau$ providing a rough measure of the effectiveness of a particular pair interaction at each instant. When $D_{jk} \ll 1$, insufficient time has elapsed for the interaction between j and k to be significant. As time passes, more couplings become sufficiently large to contribute beyond first order, and the number of admissible product operators increases. The contribution of these high-order terms to the power series becomes more and more important with time, eventually competing with the large couplings that determine the early time development. In contrast to finite systems, though, the coupling network in a solid is unbounded, and long periods of free evolution thus enable ever larger groups of spins to communicate via the dipole–dipole interaction. In the absence of other spin–spin or spin–lattice interactions the coupling network so developed increases continually, and ultimately extends to every spin present, including even the most distant.

Hence it is clear that dipolar dephasing, the principal agent of transverse relaxation in a solid, is associated not so much with the loss of some essential quality of the spin system, but simply with a redistribution of the initial condition into other modes. Indeed, since the norm of the density operator is left unchanged by the unitary transformation of Eq. (173), it is theoretically possible to restore the initial condition long after the free induction has decayed by forcing the spins to evolve under a Hamiltonian of opposite sign. Such an apparent reversal of time has in fact been realized experimentally using coherent averaging methods.[68–71] The question now turns to whether the development of complex correlations among large numbers of spins can be visualized explicitly via multiple-quantum spectroscopy.

b. Multiple-Quantum Coherence in Solids—Experimental Methods and Examples. When the system has evolved to the point where many spins are able to act collectively, application of a second pulse can transform a term like $I_{zq} \cdots I_{z2} I_{x1}$ into terms representing coherences involving up to the full complement of q spins. If multiple-quantum signals of high order are actually to be observed in a solid, however, time reversal excitation must be used to ensure that all spectral contributions of a given order are generated with same phase. Otherwise destructive interference among the countless lines within an order will severely attentuate the signal.[67]

This important experimental consideration dictates that the zero- and double-quantum dipolar Hamiltonian H_{xx} existing under a two-pulse preparation sequence must be bypassed in favor of a form better adapted to time reversal mixing. One such Hamiltonian[130] is

$$H_{(yy-xx)} = \tfrac{1}{3}(H_{yy} - H_{xx}) = -\tfrac{1}{2} \sum D_{jk}(I_{+j} I_{+k} - I_{-j} I_{-k}), \qquad (174)$$

which because it contains only double-quantum terms changes sign in response to an overall phase shift of $\pi/2$, namely,

$$\begin{aligned} &\exp[-i(\pi/2)I_z] \sum (I_{+j} I_{+k} - I_{-j} I_{-k}) \exp[i(\pi/2)I_z] \\ &\quad = -\sum (I_{+j} I_{+k} - I_{-j} I_{-k}). \end{aligned} \qquad (175)$$

Where zero-quantum $I_{+j}I_{-k} \pm I_{-j}I_{+k}$ terms are present, as in H_{xx} or H_{yy} alone, their invariance to all z rotations impedes simple time reversal. With $H_{(yy-xx)}$, however, the requisite matched pair of average Hamiltonians is easily generated during a preparation period constructed from an integral number of cycles of eight $\pi/2$ pulses, with phases and delays as diagrammed in Fig. 32, and a mixing period in which the phase of every pulse is advanced by $\pi/2$.

A set of phase-sensitive ^{1}H multiple-quantum spectra obtained from a sample of polycrystalline hexamethylbenzene with time reversal excitation is shown in Fig. 33.[131] This series illustrates how the distribution of spectral intensity over the different orders changes for preparation times ranging from 66 to 792 μs, with coherences of very high order clearly developing under prolonged excitation. The effect is particularly striking for $\tau = 792$ μs, where there are strong signals extending out to and beyond $n = 64$. This experimentally observed redistribution of spectral intensity into high-order coherences is a tangible, if qualitative, manifestation of the growth of multiple-spin correlations induced by the pairwise couplings activated during the preparation period. Subspectral structure and line widths are determined by the *response* of the prepared system to the local field of all the other spins during the subsequent evolution period. Here, as in most protonated solids, a spectrum of broad, featureless lines arises from the almost continuous distribution of eigenfrequencies in a sample containing virtually an infinite number of spins.

c. *Extension of the Statistical Model.* An exact calculation of multiple-quantum dynamics in a strongly coupled solid is of course precluded by the complexity of the energy level structure, but recourse still may be taken in the statistical model introduced earlier for systems of finite size. In extending the

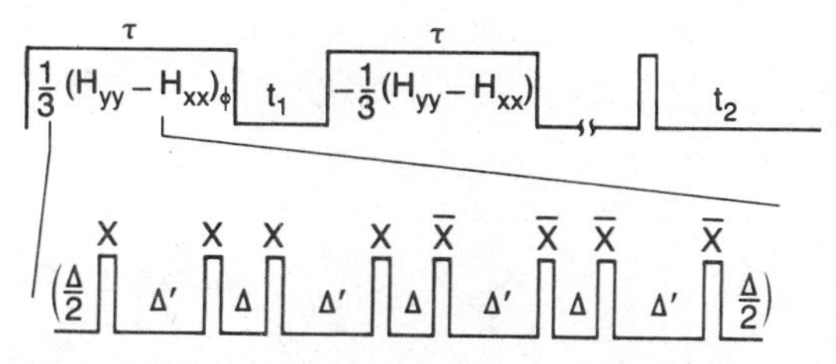

Figure 32. Pulse sequence for multiple-quantum NMR in solids, with time-reversal mixing to generate pure absorption-mode signals. The preparation and mixing propagators, generated by cycles of eight $\pi/2$ pulses with durations t_p and spacings of Δ and $\Delta' = 2\Delta + t_p$, produce average Hamiltonians of $(H_{yy} - H_{xx})/3$ for radiofrequency phases x and $-x$ and $-(H_{yy} - H_{xx})/3$ for radiofrequency phases y and $-y$. Orders are separated by time-proportional phase incrementation. A variety of detection methods, here symbolized by a single pulse, may be used during t_2. The delay before detection is inserted as a practical measure in order to allow transients to decay.

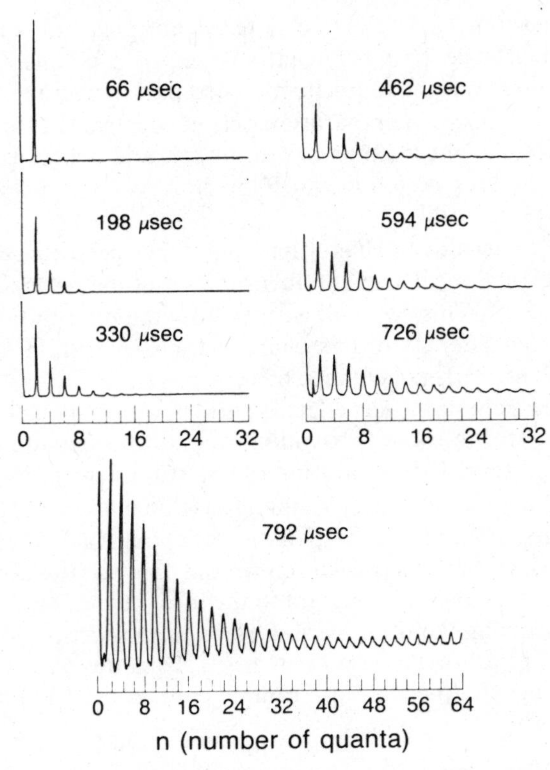

Figure 33. 360-MHz ^{1}H multiple-quantum spectra of polycrystalline hexamethylbenzene recorded with the sequence of Fig. 32 for τ = 66 to 792 μs. The basic cycle time t_c in these experiments is 66 μs (t_p = 3 μs, Δ = 2.5 μs, and Δ' = 8 μs). For τ = 66 to 462 μs the t_1 increment is 100 ns and the phase increment is $2\pi/64$; this separates each order by 156.25 kHz. For $\tau \geqslant 528$ μs the t_1 and phase increments are 50 ns and $2\pi/128$, respectively. The distribution of spectral intensity over the coherence orders broadens continuously as the preparation time increases. The lowermost trace, an expanded view of the spectrum obtained for τ = 792 μs, emphasizes the highest orders of coherence observed. (Reproduced from Ref. 131 with permission. © 1985 American Institute of Physics.)

model to the solid state, it is necessary to assume first that after a sufficiently long time the infinite system can be realistically approximated by a finite subsystem of spins among which coupling has been established; and second that all possible coherences within this group have been excited with equal probability. With regard to the first assumption, it must be borne in mind that the frontier between those spins inside a coupling network and those outside is usually blurred in a solid. If we are prepared to accept the limitations inherent in any attempt to quantify the extent of the system at a given time, then the combinatorial arguments appropriate for a well-defined group of N spins can be carried over directly. In the picture for a solid, spectral intensity versus order is again described by a Gaussian distribution with variance $\sigma^2 = N(\tau)/2$, except that $N(\tau)$ is now a *time-dependent* effective size. Best regarded as a parameter of a particular system, $N(\tau)$ grows monotonically under the influence of the bilinear spin couplings, as noted. Its change with time is determined by the detailed structure of the solid, and is influenced by factors such as the relative magnitudes of intramolecular and intermolecular dipolar interactions. Values of $N(\tau)$ versus τ for solid hexamethylbenzene, together with values similarly obtained for adamantane and squaric acid, are plotted in Fig. 34 to illustrate the kind of time dependence typically observed in fully protonated solids.[131]

One strategy for modeling the increase of $N(\tau)$ with τ is to adopt a purely stochastic view of the time development for long excitation times, treating the problem as a random walk in Liouville space. In this limit, reached perhaps after three or four dipolar correlation times, the orthogonal components of the density operator are least likely to appear correlated, and a simple time dependence may become apparent. In the view of the excitation dynamics described, the steady expansion of the effective spin size must be matched by a similar expansion of the space of basis operators needed to

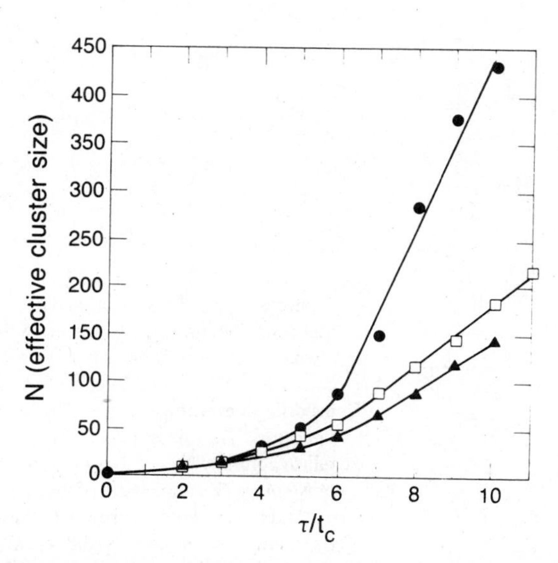

Figure 34. Effective correlated spin cluster size N versus τ/t_c for solid adamantane (●), hexamethylbenzene (□), and squaric acid (▲). The smooth curves through the points emphasize the continuous expansion of the effective size of the unbounded spin systems. (Reproduced from Ref. 131 with permission.)

represent $\rho(\tau)$ by an expression such as

$$\rho(\tau) = \sum b_i(\tau)B_i(q, N), \tag{176}$$

in which $B_i(q, N)$ is a q-spin product operator in a system of N spins. If in the statistical limit elements of the density operator at different times are truly uncorrelated, then the time development of the system can be treated as a random walk over this space of product operators representing all the possible multiple-spin states. The probability that an operator $B(q, N)$ at τ will "jump" to an operator $B(q', N')$ at τ' is then simply the product of the degeneracies of the two operator manifolds. The basic assumption is that all elements of ρ are equally accessible under the Hamiltonian, which is not unreasonable given the complexity of the dynamics and the long times involved.

A random walk over Liouville space using the product operator basis is illustrated schematically in Fig. 35. In this model (one of many possibilities) one set of terms, the q-spin operators $B(q, N)$, is monitored as it moves through the expanding operator space. Nine options are provided for

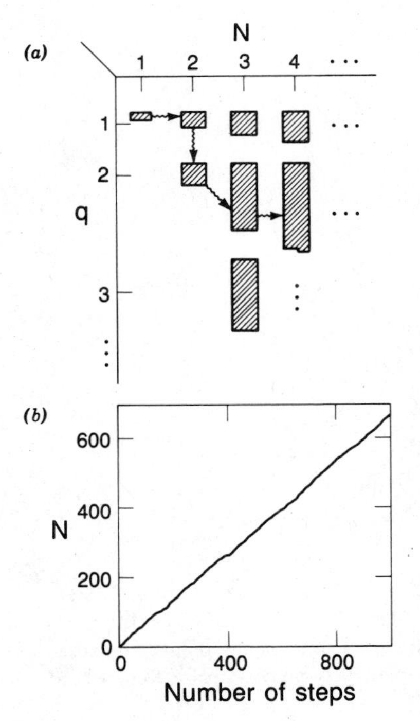

Figure 35. Stochastic approach to multiple-quantum dynamics in solids. (*a*) Schematic representation of a random walk over a Liouville space consisting of $2^{2N} - 1$ linearly independent q-spin product operators ($1 \leqslant q \leqslant N$), excluding the unit operator. The sizes of the blocks illustrate the relative numbers of q-spin operators in a manifold of N spins. The dimension of the space is allowed to expand to simulate the growth of the effective size of the spin system. (*b*) Result of a Monte Carlo simulation of this random walk, with $\Delta N = 0, \pm 1$ and $\Delta q = 0, \pm 2$. The curve is the average of 100 independent trials of 1000 steps. (Reproduced from Ref. 131 with permission.)

$B(q, N)$ to jump to $B(q', N')$; these are

$$(q, N) \rightarrow \begin{bmatrix} q-2, N-1 \\ q \quad\ , N-1 \\ q+2, N-1 \end{bmatrix} \quad \begin{bmatrix} q-2, N \\ q \quad\ , N \\ q+2, N \end{bmatrix} \quad \begin{bmatrix} q-2, N+1 \\ q \quad\ , N+1 \\ q+2, N+1 \end{bmatrix}$$

with the restriction $0 \leqslant q' \leqslant N'$. The pathways have been selected as a rough approximation to the operators actually accessible under the even-quantum Hamiltonian $H_{(yy-xx)}$. The number of q-spin operators $B(q, N)$ at any step is given by

$$\zeta_{qN} = \frac{N!}{q!(N-q)!} 3^q, \tag{89}$$

so the normalized probability for the move from $B(q, N)$ to $B(q', N')$ is simply

$$P_{qNq'N'} = \frac{\zeta_{qN}\zeta_{q'N'}}{\sum_{q'N'} \zeta_{qN}\zeta_{q'N'}}. \tag{177}$$

The trajectory is ascertained by comparing the statistical probabilities at each step to a random number, in the spirit of a Monte Carlo calculation. The time development of N, the effective cluster size, is thus directly available, and can be compared to the time dependence observed in the experimental spectra.

A typical simulation is shown in Fig. 35*b*. This plot of N versus number of steps clearly indicates a linear dependence in the statistical limit. On average, the effective size of the system increases by 1 in approximately 73% of the moves, is unchanged in approximately 21% of the moves, and decreases by 1 for the remaining 6%. These probabilities provide an overwhelming impetus for the system size to grow, apparently monotonically, with time. Although other, perhaps more realistic, pathways can be selected in a model of this type, the basic time dependence remains unchanged. The data plotted in

Fig. 34 suggest that the onset of statistical behavior occurs after approximately 300 μs under $H_{(yy-xx)}$ in these particular systems.

F. *n*-Quantum Selective Excitation

Though the number of lines in a multiple-quantum spectrum decreases dramatically as the number of quanta increases, any simplification that results is usually accompanied by a significant reduction in overall spectral intensity. Since under nonselective excitation most coherence magnitudes ultimately fall within a narrow range of limiting values, it is difficult for the components of high order to overcome the intrinsic signal-to-noise advantage of the more numerous components of low order. Clearly, if the spectrum of just one order of coherence is sought, it is better to excite only that subsystem originally rather than to accept a statistical distribution in which most of the spectral intensity is wasted.

The structure of the excitation period Hamiltonian, and through it the structure of the excitation period propagator, determines the orders of coherence available to the system. While evolving under an n-quantum Hamiltonian, which connects states where the Zeeman quantum numbers differ by $\pm n$, the density operator develops coherences of orders $[0, \pm n, \pm 2n, \ldots]$ or $[\pm 1, \pm 1 \pm n, \pm 1 \pm 2n, \ldots]$ depending upon whether the initial condition is a longitudinal (zero-quantum) mode proportional to I_z or a transverse, single-quantum mode proportional to I_x or I_y. These restrictions on coherence pathways lead to even–odd selectivity in systems governed by double-quantum bilinear Hamiltonians.[79] The more general problem is to construct an *n-quantum* propagator that will convert the initial condition of the system exclusively into nk-quantum coherences, directing spectral intensity away from undesired orders and into, perhaps, just one high order.

Coherent averaging methods may be used to generate the n-quantum Hamiltonian needed for order-selective excitation according to the experimental design laid out in Fig. 36*a*.[130,132–135] The basic unit of the excitation period is a subcycle of duration $\Delta\tau_p$, during which an effective Hamiltonian H_0 is operative. There need be no restriction on the matrix elements of H_0, which acting alone may be able to produce arbitrary modes of coherence. Selectivity is created by concatenating *phase-shifted* subcycles to form a full cycle wherein the effective Hamiltonian is a pure nk-quantum operator up to some desired order in the Magnus expansion. The complete excitation periods, mixing as well as preparation, result from the application of an integral number of cycles.

The basic subcycle Hamiltonian H_0 arises from some particular combination of radiofrequency pulses and delays. Advancement of the phase of

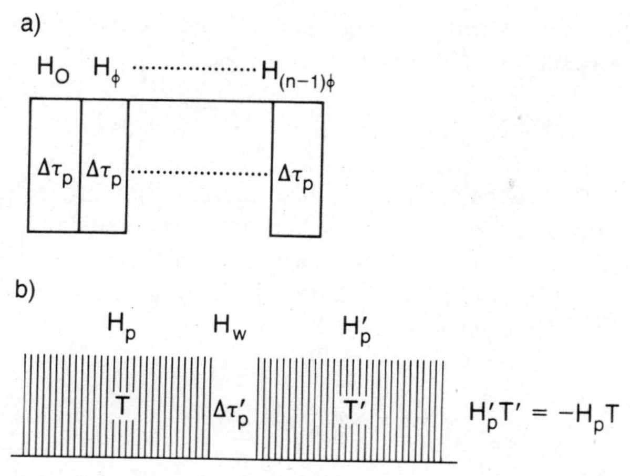

Figure 36. Pulse sequence design for selective n-quantum excitation. (a) Phase cycling is used to create an nk-quantum sequence, selective to lowest order according to average Hamiltonian theory. The selectivity arises through the concatenation of n subcycles in which the overall phase of the radiofrequency is progressively advanced in steps of $2\pi/n$. (b) Time-reversed sequence used as a subcycle in the phase cycling scheme. The conjugate propagators surrounding the window $\Delta t'_p$ allow high orders of multiple-quantum coherence to develop before the observable signal is lost to dipolar dephasing. (Reproduced from Ref. 130 with permission. © 1980 American Institute of Physics.)

every pulse through an angle ϕ is perceived as a rotation about the z axis, which yields a new operator

$$H_\phi = \exp(-i\phi I_z)H_0 \exp(i\phi I_z), \tag{178}$$

so that when a full cycle is constructed from n subcycles with overall radiofrequency phases of 0, $2\pi/n$, $4\pi/n, \ldots, 2\pi(n-1)/n$, the average Hamiltonian

$$\bar{H}^{(0)} = \frac{\Delta\tau_p}{t_c} \sum_{m=0}^{n-1} \exp[-i(2\pi m/n)I_z]H_0 \exp[i(2\pi m/n)I_z] \tag{179}$$

is simply an evenly weighted sum of the rotated Hamiltonians. Whether or not the Hamiltonian over the full cycle is any more selective than that over the subcycle alone depends on the set of matrix elements

$$\langle M_r|\bar{H}^{(0)}|M_s\rangle = \sum_{m=0}^{n-1} \langle M_r|H_{\phi=2\pi m/n}|M_s\rangle \tag{180}$$

connecting the various pairs of eigenstates. The matrix elements of the individual subcycle Hamiltonians

$$\langle M_r | H_\phi | M_s \rangle = \exp[-in_{rs}(2\pi m/n)] \langle M_r | H_0 | M_s \rangle, \qquad n_{rs} = M_r - M_s, \quad (181)$$

differ only in the order-dependent phase factors that are acquired as a result of the rotation by ϕ; but these differences are sufficient to restrict the effective Hamiltonian to nk-quantum terms, at least to lowest order in the Magnus expansion. Only for those pairs of states where $n_{rs} = nk$ does the corresponding matrix element

$$\langle M_r | \bar{H}^{(0)} | M_s \rangle = \frac{\Delta\tau_p}{n\Delta\tau_p} \sum_{m=0}^{n-1} \exp(-i2\pi m) \langle M_r | H_0 | M_s \rangle = \langle M_r | H_0 | M_s \rangle \quad (182)$$

survive the phase cycle. In all other cases the n contributions sum vectorially to zero in the complex plane.

Thus the density operator emerges from the excitation periods as if its trajectory had been determined by the n-quantum propagator

$$U(\tau) = \exp(-i\bar{H}^{(0)}\tau), \quad (183)$$

having forgotten any pathways through modes other than nk-quantum coherence.[130,135] That excursions away from nk-quantum coherence do occur during the excitation is clear from the propagator for each subcycle, but these are closed circuits that produce no net change after an integral number of cycles (Fig. 37). In principle, nothing is lost; the transformation,

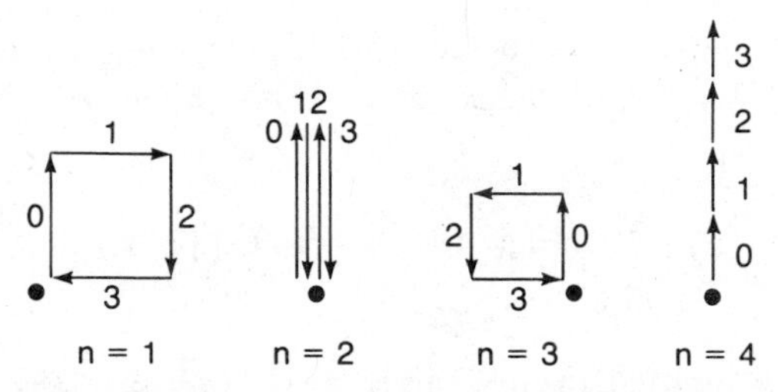

Figure 37. Symbolic representation of trajectories in Liouville space under irradiation by a $4k$-quantum selective pulse sequence. The radiofrequency phase in units of $\pi/2$ for each of the four subcycles in the sequence is indicated by the number on the arrow, and the coherence magnitude at any point on the trajectory is given by the distance from the origin (near the dot). Only the 4-quantum coherence makes a net excursion; all other components cancel owing to destructive interference over the course of the full cycle. (Reproduced from Ref. 28 with permission. © 1983 Academic Press, Inc.)

which remains unitary throughout, conserves the norm of ρ so that the nk-quantum coherences acquire amplitude originally destined for other orders.

An effective subcycle must be long enough to enable multiple-quantum operators of high order to develop, but at the same time be short enough to minimize the deleterious effects of error terms in the Magnus expansion. Apparently contradictory, these two demands nonetheless can be reconciled in a subcycle employing the principle of time reversal. The basic design, illustrated in Fig. 36*b*, calls for a sequence of three Hamiltonians—H_p, H_w, and H'_p—acting over intervals of length T, $\Delta\tau'_p$, and T'. The outermost Hamiltonians, H_p and H'_p, must differ in their algebraic signs, and must satisfy the criterion for time reversal, namely,

$$H_p T = -H'_p T'.$$

Under this arrangement the subcycle propagator

$$U = \exp(-iH_p T)\exp(-iH_w \Delta\tau'_p)\exp(iH_p T) \tag{184}$$

describes a unitary transformation of the central propagator $U(H_w, \Delta\tau'_p)$ in which whatever is done to the system during the first segment is undone in the third and final segment; and of course where no change at all is apparent when the window $\Delta\tau'_p$ disappears. Consequently, proper choice of H_p, H_w, and H'_p allows the time reversal "sandwich" to create high-order coherence through $U(H_p, T)$ while consuming only a small net amount of dynamic time $\Delta\tau'_p$. For example, one frequently adopted combination of Hamiltonians,

$$[\tfrac{1}{3}(H_{yy} - H_{xx}); T] - [H_{zz}; \Delta\tau'_p] - [-\tfrac{1}{3}(H_{yy} - H_{xx}); T],$$

is obtained by surrounding a period of free evolution with the time-reversible eight-pulse cycles detailed in Fig. 32. The free period $\Delta\tau'_p$, necessarily much shorter than the transverse relaxation time, nevertheless supports high-order coherence owing to the prolonged action of the two double-quantum propagators upon H_{zz}.

The effects of order-selective excitation are apparent in Fig. 38, where a $4k$-quantum spectrum of oriented benzene, recorded with cycles of four time-reversed $(H_{yy} - H_{xx})$ subcycles in which $\phi = m\pi/2$, is compared with the nonselective spectrum previously obtained with the basic three-pulse even–odd sequence. Both spectra have been averaged over four values of the key excitation parameters (τ for the nonselective experiment, $\Delta\tau_p$ and T for the selective experiment) to minimize any variations in excitation efficiency over the range of coupling constants. The selective excitation sequence

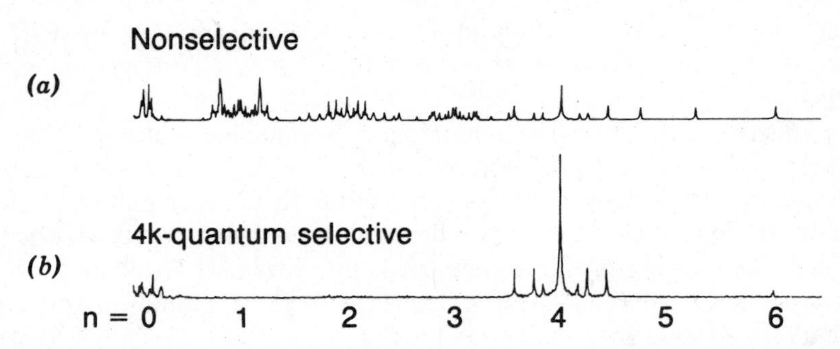

Figure 38. Spectra of oriented benzene, showing the effect of order-selective excitation. (*a*) Nonselective spectrum obtained with the simple pulse sequence of Fig. 20*a*, supplemented by a refocusing π pulse to remove linear spin interactions from the evolution period. (*b*) Spectrum obtained with a 4-quantum selective irradiation sequence constructed from phase-shifted and time-reversed subcycles, as in Fig. 36. The integrated intensity of the $n = 4$ subspectrum is increased significantly at the expense of the other components. (Reproduced from Ref. 133 with permission. © 1981 American Institute of Physics.)

effectively suppresses all orders except for $n = 0$ and $n = 4$, while enhancing the integrated spectral intensity of the $n = 4$ transition nearly fourfold. The line positions remain the same. Modifications of the experiment that incorporate additional levels of phase cycling to discriminate against the zero-quantum coherence are also availiable.[130]

The possibility of *attenuating* the spin–spin couplings, rather than reversing their dynamic sense, suggests an alternative method for the construction of subcycles, which although somewhat less flexible is perhaps easier to implement experimentally since it contains fewer pulses.[137] In "stretched" line narrowing sequences the requisite high-order operators are generated by a cycle of pulses designed to eliminate H_{zz}, but applied with the cycle time intentionally misadjusted in order to introduce significant correction terms to the average Hamiltonian. The error terms, discussed briefly in Section II.E, contain nested commutators involving the toggling frame Hamiltonians at different points in the cycle; from these commutators arise the desired multiple-spin operators.

The magnitude of the residual interaction must be adjusted critically for selective multiple-quantum excitation to be achieved. If the bilinear couplings are averaged with maximal efficiency, as they are when the cycle time is short, then the formation of multiple-quantum coherence is inhibited. Yet if the cycle is stretched far enough to interfere seriously with the averaging of H_{zz}, the observable single-quantum coherences may disappear completely before multiple-quantum coherences even develop. Proper balance is needed.

An important result of average Hamiltonian theory requires all odd-order terms in the Magnus expansion to vanish when the toggling frame Hamiltonian is symmetric about the midpoint of the cycle.[21] Consequently the leading dipolar error term in WHH-4, the symmetrized four-pulse sequence described in Section II.E, is second order in the cycle time. Given by

$$\bar{H}^{(2)} = \frac{1}{648} t_c^2 [H_{xx} - H_{zz}, [H_{xx}, H_{yy}]], \tag{185}$$

this residual coupling is a convenient source of four-quantum operators. For this reason an excitation period composed of cycles of four stretched WHH-4 sequences, throughout which the overall radiofrequency phases of the subcycles are progressively advanced by $\pi/2$, becomes $4k$-quantum selective within some range of cycle times. The effect is demonstrated in Fig. 39, which contains multiple-quantum spectra obtained in this fashion from oriented benzene.[137] Of the three examples shown, one WHH subcycle time, $t_c = 600\ \mu s$, is clearly optimal. When t_c is halved, multiple-quantum coherence

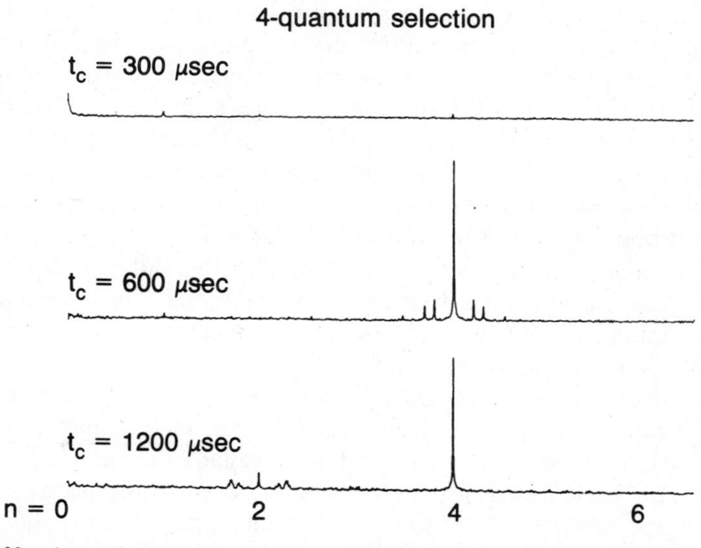

Figure 39. 4-quantum selective excitation as in Fig. 38, but with WHH-4 subcycles in place of time-reversal sandwiches. High orders of coherence are generated by the error terms in the Magnus expansion for the dipolar Hamiltonian, and then selected by phase cycling in the usual way. The excitation efficiency is controlled by the duration of the WHH-4 subcycle t_c, which must be carefully "misadjusted" to create the requisite correction terms while still averaging the homonuclear dipolar Hamiltonian to zero to lowest order. (Reproduced from Ref. 137 with permission. © 1982 Elsevier Scientific Publishing Co.)

does not develop at all; when it is doubled, the selectivity is significantly degraded.

V. EXAMPLES AND APPLICATIONS

A. Multiple-Quantum Filtering in High-Resolution NMR

A complicated solution-state NMR spectrum frequently is a superposition of individual subspectra, each originating from a small, effectively closed sybsystem. Even within a simple molecule, the limited range of the scalar interaction defines small groups of coupled spins able to respond independently of other such groups; in a more complex system, such as a mixture of species or a macromolecule, there may also be another level of organization at which each component of the mixture or each monomeric subunit is responsible for its own subspectrum. In these circumstances the behavior of multiple-quantum coherence, highly sensitive to the number and disposition of nuclei within a coupling network, may be used to simplify the overall spectrum by discriminating among the different subsystems.

The first principle of multiple-quantum filtering is simple. Since coherences no higher than n can develop among n spin-$\frac{1}{2}$ nuclei, any single-quantum component generated from an n-quantum mode must be derived from a group of *at least n* spins. The establishment of a coherence transfer pathway through a particular order of coherence thus creates a "high pass" filter in the spin number domain, eliminating signals from groups too small to support the designated order.[138–141] We should note, however, that the mere presence of n spins does not guarantee the viability of n-quantum coherence, and that a group may in fact be "too small" even if it contains the requisite minimum number of spins. Such is the case for the isolated clusters of *equivalent* spins that are found in many solvent molecules: these support no multiple-quantum coherences at all, and consequently can be effectively filtered out of a spectral mixture.[140]

In the absence of effective methods of order-selective excitation in isotropic phases, multiple-quantum filtering techniques usually employ phase cycling and order-selective detection to restrict the available coherence transfer pathways. This basic design feature has already been incorporated into a number of experiments in order to unravel coupling patterns, simplify one-dimensional and two-dimensional spectra, observe selected species, and suppress strong solvent signals.[117,138–163] The possibilities are varied, and far too numerous to review here. Nevertheless, the earliest demonstration of the filtering method, intended for ^{13}C spectroscopy, still remains one of its most useful applications.[138,142–146] We describe this experiment first to illustrate the essential features of multiple-quantum filtering, before discussing additional details and refinements needed for more general applications.

1. *Double-Quantum Filtering in ^{13}C NMR*

Naturally abundant at 1%, a ^{13}C nucleus in an isotropic phase encounters another ^{13}C coupling partner so infrequently that such a pair of nuclei almost always behaves as an isolated subsystem of two spins, provided of course that heteronuclear interactions are removed by decoupling. These dilute pairs are distinguished from the majority species of uncoupled ^{13}C nuclei by their ability to sustain double-quantum coherence, a distinction which when exploited in a double-quantum filtering experiment renders weak signals from the pairs visible against an intense background. The basic filtering experiment uses a $(\pi/2)_x$–$\tau/2$–π_y–$\tau/2$–$(\pi/2)_x$ sequence for the initial excitation, thereby creating double-quantum coherence in the ^{13}C–^{13}C subsystems while simply inverting the magnetization associated with the uncoupled ^{13}C units.[138] The coherences so generated are immediately converted to observable transverse magnetization by a mixing pulse whose phase relative to the pulses of the preparation period is cycled through the values $\phi = 0, \pi/2, \pi$, and $3\pi/2$ in a series of experiments. Under this treatment the phases of the strong transverse magnetization components from the uncoupled ^{13}C nuclei rotate in step with and in the same sense as the phase of the mixing pulse, whereas those components originating from double-quantum coherence on ^{13}C–^{13}C pairs rotate twice as rapidly and in the opposite direction. When signals obtained in this fashion are recorded with the phase of the receiver running through the sequence $0, 3\pi/2, \pi$, and $\pi/2$, what remains are only the magnetization components derived from ^{13}C–^{13}C pairs; all other components—zero-quantum coherences, population terms, and single-quantum coherences arising from imperfections in the pulse sequence—are eliminated by the phase cycling, which can become quite complex if need be. Thus is revealed the single-quantum spectrum of the pairs alone, labeled and sorted according to the phase history of the coherence. Examples of several ^{13}C satellite spectra, filtered out of a spectrum containing far stronger single lines from the uncoupled nuclei, are provided in Fig. 40 to illustrate the effect.

The optimal excitation time for a *strongly* coupled system depends upon the ratio of the scalar coupling to the chemical shift difference of the two nuclei.[143] In the limit of weak coupling, however, the double-quantum filter may be "tuned" to one or more coupling constants simply through adjustment of the excitation time to satisfy the condition

$$\tau = (2m + 1)\pi/J, \qquad m = 0, 1, 2, \ldots,$$

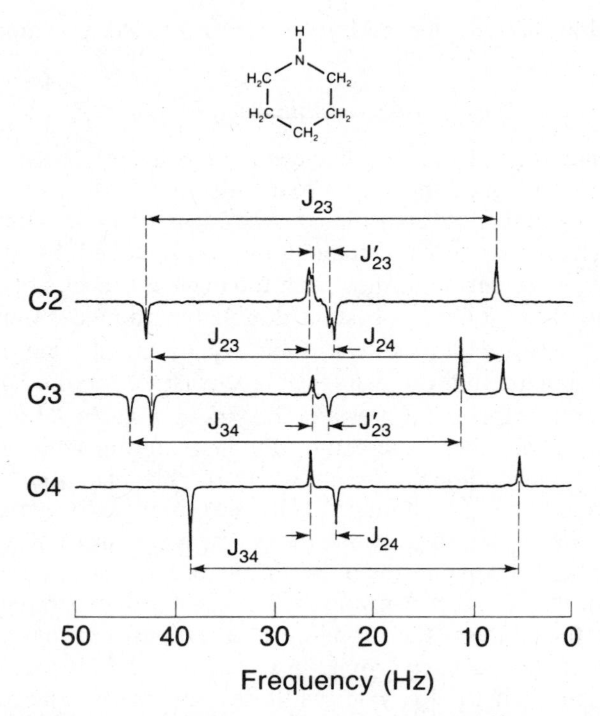

Figure 40. Sections from a natural abundance high-resolution ^{13}C spectrum of piperidine, filtered through double-quantum coherence to suppress signals from uncoupled carbon nuclei and to reveal just the direct and long-range carbon–carbon scalar couplings. The antiphase doublets are characteristic of the double-quantum experiment. Displacements of the centers of the doublets and asymmetries in the intensities are due to strong coupling effects. (Reproduced with permission from A. Bax et al., *J. Am. Chem. Soc.*, **102**, 4849. © 1980 American Chemical Society.)

for full conversion of longitudinal magnetization into double-quantum coherence.[142] The sinusoidal excitation dynamics common to all the spin pairs permit one value of τ to serve for all odd multiples of a coupling constant for which $J = \pi/\tau$. Other approaches to uniform coverage in double-quantum filtering incorporate composite pulses into the excitation period or allow τ to become a variable parameter in the experiment.[141,142,151]

A complete two-dimensional spectrum, in which the double-quantum frequencies present during an *evolution period* are correlated with the single-quantum frequencies observed during the detection period, helps to make clear the physical connectivity of the carbon atoms in a complex organic molecule. The interpretation of such a two-dimensional frequency map, an example of which is shown in Fig. 41, is straightforward. When

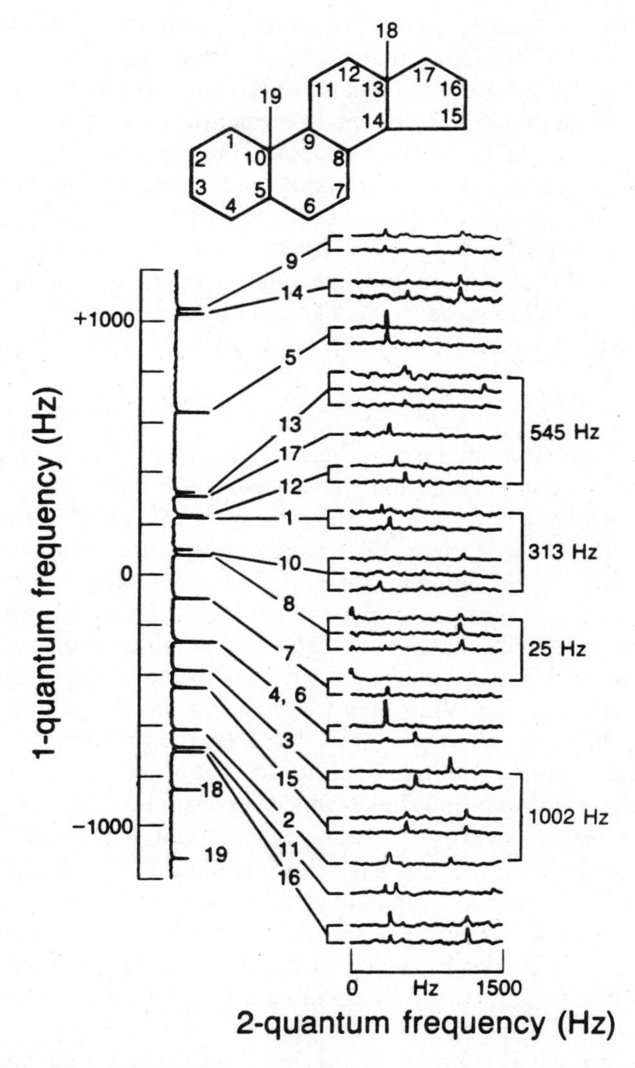

Figure 41. Two-dimensional spectrum showing the correlation between double-quantum and single-quantum signals in 5α-androstane. Every single-quantum resonance detected during t_2 originates from a double-quantum mode involving two scalar coupled carbons. As a result, the members of each pair oscillate at the same double-quantum frequency, thereby revealing the carbon framework of the molecule. A given carbon atom may be directly coupled to one, two, three, or four others. Here the connections between carbons 2 and 3, 7 and 8, 10 and 1, and 12 and 13 are indicated on the spectrum. (Reproduced with permission from A. Bax et al., *J. Am. Chem. Soc.*, **103**, 2102. © 1981 American Chemical Society.)

allowed to evolve freely, a double-quantum coherence localized on an isolated pair of ^{13}C nuclei oscillates at the sum of the chemical shifts of the two nuclei. This *one* double-quantum frequency is associated with the single-quantum spectrum subsequently detected from each member of the pair, and serves to label the two coupled spins in the two-dimensional spectrum. The appearance of a common sum frequency in the double-quantum, or $\omega^{(1)}$, dimension immediately identifies a pair of ^{13}C nuclei, while the characteristic coupling frequency that appears in the single-quantum, or $\omega^{(2)}$, dimension indicates the number of chemical bonds separating the coupling partners. Analysis of all the pairwise connectivities reveals the carbon skeleton of the molecule, atom by atom.[144-147,149,150]

2. *n-Quantum and n-Spin Filtering*

Generalized n-quantum filtering, by which we denote the selective detection of magnetization components that have passed through a transient state of n-quantum coherence, is accomplished with the phase cycle $\phi = 0$, π/n, $2\pi/n$, $\ldots(2n-1)\pi/n$ throughout which signals are alternately added and subtracted.[63] Although this kind of filter only blocks signals from subsystems with total spin less than n, a slight modification of the procedure sharpens the selectivity further by establishing an approximate *n-spin* bandpass. The trick is to allow the system to evolve for a time t_1 before the mixing period commences, and then to average a series of filtered spectra obtained for different values of t_1.[117,153-155] By the time the mixing pulse is applied, the different multiple-quantum components will have acquired distinctive phase and amplitude labels which vary with time and consequently may average to zero over t_1. Nevertheless the total spin coherence for a subsystem, oscillating at the sum of the chemical shifts of the n nuclei, remains recognizable. This "trivial" modulation can be removed by applying a refocusing π pulse at $t_1/2$, or if one specific group is to be selected, by setting the radiofrequency carrier frequency at the center of the desired multiplet, as in Fig. 42.[153-155] In this way signals from n-spin clusters survive the averaging process while signals from larger clusters are suppressed. The bandpass is not perfect, however, because spin-inversion coherences are also not modulated by J coupling, and cannot be blocked in this fashion.[117]

3. *Spin Topology Filtration*

Spin topology filtration, the recognition of specific coupling patterns, offers an additional level of spectral discrimination in high-resolution NMR.[162,163] Formulated in the spirit of selective excitation rather than selective detection, these methods further narrow the bandpass of a spin filter by isolating signals from subsystems with designated distributions of

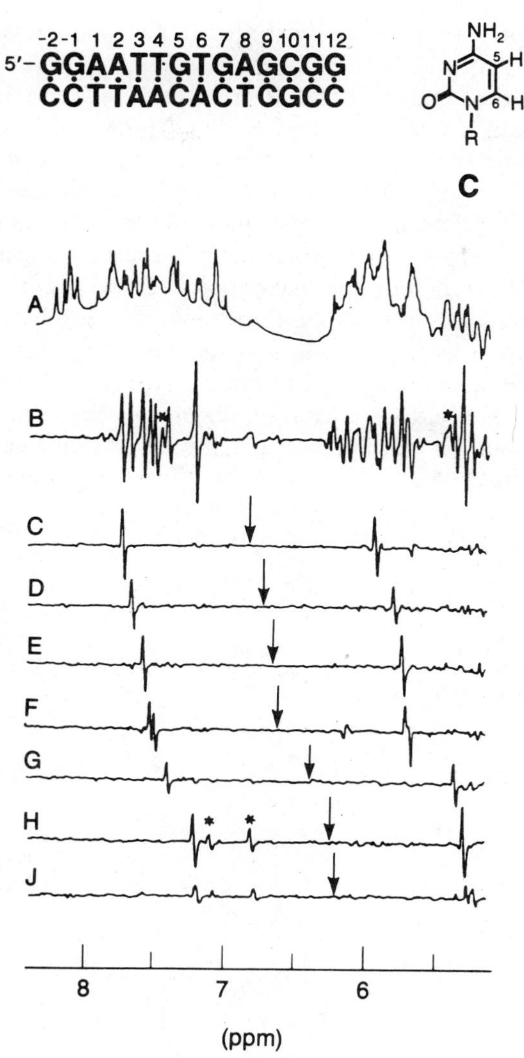

Figure 42. ^{1}H spectra of a 14-base-pair fragment of double-stranded DNA, showing the enhancement in resolution possible with two-spin filtering. (*A*) Conventional high-resolution spectrum at 360 MHz. The resolution is degraded considerably by the large numbers of resonances crowding the spectrum. (*B*) Spectrum obtained with double-quantum filtering to select doublets arising from pairs of coupled spins on the cytosine bases. Single-spin and single-quantum signals from the other nitrogen bases are suppressed. (*C*) to (*J*) Cytosine spectra, filtered through double-quantum coherence and averaged over different values of a fixed evolution time. The irradiation frequency, indicated by the arrows, is adjusted to the mean resonant frequency in each coupled pair in order to average away signals from other two-spin groups. Asterisks denote impurities. (Reproduced from Ref. 154 with permission. © 1982 Academic Press, Inc.)

spin–spin couplings, such as those illustrated in Fig. 43. Thus can be distinguished groups with identical numbers of spins that would otherwise pass through a multiple-quantum filter. For example, a four-spin system of the type AX_3, in which one spin A is weakly coupled to three equivalent spins X, can be made to look different from a four-spin system of the type A_2X_2, in which two equivalent spins are weakly coupled to two other equivalent spins. No reference need ever be made to any actual *structure*, for it is the set of nonzero couplings, not the geometric arrangement of the nuclei, that defines the topology and makes possible the selectivity.

The topologically selective excitation methods that have been developed so far are designed for weakly coupled, principally ^{1}H, systems where the spread of values among the coupling constants is narrow. Although different for each topology, the various excitation sequences share as a common building block the familiar $\pi/2$–$\tau/2$–π–$\tau/2$–$\pi/2$ pulse cluster, which refocuses the chemical shift interaction and directs the bilinear coupling interaction along the x, y, or z direction, as desired. Even when just one of these three-pulse units is used to generate a Hamiltonian such as

$$H_{xx} = -\sum J_{jk} I_{xj} I_{xk},$$

some measure of topological sensitivity is achieved. Acting on each initial component I_{zj}, the propagator

$$U = \prod_{j<k} \exp(-iJ_{jk} I_{xj} I_{xk} \tau),$$

a b c

d e f

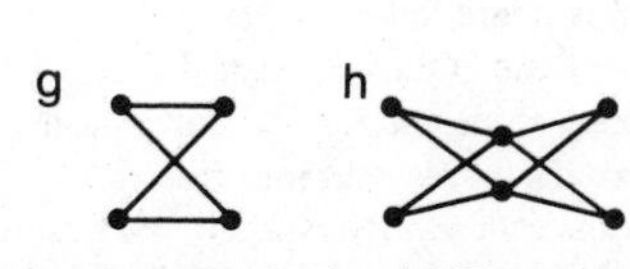

Figure 43. Common spin coupling topologies. (*a*) Two-spin. (*b*) Linear three-spin. (*c*) Linear four-spin. (*d*) AX_m with m even. (*e*) AX_m with m odd. (*f*) Triangular three-spin. (*g*) A_2X_2. (*h*) $A_2M_2X_2$. The solid lines denote nonzero pairwise couplings. Unconnected spins are equivalent. (Reproduced from Ref. 163 with permission.)

factored into a product of commuting operators, generates a starlike coupling network that includes only those spins directly coupled to the central spin j. Hence in a system of three spins with couplings J_{12}, J_{13}, and J_{23}, we effectively have (for spin 1)

$$\rho(\tau) = \exp(iJ_{13}I_{x1}I_{x3}\tau)\exp(iJ_{12}I_{x1}I_{x2}\tau)I_{z1}\exp(-iJ_{12}I_{x1}I_{x2}\tau) \times \exp(-iJ_{13}I_{x1}I_{x3}\tau), \tag{186}$$

since the operator $J_{23}I_{x2}I_{x3}$ always commutes with the evolving density operator as it is transformed first to $I_{y1}I_{x2}$ by the 1–2 coupling and then to $I_{z1}I_{x2}I_{x3}$ by the 1–3 coupling. Alternatively we may simply remove the 2–3 propagator and its inverse at the outset, recognizing that these operators commute both with I_{z1} and with all the other propagators. Recall that this behavior emerged in Section IV.E.1 in a comparison of the effects of strong and weak coupling. Now more complex sequences are to be used to shape the coupling network more specifically, while preferentially exciting high yields of *spin-inversion* coherences.

Shown in Fig. 44 are two experiments designed for AX_m filtration: sequence a for m even and sequence b for m odd.[162,163] Schemes for the triangular three-spin topology ($J_{12} = J_{13} = J_{23}$) and for the six-spin $A_2M_2X_2$ topology have also been reported.[163] Each of these experiments combines a topologically selective preparation period, which directs the system largely into spin-inversion coherences, with a conjugate mixing

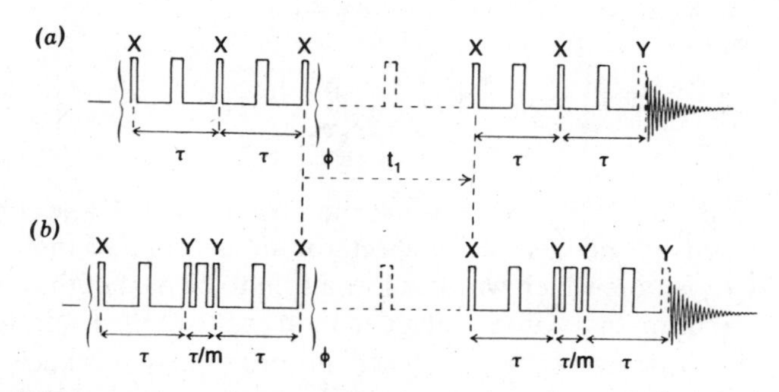

Figure 44. Pulse sequences designed to select the AX_m topology both for m even (a) and m odd (b). Flip angles are $\pi/2$ for the short pulses and π for the long pulses; all delays τ are set equal to π/J, where J is the coupling constant in radians per second. Once signals from the designated topology have been excited, the phase of the excitation sequence ϕ may be varied to select coherences of a particular order, and an optional modulation period may be added to suppress unwanted components still further. (Reproduced from Ref. 162 with permission. © 1983 Elsevier Scientific Publishing Co.)

period to convert these coherences to transverse magnetization most efficiently.[28] The mixing period, derived from the preparation period by inverting the phase of each pulse and reversing the order of application, serves two purposes. First, even though it does not effect a reversal of time, the conjugate mixing converts the spin inversion coherences to single-quantum lines with the same phase, and in so doing maximizes the observed signal intensity. Second, by using a series of pulses rather than a single pulse, the mixing sequence bypasses the selection rules discussed earlier in Section IV.C.3, which would otherwise restrict the transfer of coherence to only those spins with sufficiently high multiplicity.[114] Optionally accompanying the selective excitation of in-phase spin-inversion coherences are the standard techniques of *n-quantum* filtration, such as phase cycling and t_1 averaging, to suppress unwanted coherences still further and to provide a *higher than statistical* yield of coherence of the desired order from the chosen topology. Note that the specially prepared spin-inversion coherences evolve independently of the spin–spin couplings, and therefore are unaffected by any evolution period.

The propagator for sequence *a*, taken as a succession of two bilinear rotations and one linear rotation, can be expressed as

$$U = \exp[i(\pi/2)I_x] \exp(-iH_{zz}\tau) \exp(-iH_{yy}\tau). \tag{187}$$

The interval τ is chosen to satisfy the condition $\tau = \pi/J$ in order to engineer $\pi/2$ rotations in the appropriate product operator subspaces. Under the influence of this propagator an AX_m system (m even, A = spin number *1*) initially in thermal equilibrium is transformed to

$$\rho = (-1)^{m/2} \left[2^m I_{y1} \prod_{k=2}^{m+1} I_{xk} + \sum_{k>1} 2^m I_{x1} I_{xk} \prod_{j \neq 1,k} I_{yj} \right], \tag{188}$$

a state consisting exclusively of spin-inversion coherences. These coherences are distinguished by the absence of spectator spins: none of the operators contains any I_{zj} factors which would allow a spin to be part of the coherence without contributing to its order. Included in the collection of spin-inversion coherences is the total spin coherence of order $m + 1$ which, with a normalized magnitude of

$$[f^{m+1}]^2 = \frac{\Sigma_{r,s}[\rho_{rs}]^2}{\mathrm{tr}[\rho^2]} = (m+1)2^{-m}, \qquad |r-s| = m+1, \tag{189}$$

is excited with maximal efficiency. Coherences of the same order originating from groups with the "wrong" coupling topology are excited with considerably less success than the corresponding topologically correct coherences. These unwanted components, not necessarily spin-inversion transitions, may be suppressed even further in the t_1 average. The desired coherence, which definitively labels the AX_m group, can then be filtered out of the mixture and converted to observable magnetization. The action of sequence *b*, and other topologically selective sequences, may be understood using similar arguments and algebraic manipulations.

The high-resolution ^1H spectra shown in Fig. 45 illustrate the effects of AX_2 and AX_3 filtration on a mixture of four components: 1,1-dichloroethane (AX_3), 1,1,2-trichloroethane (AX_2), 1,3-dibromobutane ($A_3MPQ'XY'$), and 1,2-dibromopropane (A_3MPQ). The complete spectrum, reproduced in Fig. 45*a*, is a superposition of overlapping multiplets from the four subspectra. AX_2 and AX_3 topological filtering disentangles the

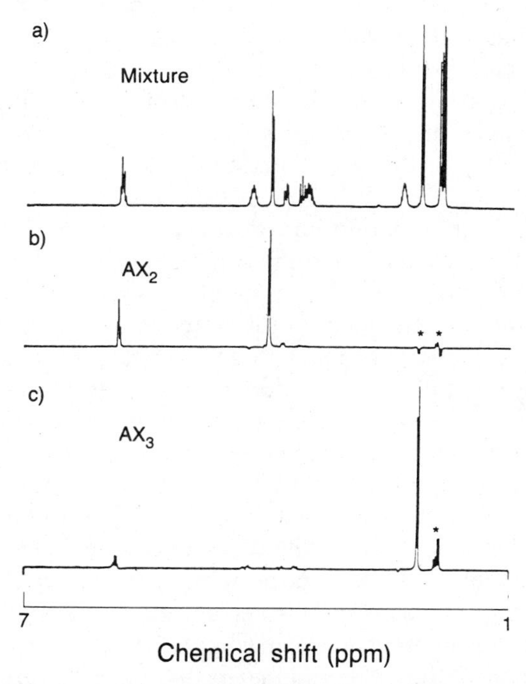

Figure 45. High-resolution ^1H spectra at 300 MHz of a mixture of 1,1-dichloroethane (AX_3), 1,1,2-trichloroethane (AX_2), 1,3-dibromobutane ($A_3MPQ'XY'$), and 1,2-dibromopropane (A_3MPQ). (*a*) Conventional spectrum, complicated by overlapping signals from the different components. (*b*), (*c*) Simplified spectra obtained with $(m + 1)$-quantum AX_m topological filtering. Incompletely suppressed peaks are marked with asterisks. (Reproduced from Ref. 162 with permission. © 1983 Elsevier Scientific Publishing Co.)

resonances, however, producing the well-resolved subspectra of 1,1,2-trichloroethane and 1,1-dichloroethane shown in Fig. 45*b* and *c*. In other reported applications of the method, coherences with orders as high as 7, normally rarely seen in high-resolution NMR, have been generated with reasonable intensities.[163]

It is hoped that filtering methods, *n*-quantum and *n*-spin as well as topological, will aid in the simplification and analysis of high-resolution spectra of biological macromolecules, which frequently are obtained with two-dimensional *single*-quantum coherence transfer methods. According to the basic method of two-dimensional correlated spectroscopy, signals oscillating at $\omega^{(1)}$ during an initial response period are associated with signals oscillating at $\omega^{(2)}$ during a second response period in order to elucidate networks of scalar couplings or dipolar relaxation pathways, and thereby to map out the molecular framework.[24,164,165] The fundamental idea is as follows: a spin unconnected to any other spin either through bonds or through space exhibits the same frequency in both dimensions, and therefore gives rise to a signal on the diagonal of the two-dimensional spectrum, $\omega^{(1)} = \omega^{(2)}$. If instead there occurs a transfer of magnetization, coherent or incoherent, a pair of *cross peaks* symmetrically disposed about the diagonal appears at the intersections of the two frequencies $\omega^{(1)}$ and $\omega^{(2)}$ associated with the two correlated spins. In this way the atom-by-atom connectivity becomes apparent.

Typically rich in information but often correspondingly complicated, a two-dimensional correlated spectrum can be simplified considerably by multiple-quantum filtering.[140,157] By blocking signals from unconnected spins, a double-quantum or higher order filter helps to reduce the intense, but largely uninformative, diagonal ridge characteristic of the correlated spectrum. Moreover, in many instances such a filter blocks large signals from solvent molecules, thus revealing weak features that would otherwise be obscured. These advantages are evident in an example of a double-quantum correlated spectrum reproduced in Fig. 46*a*. Obtained from basic pancreatic trypsin inhibitor[166] (molecular weight ≈ 6500) in D_2O, the spectrum clearly shows the cross peaks against a reduced set of diagonal peaks, including a greatly attenuated solvent peak at the position marked HDO.[163]

The analysis is simplified even further by the incorporation of an AX_3 topological filter, which limits the signals mainly to the six alanine residues in the protein, isolated from other residues by the deuterated amide backbone. Shown in Fig. 46*b*, the AX_3-selective four-quantum filtered spectrum contains all the alanine resonances plus two resonances from a threonine, which under some circumstances may behave as an AX_3 subsystem as well. Only one undesired peak, originating from leucine, passes through the filter.[163]

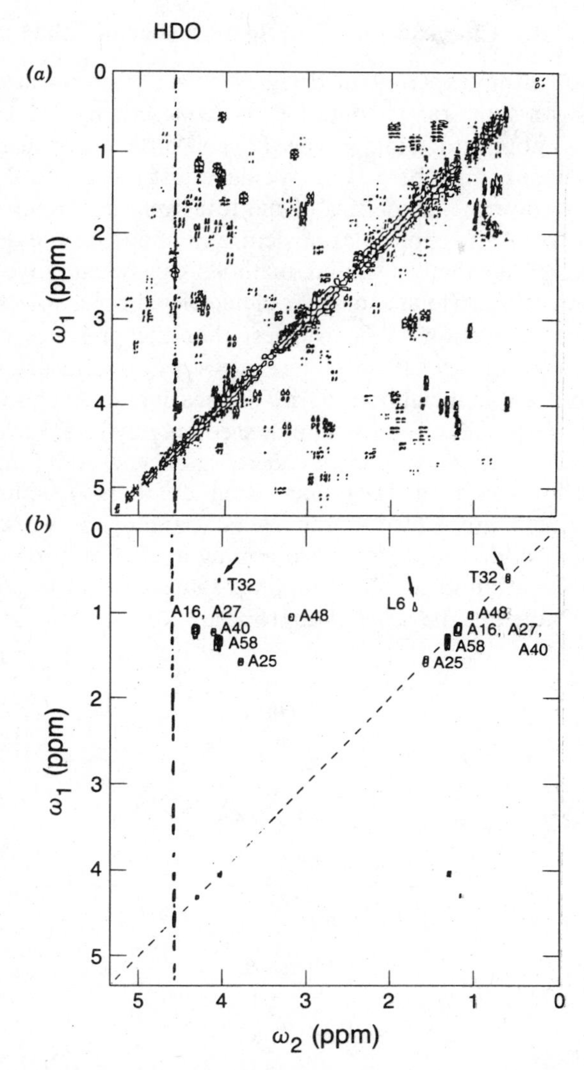

Figure 46. Two-dimensional correlated ^{1}H spectra of the aliphatic region of basic pancreatic trypsin inhibitor in D_2O. (*a*) Spectrum simplified by double-quantum filtering. The uncorrelated peaks on the diagonal, which include a solvent resonance at the position marked HDO, are attenuated considerably compared to the conventional spectrum. (*b*) Equivalent spectrum excited with AX_3 topological filtering and detected with a single $\pi/2$ pulse contains peaks originating mainly from six alanine residues and one threonine residue. Coherence transfer selection rules (Section IV.C.3) predict that the most important detection pathway is from the X spins to the A spin via 4-quantum coherence. Thus the favored A single-quantum signals, which appear in the upper left-hand corner of the spectrum, are correspondingly strong, while those of the X spins, which appear in the right-hand half of the spectrum, are correspondingly weak. (Reproduced from Ref. 163 with permission. © 1985 American Institute of Physics.)

B. Characterization of Spin Clusters in Solids

Multiple-quantum experiments designed to ascertain the size of a system generally rely on short-range couplings to define the number of interacting spins sharply enough to establish a *maximum* order of coherence, beyond which the system cannot be driven. We have just seen several examples of techniques that use the properties of the total spin coherence to elucidate coupling networks and molecular structure in solution; the question now, posed in Fig. 47, is whether similar methods can be employed to identify and characterize dilute isolated clusters of nuclei in dipolar solids.

The results of Section IV.E.3 suggest that the presence of an almost continuous range of dipolar couplings in a typical solid usually makes it unrealistic to draw artificial boundaries delineating molecules or functional groups. Its pattern of development influenced only by internuclear distances and angles, not by chemical bonds, coherence excited under an anisotropic dipolar Hamiltonian extends to increasing numbers of spins as distant coupling partners are added to an ever-expanding network. Clearly, if coherence in a solid is to be localized among a small group of spins, there must exist a variation in dipolar couplings sharp enough to create isolated clusters on whatever time scale is appropriate for the experiment at hand.

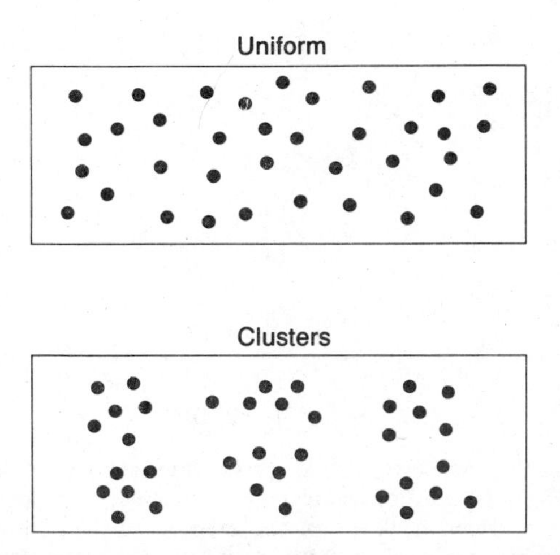

Figure 47. Use of multiple-quantum NMR to study spin clustering in solids. The single-quantum free induction decay in a dipolar solid reflects the response of independently excited spins to a collective local field. The homogeneously broadened spectrum so obtained usually lacks sufficient fine structure to distinguish a situation in which the nuclei are clustered in small groups from one in which they are distributed uniformly throughout the solid. An alternative approach to the problem uses collective multiple-quantum excitation patterns as a measure of the spatial distribution of the nuclei.

The notion of isolation therefore has no absolute meaning in a solid, and can only be defined operationally with respect to some particular multiple-quantum excitation scheme. The basic requirement is for the couplings between proximate spins within a cluster to be large enough to inhibit any communication with distant spins during the preparation period. In this situation the number of interacting spins is limited, at least temporarily, to the small, tightly coupled group; and $N(\tau)$, the time-dependent effective size of the system, will grow either very slowly or not at all once the largest couplings have been activated.

To assess the value of multiple-quantum NMR as a "spin counting" tool in studies of the dynamics of clustering in solids, we consider first a simple model system: the six methyl hydrogens on a molecule of 1,8-dimethylnaphthalene in which each of the ring positions is deuterated. In the monoclinic crystal structure assumed by dimethylnaphthalene, the methyl groups on adjacent molecules face each other in pairs, with the shortest intermolecular contact approximately equal to 2.0 Å.[167] Since this distance is less than the 2.93-Å separation between intramolecular methyl substituents, the dimer apparently supports a twelve-spin system, which is buffered from other groups of twelve spins by the large, magnetically inert aromatic rings that intervene. But the disparity between intracluster and intercluster dipolar couplings is not great enough to bound the development of coherence under the time-reversed $H_{(yy-xx)}$ excitation sequence. The growth of the spin system with time in a polycrystalline sample is reflected in the ^1H multiple-quantum spectra, shown in Fig. 48, by a steady growth of the number of orders observed and of the parameter $N(\tau)$ used to characterize the distribution. The pattern changes abruptly, however, when the molecule is incorporated into a perdeuterated lattice at a level of 5 mol-%. The spectra obtained from the solid solution, some of which are shown in Fig. 48 as well, never extend beyond $n = 6$, and $N(\tau)$ increases only gradually from 5 to 8 over the range $\tau = 66$ to 960 μs (Fig. 49). The conclusion, arrived at empirically, is that the additional dilution by the bulky perdeuterated molecules creates the physical, and therefore temporal, isolation necessary to excite just the pairs of methyl groups on the individual molecules. The slow increase in effective size above 6 arises from the inevitable influence of long-range couplings as the excitation time is increased.[131]

The example of dimethylnaphthalene suggests that multiple-quantum spectroscopy may be a useful method for characterizing homonuclear clusters in the solid state, provided that the separation between groups is sufficiently large. There are in fact many important systems in which the species of interest is present in a concentration low enough to meet even the most stringent standards of isolation. Among the most likely canidates are reactive molecules or intermediates either trapped in an inert matrix or

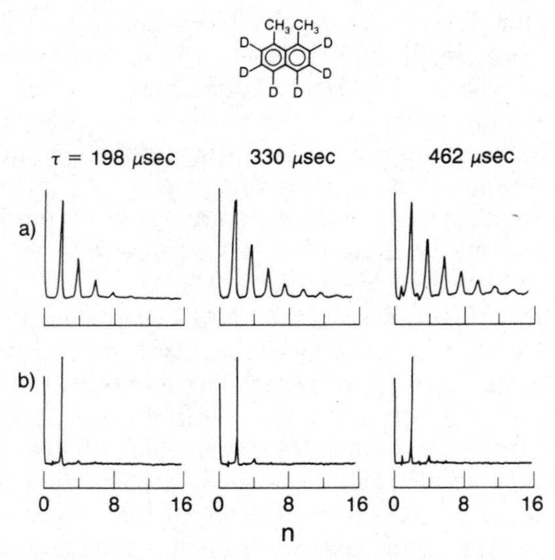

Figure 48. Multiple-quantum ^{1}H spectra of polycrystalline dimethylnaphthalene recorded at 360 MHz with the pulse sequence of Fig. 32. (*a*) Results for neat dimethylnaphthalene, with the ring positions deuterated as shown. (*b*) Results for a 5-mol-% solid solution of the same species in a perdeuterated host. The two sets of spectra highlight the different multiple-quantum excitation pathways possible in bounded and unbounded spin systems. Only low-order coherence can develop among the six isolated spins in the dilute system. Consequently the distribution of multiple-quantum spectral intensity changes very little over the range of preparation times shown. By contrast, the instantaneous effective size of the neat material increases continuously over the same range of times. (Reproduced from Ref. 131 with permission. © 1985 American Institute of Physics.)

adsorbed onto a surface at low coverage. When combined with other structural data, perhaps obtained also by NMR, knowledge of the *number* of atoms of a particular element may be all that is needed for a complete elucidation of the structure.

A striking demonstration of the utility of multiple-quantum spin counting is provided in a study of the intermediate state of acetylene adsorbed onto small particles of platinum.[168] In this work the amplitudes of single-, double-, and triple-quantum signals were used to discriminate among three proposed models: (1) 77% CH_2 and 23% HCCH, (2) 50% CH_3 and 50% CCH, and (3) 100% CCH. Statistical interpretations of the coherence distributions are of limited value for clusters this small, but it is meaningful to compare the experimental data to data obtained from simple systems of known structure under identical conditions. Following the analysis of the original paper, we note first that the mere existence of a triple-quantum signal rules out the

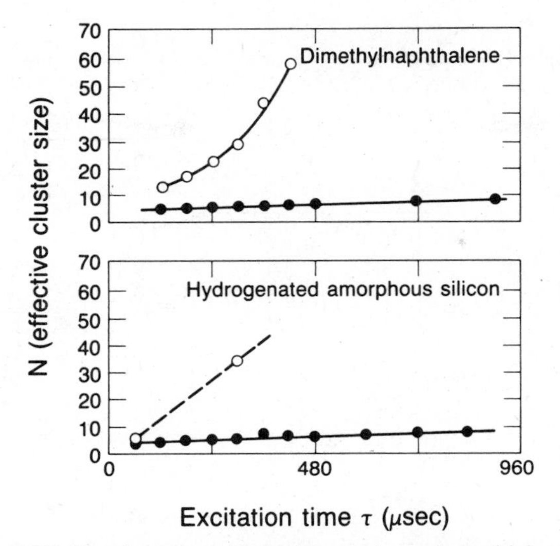

Figure 49. Plots of the time-dependent effective cluster size under multiple-quantum excitation for neat and dilute systems of dimethylnaphthalene and for 50-atom-% and 8-atom-% preparations of hydrogenated amorphous silcon. Smooth curves are drawn through the experimental points. In each set, open circles denote data for the extended spin systems and filled circles denote data for the limited spin configurations. The similarity between the results for the 8-atom-% hydrogenated amorphous silicon, a device-quality sample, and the dilute dimethylnaphalene system suggests that clusters of approximately six spins exist in the hydrogenated silicon. (See Ref. 170.)

third possibility, 100% CCH, a species which even when it occurs as a dimer is incapable of supporting triple-quantum coherence. The remaining choices, CCH_2 and CCH_3, are approximated physically by dilute frozen solutions of CH_2Cl_2 and CH_3CCl_3. The multiple-quantum spectral intensities for these model systems and for acetylene are reproduced in Table II. From these values it is clear that the surface-adsorbed species looks more like CH_2Cl_2 than CH_3CCl_3. The correct model for the intermediate therefore is the mixture of the two species CCH_2 and HCCH. Given these configurations, the triple-quantum coherence develops not from single units of acetylene, but from *pairs* of molecules interacting on the surface. The probability for two molecules to be close enough to interact is proportional to the concentration of the adsorbed species at low coverage, and is reflected in the spectra by the amplitude of the triple-quantum signal relative to that of the double-quantum signal.

Two possible structures for the CCH_2 species are sketched in Fig. 50*a*. With other NMR evidence showing that the carbon–carbon bond length is 1.44 Å, midway between a single and a double bond, it is likely that the

TABLE II
Amplitudes of Multiple-Quantum Coherences for Adsorbed Acetylene, Diluted CH_2Cl_2, and Diluted CH_3CCl_3.[a]

	$n = 1$	$n = 2$	$n = 3$	$\frac{n=3}{n=2}$
Acetylene				
19% coverage	3.2	3.0	0.44 ± 0.02	$(15 \pm 0.7)\%$
11% coverage	1.8	1.9	0.18 ± 0.02	$(9.5 \pm 1.0)\%$
CH_2Cl_2	18.6	15.4		
20% mole fraction				11%
11% mole fraction				6%
CH_3CCl_3	28.8	13.3	6.0	45%

[a]From Ref. 168.

correct structure lies somewhere between these extremes, similar perhaps to the pattern of chemical bonds in the stable molecule shown in Fig. 50*b*.[168]

Multiple-quantum spin counting techniques have also been applied to thin films of hydrogenated amorphous silicon in an effort to characterize the distribution of hydrogen in these materials. The basic material, amorphous silicon, acquires the properties of a semiconductor only after the incorporation of hydrogen, which is thought to exist in at least three distinct forms and environments: in molecular form as H_2, dispersed throughout the film; as well-separated silicon monohydride units; and as well-separated *groups* of hydrides. These different distributions are reflected in the conventional ^1H NMR spectrum by two featureless components—a 4-kHz Lorentzian signal arising from the dilute spins and a 24-kHz nearly Gaussian signal arising from the groups—from which it is difficult to deduce the average number of nuclei in a cluster.[169] By contrast, analysis of the growth of the effective size of the system under time-reversed multiple-quantum excitation (Fig. 49) suggests that the broad component of the spectrum originates from groups of five or six hydrogens, exhibiting a pattern of development similar to that observed in dimethylnaphthalene.[170]

An important technical point to note here is that since only the intensities of the multiple-quantum resonances are used to extract $N(\tau)$, information about the line widths need not be gathered. In principle, the integrated intensity of the signal from each order is contained in the first point of the interferogram. Accordingly, the evolution period may be omitted from the pulse sequence if only the intensities of the frequency components are sought. In this modified experiment, normal time-proportional phase incrementation serves to separate the signals from the different orders, which upon Fourier

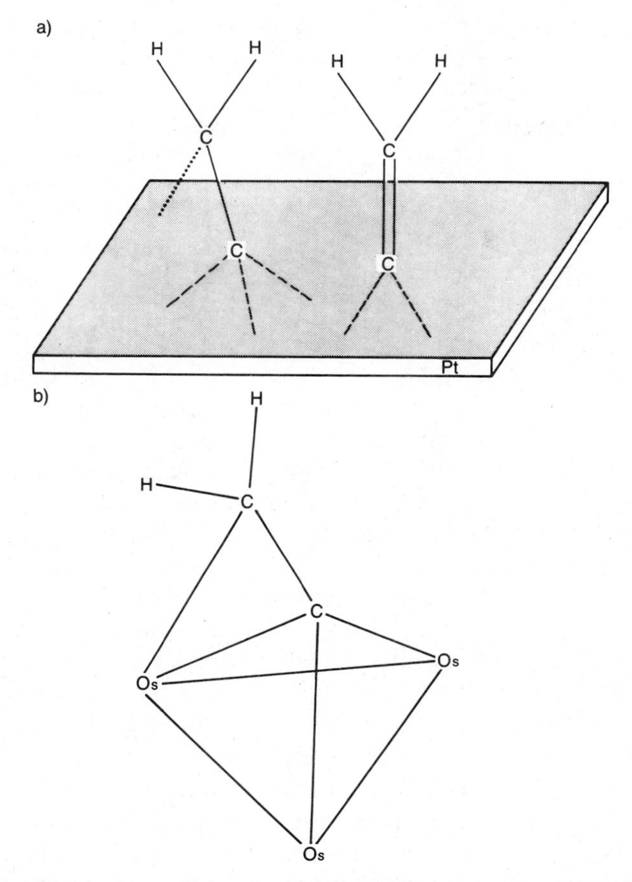

Figure 50. Possible structures of the acetylene-like CCH_2 species on a platinum surface as determined by multiple-quantum and conventional NMR methods. The actual carbon–carbon bond length is 1.44 Å, intermediate between a single and a double bond. The pattern of bonding in the stable compound $H_2Os_3(CO)_9$ has been suggested as a model for the attachment of acetylene to the surface of the metal. (Part *b* reproduced from Ref. 168 with permission.)

transformation appear as δ functions in the frequency domain. Both ease of operation and more accurate intensity distributions are gained in return for the loss of information concerning fine structure and line widths.[171]

C. Heteronuclear Coherence Transfer

Although seemingly omnipresent, hydrogen nuclei are neither omnipotent nor omniscient in their role as NMR probes of geometric and electronic structure. Information from other nuclei, taken either alone or together with what is obtained from ^{1}H spectra, is often needed to provide a complete picture of a molecular system, particularly in a liquid phase. Just a partial list of the many heteronuclei useful for NMR might include common spin-$\frac{1}{2}$ and spin-1 isotopes such as ^{13}C, ^{15}N, ^{31}P, ^{29}Si, ^{2}H, and ^{14}N, as well as more exotic species such as the metals ^{113}Cd, ^{119}Sn, and ^{199}Hg. The obvious problems of low gyromagnetic ratios and low natural abundances notwithstanding, the reduced sensitivity of these magnetically dilute, or "rare," spins frequently is offset by the benefits of a broad range of chemical shifts and large scalar couplings to hydrogen. Moreover, since the heteronuclear scalar coupling usually falls off rapidly when more than one chemical bond separates a pair of nuclei, a heteronucleus S typically finds itself at the center of a small well-defined subsystem I_mS that can be manipulated with a variety of coherence transfer techniques.[111,172,173] Numbering among the diverse applications of heteronuclear coherence transfer are experiments designed to move polarization from the abundant to the rare species, and at the same time, perhaps, isolate signals from I_mS groups of selected size; detect the spectrum of the rare spins via the more sensitive response of the abundant spins; obtain high-resolution spectra in inhomogeneous magnetic fields; and decouple homonuclear interactions in isotropic phases.[124,125,174–209] Through these experiments are combined the complementary advantages of high sensitivity and high resolution offered by the two species of spins. The various methods, though differing perhaps in the details of their operation, all exploit the properties of either multiple-quantum or multiple-spin modes, and are treated most conveniently as a group.

For the sake of simplicity we neglect interactions among both the abundant I spins and the dilute S spins—an approximation usually valid for $I = {}^1$H, where homonculear couplings are generally small compared to heteronuclear couplings, as well as for most S spins present in low concentrations. What remains is the interaction of just the unlike nuclei, which with a Hamiltonian

$$H_J = -\sum_{j=1}^{m} J_{jS} I_{zj} S_z \tag{190}$$

is formally identical to weak homonuclear coupling; it may even be argued that "like" nuclei become "unlike" as soon as their resonant frequencies diverge enough to short-circuit energy-conserving flip-flops in the rotating frame. For this reason the principal distinction between homonuclear and heteronuclear coherence transfer methods lies not in the formalism but in the experimental design, with heteronuclear experiments generally calling for the independent generation and application of pulses at both Larmor frequencies. Thus in its heteronuclear version, the refocused

$\pi/2$–$\tau/2$–π–$\tau/2$–$\pi/2$ sequence is supplemented by a simultaneous central π pulse and a final $\pi/2$ pulse at ω_{0S}, as shown in Fig. 51*a*.[111] Subjected to this array of five pulses, the system, initially described by the density operator

$$\rho(0) = \beta_I I_z + \beta_S S_z, \tag{191}$$

is transformed in essentially three stages. The first pulse, whose radiofrequency phase we arbitrarily take to be $y(\pi/2)$, acts only upon the I spins,

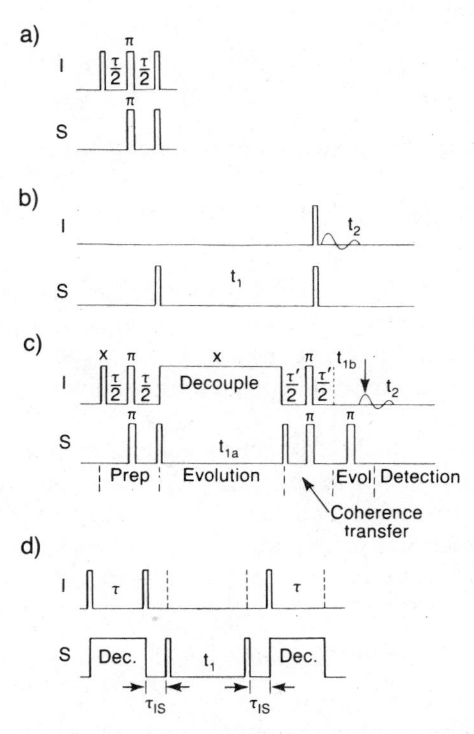

Figure 51. Pulse sequences for heteronuclear coherence transfer experiments. Unmarked pulses have flip angles of $\pi/2$. (*a*) Heteronuclear version of the refocused multiple-quantum excitation scheme of Fig. 20*b*. The radiofrequency phase difference between the I $\pi/2$ pulses determines whether the sequence transfers polarization from the I species to the S species or creates heteronuclear double-quantum coherence. (*b*) Experiment designed to correlate S chemical shift frequencies present during t_1 with I chemical shift frequencies present during t_2. (*c*) Experiment designed to eliminate inhomogeneous broadening due to magnetic field inhomogeneity while preserving information about the chemical shifts. (*d*) Excitation scheme based on the homonuclear I–I couplings. Homonuclear multiple-quantum coherence of order n_I is first created in the I subsystem by the two strong pulses while the S subsystem is decoupled. An additional $\pi/2$ pulse after a period τ_{IS} in the presence of the heteronuclear coupling adds an S coherence component of order $n_S = \pm 1$ to the existing coherence of order n_I.

rotating I_z to I_x. With the π pulses refocusing the linear interactions of both the I and the S spins while preserving the bilinear coupling between them ($I_zS_z \rightarrow [-I_z][-S_z]$), the density operator then develops into the combination

$$\rho(t^-) = \beta_I \sum_j [I_{xj} \cos(J_{jS}\tau/2) + I_{yj}S_z \sin(J_{jS}\tau/2)] + \beta_S S_z \qquad (192)$$

over the interval τ. The last pair of pulses generates either the odd-order (single-quantum) condition

$$\rho(t^+) = \beta_I \sum_j [I_{xj} \cos(J_{jS}\tau/2) - I_{zj}S_x \sin(J_{jS}\tau/2)] + \beta_S S_x, \qquad (193)$$

or the heteronuclear zero-quantum and double-quantum condition

$$\rho(\tau^+) = \beta_I \sum_j [-I_{zj} \cos(J_{jS}\tau/2) + I_{yj}S_x \sin(J_{jS}\tau/2)] + \beta_S S_X, \qquad (194)$$

for each *pair* of spins I_jS, depending upon whether the phase of the I pulse is x or y.[111] Little is changed from the homonuclear operator treatment except that the long suppressed constants $\beta_I = \hbar\omega_{0I}/kT$ and $\beta_S = \hbar\omega_{0S}/kT$ must be recalled to label the different equilibrium polarizations of the I and S spins.

This basic scheme of heteronuclear coherence transfer runs through many different experiments as pulses are added and deleted, phases and flip angles varied, and intervals adjusted to accomplish the desired ends. In the following we consider some of the more well-known examples, omitting discussion of the many improvements and refinements that have been proposed in nearly every case.[210]

1. *Polarization Transfer and Spectral Editing**

A complete conversion of $\beta_I I_z$, the initial order of the I spins, into the component $\beta_I I_z S_x$ occurs when the coupling interval is adjusted to $\tau = \pi/J$ and terminated with an x pulse at the I Larmor frequency. In the continued presence of the heteronuclear coupling this antiphase magnetization develops further into *observable* S magnetization, $\beta_I S_y \sin(Jt/2)$, its amplitude enhanced by a factor of β_I/β_S compared to the transverse magnetization $\beta_S S_x$ which is obtained following the application of a single $\pi/2$ pulse directly to the S spins. The enhancement, directly proportional to the ratio of the gyromagnetic constants γ_I/γ_S, is approximately equal to 4 in systems where

†See Refs. 174–188.

$I = {}^1H$ and $S = {}^{13}C$, and correspondingly greater for weaker heteronuclei. The result is an apparent transfer of the usually larger population difference of the I spins to the less sensitive S spins, but a *net* transfer of polarization does not occur, for the two enhanced frequency components are equal in intensity but opposite in phase.[174] All multiplet components can be generated with the same phase, however, if an additional refocusing period

$$\begin{array}{ll} I & \tau'/2\text{–}\pi\text{–}\tau'/2 \\ S & \quad\;\; \pi \quad \pi \end{array}$$

is inserted before acquisition of the S-free induction decay. During this time in-phase transverse magnetization S_x develops from the component I_zS_y, growing in as

$$S_x(\tau'; IS) = S_x \sin(J\tau'/2), \tag{195a}$$

$$S_x(\tau'; I_2S) = S_x \sin(J\tau'), \tag{195b}$$

and

$$S_x(\tau'; I_3S) = S_x[\sin(J\tau'/2) + \sin(3J\tau'/2)] \tag{195c}$$

with enhancement factors of γ_I/γ_S, γ_I/γ_S, and $3\gamma_I/4\gamma_S$ for IS, I_2S, and I_3S groups, respectively.[175] The transfer functions in Eq. (195) have been simplified by assuming that all coupling constants are equal. With this approximation, the sum and difference frequencies follow directly from the transformation

$$\rho(\tau') = \sum_{j=1}^{m} \prod_{k=1}^{m} \exp(-iI_{zk}S_zJ\tau')I_{zj}S_y \exp(iI_{zk}S_zJ\tau'), \tag{196}$$

and the differences among them permit the three subsystems to be distinguished. The refocused polarization transfer experiment therefore can be used as a spectral editing tool, capable of isolating the response of a subsystem when spectra obtained with different values of τ' are properly combined.

Also useful for polarization transfer and spectral editing is the "distortionless" sequence

$$\begin{array}{lll} I & (\pi/2)_y\text{–}\tau/2\text{–}\pi\text{–}\tau/2\text{–}\theta_x\text{–}\tau/2 & \text{decouple (optional)} \\ S & \qquad\quad (\pi/2)_y \qquad \pi & \text{acquire FID,} \end{array}$$

which for $\tau = \pi/J$ acts selectively on I_mS groups depending on the flip angle θ.[178-180] Since polarization enhancements of $(\gamma_I/\gamma_S) \sin\theta$, $(\gamma_I/\gamma_S) \sin 2\theta$, and $(3\gamma_I/4\gamma_S)(\sin\theta + \sin 3\theta)$ are obtained for $m = 1$, 2, and 3, respectively, flip angles of $\pi/4$, $\pi/2$, and 0.196π may be used to optimize the transfer in either the IS. I_2S, or I_3S subsystems. Individual subspectra now are generated by combining spectra obtained with different values of θ—not τ, which is chosen to reflect an average coupling and then held constant throughout the series of experiments. Thus are eliminated many of the problems arising from a distribution of coupling constants. Moreover, the sequence brings about a net transfer of polarization while preserving both the phases and the relative intensities of all components in each multiplet, thereby permitting a conventional analysis of the enhanced spectra. An example of a high-resolution ^{13}C spectrum, enhanced and then decomposed into CH, CH_2, and CH_3 subspectra, is provided in Fig. 52.[179]

The simpler pulse sequence

$$
\begin{array}{lll}
I & \theta_x & \\
S & (\pi/2)_x\text{–}\tau/2\text{–}\pi_x\text{–}\tau/2 & \text{– acquire FID}
\end{array}
$$

selects S, IS, I_2S, and I_3S groups without effecting a polarization transfer. In this experiment the central S spin, by acquiring multiple coupling partners over the excitation interval, acts as a seed nucleus for the formation of coherence. The radial pattern, through which may appear relatively high orders of heteronuclear multiple-quantum coherence, is to be contrasted with the sum of independent IS coherences that develops from an initial condition proportional to I_x or I_y. When τ is fixed at its usual value of π/J, the variable flip angle effectively controls the multiplicity by relegating signals from undesired subsystems to unobservable modes of coherence.[187,188]

2. *Chemical Shift Correlation and Indirect Detection**

Chemical shift information is introduced into a heteronuclear experiment when refocusing π pulses are removed. This option may be exercised separately for each species, and makes possible a number of additional useful manipulations. Consider, for example, the pulse sequence

$$
\begin{array}{lll}
\mathrm{I} & (\pi/2)_y & \text{– acquire FID} \\
S & (\pi/2)_y\text{–}t_1\text{–}(\pi/2)_y & -t_2
\end{array}
$$

†See Refs. 124, 172, 189–196.

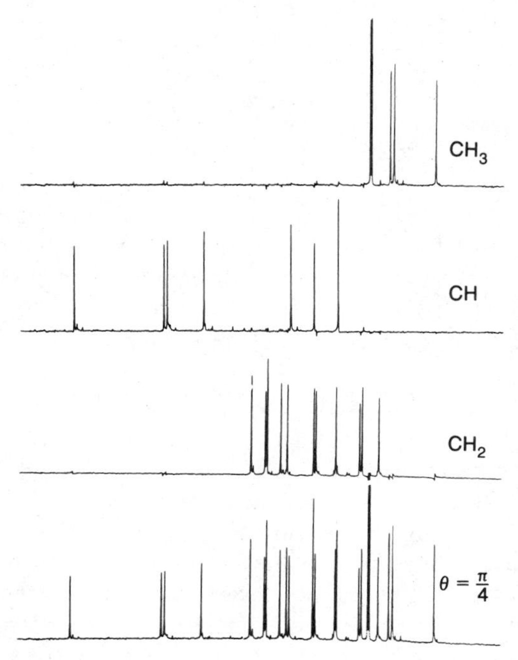

Figure 52. High-resolution ^{13}C spectra of cholesterol obtained with the distortionless polarization transfer method. At the bottom is a spectrum recorded with $\theta = \pi/4$, containing enhanced signals from all CH_m groups. The lines within each multiplet appear with the correct phases and with the correct relative intensities. Above are subspectra generated by combining results for $\theta_1 = \pi/4$, $\theta_2 = \pi/2$, and $\theta_3 = 3\pi/4$ according to the rules (CH) $[\theta_2 - z(\theta_1 + x\theta_3)]$, ($CH_2$) $[(\theta_1 - x\theta_3)/2]$, and ($CH_3$) $[(\theta_1 + x\theta_3)/2 - y\theta_2]$. The theoretical values of the parameters x, y, and z are 1.00, 0.71, and 0.00, respectively; in practice they are optimized experimentally. (Reproduced from Ref. 179 with permission. © 1982 Academic Press, Inc.)

shown in Fig. 51*b*. This scheme, which is similar to an early demonstration of heteronuclear coherence transfer by radiofrequency pulses, is designed to measure chemical shift interactions of the rare spins in one interval through their effect on the free induction decay of the abundant spins in a later interval.[172,189] Under the sequence, transverse S magnetization created by the first $\pi/2$ pulse evolves into components proportional to $I_zS_y \sin(Jt_1/2) \cos(\Omega_S t_1)$ and $I_zS_x \sin(Jt_1/2) \sin(\Omega_S t_1)$ while responding to both the chemical shift and the scalar coupling during t_1. The final $\pi/2$ pulses transform the antiphase S magnetization into antiphase I magnetization, which becomes observable during t_2 at the higher Larmor frequency. Fourier transformation of this signal, obtained point by point, reveals the spectrum of the S nuclei.

The sensitivity advantage inherent in a higher measuring frequency is pressed even further when the indirect detection scheme is preceded by a polarization transfer from I to S. A later version,

I $\quad (\pi/2)_y\text{–}\tau/2\text{–}\pi\text{–}\tau/2\text{–}(\pi/2)_x\text{–}t_1/2\text{–}\pi\text{–}t_1/2\text{–}\pi/2\text{–}\tau/2\text{–}\pi\text{–}\tau/2$ $\quad$ –acquire FID

S $\quad \pi \quad \pi/2 \quad \pi/2 \quad \pi$ $\quad$ –decouple

replaces the initial $\pi/2$ pulse on the S nuclei with the standard $I \rightarrow S$ polarization transfer sequence to produce the initial condition $\beta_I I_z S_x$. This enhanced mode evolves for a time t_1, during which the unmatched π pulse on the I spins removes both the I chemical shift interaction *and* the heteronuclear coupling, developing eventually into the combination

$$\rho(t_1) = 2\beta_I[\cos(\Omega_S t_1)I_z S_x + \sin(\Omega_S t_1)I_z S_y]. \tag{197}$$

The concluding polarization transfer sequence generates in-phase transverse I magnetization, modulated only by the S chemical shifts.[192] In this manner a simpler and more sensitive spectrum of the rare spins is obtained.

The single-quantum spectrum of the abundant spins can also be correlated with the sum and difference frequencies exhibited by the heteronuclear double-quantum and zero-quantum components generated with a pulse sequence such as

I $\quad (\pi/2)_y\text{–}\tau/2\text{–}\pi\text{–}\tau/2\text{–}(\pi/2)_y\text{–}t_1/2\text{–}\pi\text{–}t_1/2\text{–}$ $\quad$ acquire FID

S $\quad \pi \quad \pi/2 \quad -\pi/2\text{–}t_2.$

This experiment retains the sensitivity advantage of the double polarization transfer technique while employing fewer pulses.[111] The two methods differ operationally only in the relative phases of the first two $\pi/2$ pulses on the I spins.

Still fewer pulses are used in the sequence

I $\quad \pi/2\text{–}(\tau = \pi/J)\text{–}$ $\quad$ –acquire FID

S $\quad \pi/2\text{–}t_1\text{–}\pi/2\text{–}t_2,$

which also correlates IS zero-quantum and double-quantum frequencies with I single-quantum frequencies, but in a manner that is less susceptible to interference from spurious I signals inadvertently generated by the various pulses.[194,195] Only one pulse is applied to the I spins in order to establish the initial transverse magnetization $\beta_I I_x$. From here the system is transformed into a combination of zero-quantum and double-quantum coherence by the period of heteronuclear coupling and the first S $\pi/2$ pulse. For pairs of spins, the zero-quantum and double-quantum modes are spin–inversion coherences, which then oscillate independently of the spin–spin coupling. The second S pulse generates antiphase magnetization components modulated at the sums and differences of the I and S chemical shift frequencies. These components may either be observed immediately or transformed into in-phase magnetization during an additional delay. This method has been applied with great success to the study of ^{15}N chemical shifts, enhancing by a factor of 1000 the sensitivity of the nitrogens at their natural abundance of 0.36%.[196] A sample two-dimensional chemical shift correlation map, presented in Fig. 53, illustrates the effect. In this example, ^{15}N resonances in gramicidin S are correlated with ^{15}N–^{1}H zero-quantum and double-quantum resonances, thereby enhancing these signals selectively while suppressing those from nitrogens not bonded to hydrogens.[195]

3. *Removal of Magnetic Field Inhomogeneity*

Although easily surmounted in conventional applications, the technical problem of spatial inhomogeneity in the external magnetic field is severe

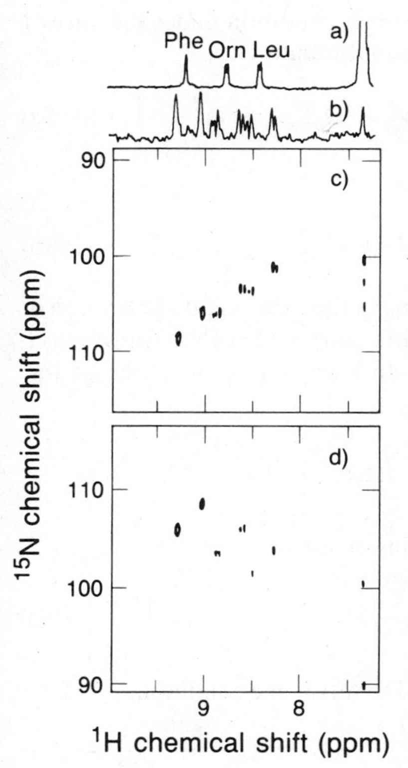

Figure 53. Indirect detection of ^{15}N chemical shifts via the ^{1}H spectrum, using heteronuclear zero-quantum and double-quantum coherence. (*a*) Imino region of the conventional high-resolution ^{1}H spectrum of gramicidin *S*. The two-dimensional spectrum correlating the double-quantum frequencies $\Omega_H - \Omega_N$ with the single-quantum frequencies Ω_H is shown in (*c*), with the projection onto the Ω_H axis directly above in (*b*). The projection contains signals only from ^{1}H nuclei directly connected to ^{15}N. An equivalent two-dimensional spectrum correlating the zero-quantum frequencies $\Omega_H + \Omega_N$ with Ω_H is shown in (*d*). The seemingly anomalous signs in the modulation frequencies arise from the negative gyromagnetic ratio of ^{15}N. (Reproduced with permission from A. Bax et al., *J. Am. Chem. Soc.*, **105**, 7188. © 1983 American Chemical Society.)

enough to degrade the resolution of spectra obtained from extremely large samples, such as those frequently encountered in topical and *in vivo* NMR experiments. Simple spin echo methods are of limited help in this situation, for in removing the effect of the field inhomogeneity they simultaneously remove the effect of the chemical shift interactions. Better suited to the problem is the coherence transfer echo, which offers more precise control over the rates of dephasing and rephasing of coherence in the external field.[122–125,211,212]

The experiment diagrammed in Fig. 51*c* is an example of such an approach.[197] Designed to obtain a high-resolution chemically shifted spectrum from a heteronucleus in an inhomogeneous magnetic field, it also exploits the presumably higher sensitivity of the coupling partner by employing the techniques of polarization transfer and indirect detection discussed in the preceding sections. This sequence differs noticeably from the others in its evolution period, however, which is broken into two distinct intervals separated by a period of coherence transfer. During the evolution period the double-quantum coherence $I_x S_y$ created during the preparation period responds first to the linear S spin interactions alone; and then, after conversion to $I_x S_z$ and ultimately to I_y, resumes evolution under the linear I spin interactions alone. The two desired Hamiltonians

$$H(t_{1a}) = -\gamma_S B_0(\mathbf{r}) S_z + \Omega_S S_z \tag{198a}$$

and

$$H(t_{1b}) = -\gamma_I B_0(\mathbf{r}) I_z + \Omega_I I_z \tag{198b}$$

are obtained by continuous-wave decoupling of the I spins and application of an S refocusing pulse, respectively. The initial amplitude of the transverse I magnetization eventually observed is modulated at the difference of the precession rates,

$$[\gamma_S B_0(\mathbf{r}) t_{1a} - \gamma_I B_0(\mathbf{r}) t_{1b}] + (\Omega_S t_{1a} - \Omega_I t_{1b}),$$

from which can be obtained the *homogeneous* frequency

$$\omega^{(1)} = \Omega_S - (\gamma_S/\gamma_I)\Omega_I \tag{199}$$

if the lengths of the subintervals are adjusted to satisfy the condition

$$\gamma_I t_{1b} = \gamma_S t_{1a}.$$

The effect of the experiment is illustrated in Fig. 54, in which a sharp spectrum of ^{13}C is correlated in two-dimensional fashion with a broad spectrum obtained in the same inhomogeneous magnetic field.

The basic idea here is to allow the magnetization to dephase on one species and rephase on another species, while ensuring that the net result is independent of nuclear positions. In this regard the method is similar to *total spin coherence transfer echo spectroscopy*, which removes the effect of field inhomogeneity in homonuclear systems by allowing n-quantum coherence to rephase through the total spin coherence of order N.[211,212] The rephasing of the N-quantum coherence under the resonance offset interaction alone leaves just the chemical shift and spin–spin couplings to determine the response.

4. *Alternative Excitation Schemes*

a. I–I Couplings. Where they exist and are of sufficient magnitude, homonuclear I–I couplings can be exploited to generate heteronuclear

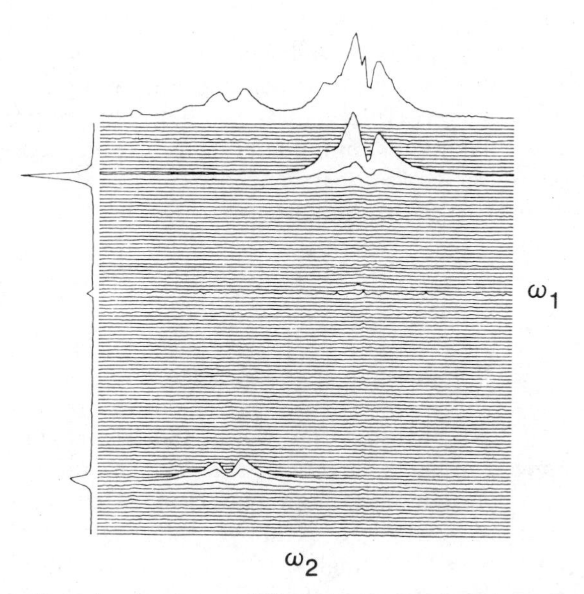

Figure 54. Two-dimensional spectrum of ^{13}C-enriched ethanol obtained with a pulse sequence incorporating heteronuclear coherence transfer echoes (Fig. 51*c*). The ^1H spectrum projected onto the ω_2 axis is broadened as the result of a severe inhomogeneity in the magnetic field (line widths ≈ 150 Hz). In contrast is the relatively narrow echo spectrum projected onto the ω_1 axis (line widths ≈ 6 Hz). Signals in the homogeneous dimension appear at the frequencies $\omega_1 = -4\omega_C/5 + \omega_H/5$, with each species contributing according to its gyromagnetic ratio. (Reproduced from Ref. 197 with permission. © 1985 Academic Press, Inc.)

multiple-quantum coherence even without the active presence of the heteronuclear interaction. Such opportunities arise most frequently in anisotropic systems, for $^1H-^1H$ dipolar couplings often are larger than the corresponding ^1H–rare-spin couplings. Since the introduction of homonuclear interactions, either strong or weak, by itself does nothing to connect the thermally isolated I and S subsystems, separate I and S Zeeman quantum numbers are retained; and the order of heteronuclear multiple-quantum coherence, $n = n_I + n_S$, is simply the sum of the two independent suborders. Consequently an n-quantum coherence just among the I spins, given by $n_I = n$ and $n_S = 0$, is properly treated as a heteronuclear coherence—as heteronuclear as the n-quantum coherence ($n_I = n - 1$, $n_S = 1$) of the same order.

Modes for which $n_I = n$ and $n_S = 0$ are realized experimentally by preparing the I spins as a group while decoupling the I–S interaction, as in Fig. 51*d*. Any of the standard homonuclear excitation sequences may be used to prepare the I spin coherence; subsequently an S $\pi/2$ pulse following a free period under the heteronuclear coupling may be applied to append a single-quantum S operator, S_x or S_y, to the n-quantum I component if desired.[125]

b. Strong Coupling Effects. Heteronuclear couplings, intrinsically weak, become strong when spins of the two species are forced to nutate at the same angular velocity in a frame rotating at both Larmor frequencies.[198] Individual spin angular momentum components I_z and S_z are no longer conserved under these conditions, which may be established via the simultaneous application of two radiofrequency fields,

$$\omega_{1I} = \gamma_I B_{1I} \cos(\omega_{0I}\tau) \tag{200a}$$

and

$$\omega_{1S} = \gamma_S B_{1S} \cos(\omega_{0S}\tau), \tag{200b}$$

equal in amplitude. Nevertheless, the *sum* of the two angular momenta, $I_z + S_z$, is quantized along a direction transverse to the static Zeeman field, and remains a constant of the motion under the recoupled Hamiltonian $\mathbf{I}\cdot\mathbf{S}$. It then becomes possible to exchange polarization and coherence, multiple-quantum as well as single-quantum, between the two species through the conservative flip-flop channels opened in the new frame of reference.[198–207] Thus the difference of two transverse angular momentum components evolves under a strong coupling interaction as

$$\begin{aligned} &\exp[-iJt\mathbf{I}_j\cdot\mathbf{I}_k](I_{xj} - I_{xk})\exp[iJt\mathbf{I}_j\cdot\mathbf{I}_k] \\ &\quad = (I_{xj} - I_{xk})\cos(Jt) + (2I_{yj}I_{zk} - 2I_{zj}I_{yk})\sin(Jt), \end{aligned} \tag{201}$$

while the difference of two longitudinal components, $I_{zj} - I_{zk}$, evolves into a zero-quantum coherence $(2I_{xj}I_{yk} - 2I_{yj}I_{xk})$.[54] Exact density operator calculations of this sort, feasible for small systems,[201-203] give way to thermodynamic descriptions in large systems such as solids.[198-200]

Weak heteronuclear couplings are also made strong when certain pulse sequences are applied simultaneously to both spin systems. These methods are designed to average the original Hamiltonian I_zS_z to $\mathbf{I}\cdot\mathbf{S}$, and can be used to effect a polarization transfer through mechanisms similar to those mentioned.[125] This kind of approach works for weakly coupled systems as well.[213,214]

D. Simplification of Spectra

1. Homonuclear Multiple-Quantum Spectroscopy

The conventional single-quantum spectrum of even one species is likely to become exceedingly complicated once the number of coupled spins grows beyond five or six.[215,216] Limitations of this sort are felt especially keenly in NMR studies of anisotropic fluids, where it is necessary to extract a full set of dipolar coupling constants from a spectrum of well-resolved lines in order to model details of the structure and conformation of a molecule.[217-220] Unfortunately there is an inevitable loss of resolution when larger and larger numbers of single-quantum transitions begin to appear over approximately the same frequency range. The trend is clearly evident in the ^{1}H spectra of Fig. 55, obtained from various small molecules oriented in liquid crystal solvents. Despite the high molecular symmetry that significantly reduces the number of distinct coupling constants in some of these systems, the spectra become increasingly complex with the addition of relatively small numbers of spins, and eventually lose all distinguishing features. At this point meaningful analysis is of course impossible.

The difficulty arises because the information carried by the single-quantum response—at most, $N(N + 1)/2$ coupling constants and N chemical shifts—is contained in a spectrum with up to $(2N)!/[(N + 1)!(N - 1)!]$ lines. In marked contrast, multiple-quantum spectra of the higher orders may convey the same information in less redundant form through a spectrum with far fewer lines. In particular, the $n = N - 1$ spectrum, which has no more than $2N$ components, and the $n = N - 2$ spectrum, which has no more than $N(2N - 1)$ components, are in many instances sufficient to yield all the parameters of the internal Hamiltonian. The advantages offered by multiple-quantum spectroscopy are suggested in Fig. 56, in which the five-, six-, and seven-quantum responses of an eight-spin system are compared with the conventional spectrum.[1,221] Although the problem of fitting the observed line positions and intensities to a set of coupling constants may still be

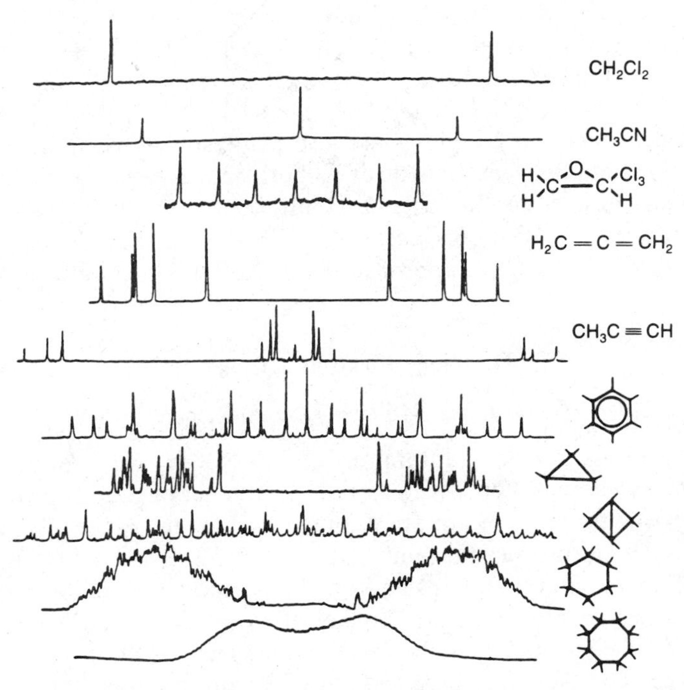

Figure 55. Conventional ^1H spectra of various molecules oriented in liquid crystal solvents. Highly informative for the smaller systems, the spectra become progressively more complex, and ultimately less informative, as the sizes of the solute molecules increase. (Courtesy of Z. Luz.)

formidable even for the comparatively few lines in a set of multiple-quantum spectra, the analysis can only be aided by stripping away as much of the redundant information as possible.

An analogy between the influence of local fields on the evolution of multiple-quantum coherence and the influence of isotopic labeling on the evolution of single-quantum coherence, though not a substitute for a proper group theoretical treatment,[215,216,220,222] is nonetheless useful for counting the number of allowed multiple-quantum resonances in certain circumstances.[223] Spins actively involved in a multiple-quantum coherence are distinguishable from those outside, and act in some respects as if they were nonmagnetic species. Thus the spectrum of an $(N - 1)$-quantum coherence may be viewed roughly as arising from the response of $N - 1$ spins-$\frac{1}{2}$ to the local field of one passive spin, up or down, through which appears two resonances for every distinct arrangement of active and passive spins. In the same way, replacement of $N - 1$ spins either by another isotope or

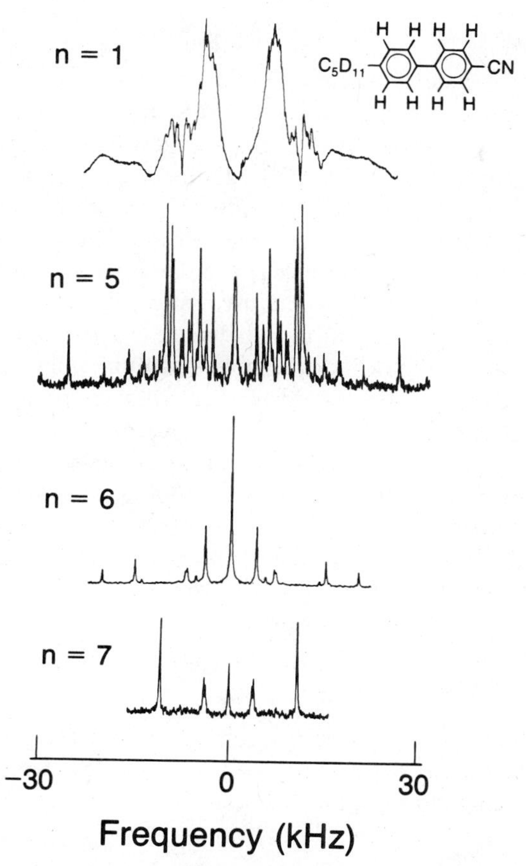

Figure 56. Comparison of the conventional single-quantum spectrum of the biphenyl portion of the liquid crystal 5CB-d_{11} with multiple-quantum spectra of orders 5, 6, and 7. In principle the 6-quantum and 7-quantum responses contain enough information to yield a full set of dipole–dipole coupling constants. (Adapted from Ref. 221. Original illustrations © 1984 Taylor and Francis, Ltd.)

chemical species reduces the number of single-quantum resonances to the number of distinguishable substituted isomers. For a highly symmetrical molecule such as benzene, where there is only one possible monoprotonated isomer, there are just two single-quantum transitions in the substituted form and two five-quantum transitions in the fully protonated form. Similar arguments concerning the $(N-2)$-quantum and lower order coherences complete the analogy between the spectral simplification realized by spin labeling on the one hand, and multiple-quantum labeling on the other.

2. *Heteronuclear Multiple-Quantum Spectroscopy**

Knowledge of the m heteronuclear coupling constants defining an I_mS grouping is important to the elucidation of the structure of the unit, sometimes proving to be as informative as knowledge of the more numerous homonuclear coupling constants. The desired information is contained in the response of the various orders of heteronuclear coherence which, once prepared, generally oscillate at characteristic arithmetic combinations of the m coupling frequencies. Homonuclear dipolar interactions, certainly a significant factor in anisotropic systems involving ^1H, may be used most expeditiously for this purpose. Employed during the excitation phase to promote the formation of high-order multiple-quantum coherence in the I_m subsystem, they are easily terminated during the evolution period to simplify the response. In practice the I–I Hamiltonian is averaged to zero during the evolution period by a multiple-pulse line narrowing sequence such as WHH-4, which at the same time reduces the I–S couplings uniformly by a factor κ.[45] With the chemical shift and resonance offset interactions controlled by refocusing pulses, elimination of the I–I dipolar coupling leaves the truncated heteronuclear interaction

$$H_{IS} = -\kappa \sum_{j=1}^{m} 2(D_{jS} + J_{jS}/2) I_{zj} S_z \tag{202}$$

and the usually small homonuclear indirect interaction $H_{J,\mathrm{II}}$ to determine the response over the interval t_1. Thus evolves an independent component $\rho^{n_I n_S}$ according to the equation of motion

$$\frac{d\rho^{n_I n_S}}{dt} = i[\rho^{n_I n_S}, H_{IS} + H_{J,\mathrm{II}}], \tag{203}$$

producing a spectrum that is usually simpler than that obtained from an equivalent order of homonuclear coherence responding to the homonuclear dipolar interactions.

Evaluation of the Liouville–von Neumann equation is straightforward in the product operator basis, for both the two single-spin raising and lowering operators

$$I_{+j} = |m_j = +\tfrac{1}{2}\rangle\langle m_j = -\tfrac{1}{2}|$$

and

$$I_{-j} = |m_j = -\tfrac{1}{2}\rangle\langle m_j = +\tfrac{1}{2}|$$

†See Ref. 125.

as well as the two single-spin zero-quantum operators

$$I_{0j}^{(+)} = |m_j = +\tfrac{1}{2}\rangle\langle m_j = +\tfrac{1}{2}|$$

and

$$I_{0j}^{(-)} = |m_j = -\tfrac{1}{2}\rangle\langle m_j = -\tfrac{1}{2}|$$

contributing to

$$I_{zj} = \tfrac{1}{2}[I_{0j}^{(+)} - I_{0j}^{(-)}]$$

are also eigenoperators of H_{IS}. Hence a coherence that commutes with $H_{J,\text{II}}$ oscillates harmonically as

$$\rho^{n_I n_S}(t_1) = \rho^{n_I n_S} \exp[i\omega(n_I, n_S)t_1], \tag{204}$$

with its frequency $\omega(n_I, n_S)$ given by the commutator

$$[\rho^{n_I n_S}, H_{IS}] = \omega(n_I, n_S)\rho^{n_I n_S}. \tag{205}$$

When the coherence fails to commute with the indirect coupling, extra lines may appear in the spectrum, depending on the order.[125]

Like their homonuclear counterparts, heteronuclear multiple-quantum spectra become increasingly complex as the order of coherence decreases but fail to provide any new information below some particular order. Once again the goal is to obtain the most information possible from the fewest spectral lines.

First, there is the combined total spin coherence ($n_I = m$, $n_S = 1$) for the $m + 1$ spin-$\frac{1}{2}$ nuclei. This component commutes with all heteronuclear and homonuclear couplings, and therefore oscillates only at the sum of the chemical shifts, which is zero by convention, and the resonance offset, which can be adjusted to zero. Its spectrum, a single line at zero frequency, can scarcely be simpler, but unfortunately conveys no information about the dipolar couplings.

One order removed, however, are the components

$$\rho^{m,0(+)} = S_0^{(+)} \prod_{j=1}^{m} I_{+j}$$

and

$$\rho^{m,0(-)} = S_0^{(-)} \prod_{j=1}^{m} I_{+j},$$

which, being total spin coherences for just the I_m subsystem, commute with $H_{J,\mathrm{II}}$ but not with H_{IS}. These coherences oscillate at the sum of the heteronuclear coupling frequencies,

$$\omega(n_I = m, n_S = 0) = \pm \sum_{j=1}^{m} (D_{jS} + J_{jS}/2), \tag{206}$$

scaled by κ if homonuclear decoupling is used during the evolution. The spectrum is a doublet with splitting $2\omega(m, 0)$.

More informative are the $2m$ coherences

$$\rho^{m-1,1} = S_+ I_{0k}^{(\pm)} \prod_{j \neq k} I_{+j},$$

which evolve under H_{IS} at individual frequencies of $\mp \kappa(D_{jS} + J_{jS}/2)$ to yield a spectrum of m doublets. Each splitting supplies the magnitude of one heteronuclear coupling constant, although not its sign. The same information is found in the response of strictly single-quantum S coherence to the 2^m static local field configurations of the m spins $I = \frac{1}{2}$, but in a more complex spectrum of 2^m lines rather than $2m$.[224-233] Moreover, unlike the multiple-quantum spectrum, which acquires no additional lines when nonvanishing homonuclear interactions are present, the single-quantum local field spectrum may be complicated considerably by the inclusion of homonuclear couplings.

Additional information is contained in the spectrum of the $4m$ coherences

$$\rho^{m-1,0} = S_0^{(\pm)} I_{0k}^{(\pm)} \prod_{j \neq k} I_{+j},$$

which oscillate at the frequencies

$$\omega(m-1, 0) = \pm \kappa \sum_{j \neq k} (D_{jS} + J_{jS}/2). \tag{207}$$

Each line in the spectrum appears at the sum of all but one of the coupling constants, different for each of the coherences, so a complete analysis of all the frequencies reveals both the magnitudes and the relative signs of the coupling constants.

The resolution advantages of heteronuclear multiple-quantum spectroscopy are well illustrated in Fig. 57, in which single-quantum S local field spectra are compared with high-order multiple-quantum spectra for the I_6S group formed by a ^{13}C nucleus in oriented benzene.[125]

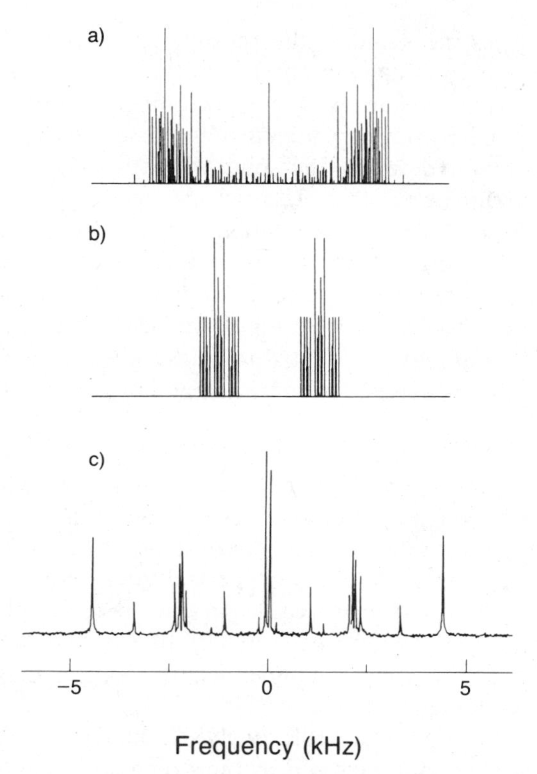

Figure 57. Demonstration of the resolution advantage inherent in heteronuclear multiple-quantum spectroscopy of dipole coupled systems. (*a*), (*b*) Simulations of the single-quantum spectrum of a ^{13}C nucleus in the local field of the six ^{1}H nuclei in oriented [1 − ^{13}C] benzene, (*a*) with and (*b*) without the involvement of the homonuclear ^{1}H–^{1}H interactions. (*c*) Experimental spectrum from the same system, but reflecting the response of the heteronuclear mode $n_I = 5$, $n_S = 0, \pm 1$ to an effectively simpler local field. No information is sacrificed, but the number and the density of the resonances in the multiple-quantum spectrum are reduced considerably. (Reproduced from Ref. 125 by permission © 1982 American Institute of Physics.)

E. Resonance Offset Effects: Applications of Multiple-Quantum NMR to Imaging and the Measurement of Diffusion Constants

Implicit in the design of most NMR experiments is the assumption that the applied magnetic field is nearly uniform throughout space, possessing at worst a small residual inhomogeneity responsible for negligible differences in Larmor frequency over the sample. The presence of a homogeneous field, now virtually guaranteed for samples of reasonable size, ensures that the resonant frequencies measured in the rotating frame originate exclusively from internal couplings and are unaffected by the macroscopic location of a nucleus. Yet there are instances where position in space is precisely the information sought, in which case the introduction of a *well-defined* inhomogeneity may be required to tag the spins with spatially varying resonant frequencies. Experiments in this class, which usually employ linear field gradients to establish the necessary relationship between resonant frequency and position, clearly can benefit from the n-fold enhancement of the offset perceived by an n-quantum coherence. To realize the n-quantum advantage, offset-dependent methods used, for example, in NMR imaging and in studies of molecular diffusion may be adapted to measure the accelerated response of multiple-quantum coherence under the same physical conditions.

1. NMR Imaging of Solids by Multiple-Quantum Resonance

The coordinates of a selected group of spins in a nonuniform magnetic field are revealed through the frequency or relaxation time of the associated NMR signal; at the same time the number of spins within the labeled volume is revealed through the corresponding spectral intensity. Imaging methods exploit these variations in spectroscopic parameters to ascertain the spatial distribution of the resonating nuclei, translating frequency and intensity into position and density. In this manner a static picture of the macroscopic structure of the material is obtained.[234–238]

The spatial resolution attainable by NMR imaging is limited by the magnitude of the spin couplings and by the strength of the gradient, for the externally imposed field within a volume must overpower any existing internal fields in order to label the region cleanly. This requirement is easily satisfied in mobile phases, where motional averaging eliminates large anisotropic interactions such as the dipole–dipole coupling. In rigid solids, however, local dipolar fields are often strong enough to degrade the resolution by competing with the additional external field supplied by the gradient. In these circumstances either the local field must be reduced with the aid of a multiple-pulse line narrowing sequence or the strength of the gradient must be increased, or both.[239–241] Since the ability to generate arbitrarily large gradients is limited by engineering considerations, any apparent intensification of the gradient afforded by multiple-quantum spectroscopy is a potentially valuable aid.

Shown in Fig. 58 are the results of a prototype solid-state imaging experiment in which even-order coherences generated by the eight-pulse $H_{(yy-xx)}$ sequence are allowed to evolve in a field gradient for a variable time t_1, and then are monitored after a time-reversed $-H_{(yy-xx)}$ mixing sequence.[242] As usual, the multiple-quantum spectrum is obtained as the Fourier transform of the interferogram $S(t_1)$. In this test the sample, composed

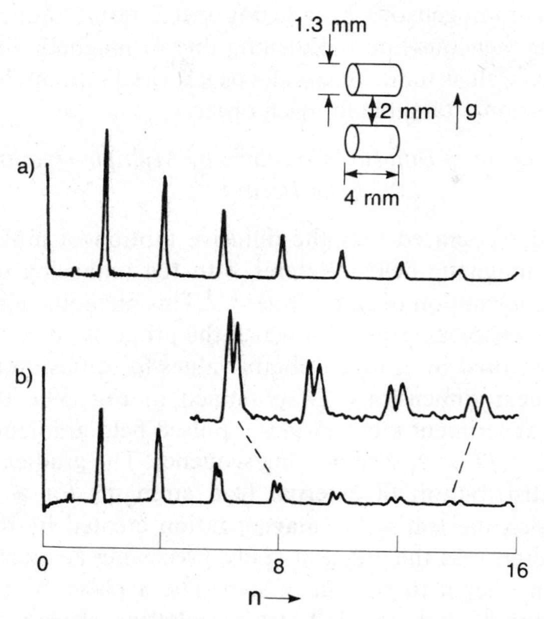

Figure 58. Application of multiple-quantum NMR to imaging in solids. Multiple-quantum ^{1}H spectra of polycrystalline adamantane (*a*) without and (*b*) with a static *z* gradient of 20 kHz/cm. Sample geometry is shown in the inset. The apparent *n*-fold intensification of the effect of the gradient on an *n*-quantum spectrum is evident in the progressive improvement in the resolution of the signals from the two plugs. (Reproduced from Ref. 242 with permission. © 1984 Academic Press, Inc.)

of two cylindrical plugs of polycrystalline adamantane separated by 2 mm, was oriented with its long axes perpendicular to a 20-kHz/cm *z* gradient and excited for 396 μs. Spectra recorded with the gradient in place and sorted according to order by time-proportional phase incrementation are displayed in Fig. 58*b*; these are to be compared with the spectra in Fig. 58*a*, obtained in a homogeneous field. Although separated multiple-quantum signals from the two plugs are not resolved in the absence of the gradient, in its presence individual peaks corresponding to the two sites are clearly evident. The signals just begin to separate at $n = 4$, and finally become well resolved at $n = 10$, where the gradient is effectively 10 times larger than for single-quantum coherence. Thus a gradient unable to distinguish the two plugs in a conventional imaging experiment proves to be sufficient when enhanced by the multiple-quantum response.

The multiple-quantum imaging experiment works because the linear increase in apparent resonance offset with coherence order outpaces any order-dependent homogeneous broadening mechanisms that may also be operative. Inhomogeneous line broadening due to magnetic field inhomogeneity, invited as well as uninvited, scales as $n\Delta\omega$ and ultimately determines the spatial resolution attainable for each order.

2. *Measurement of Diffusion Constants by Multiple-Quantum Spin Echo Decays*

It has long been recognized that the diffusive motion of molecules in an inhomogeneous magnetic field interferes with the rephasing of coherence necessary for the formation of spin echoes.[243] This phenomenon can be put to work in a spin echo experiment in which the progressive deterioration of the signal is monitored in order to obtain values for diffusion constants in one or more linear dimensions. Diagrammed in Fig. 59*a*, the standard single-quantum experiment incorporates a pulsed field gradient of variable length into a $\pi/2$–$t_1/2$–π–$t_1/2$ refocusing sequence. The gradient establishes a nonuniform distribution of external field strengths for a well-defined time, during which the transverse magnetization created by the $\pi/2$ pulse loses phase coherence as the different spins, precessing at rates determined by their positions, begin to get out of step. The π pulse, by reversing the *sense* of precession at each site, initiates a rephasing process, which to be effective must occur while all the molecules remain in place. A spin transported from its original location to a position where the magnetic field is different is unable to reverse its precessional history completely, and therefore fails to contribute fully to the echo.[244]

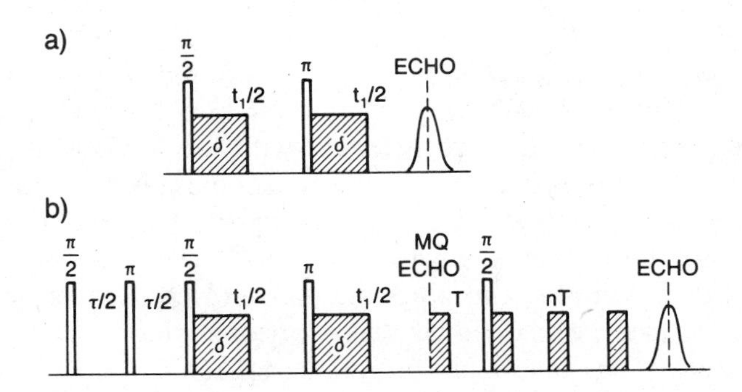

Figure 59. (*a*) Single-quantum pulsed gradient echo experiment for measuring diffusion constants. (*b*) Multiple-quantum adaptation. The amplitude of the echo is recorded for different durations of the gradient pulse δ at a fixed value of t_1. Cross-hatched areas denote gradient pulses. The detection method illustrated employs coherence transfer echoes to observe n-quantum signals selectively. Other techniques may be used as well. (Reproduced from Ref. 246 with permission. © 1983 American Institute of Physics.)

The intervals between pulses in the diffusion experiment are fixed, so that the elapsed time between the initial $\pi/2$ pulse and the appearance of the echo is always t_1, regardless of the duration of the gradient. As a result, homogeneous dephasing processes simultaneously present uniformly reduce the amplitude of the echo by a transverse relaxation factor $\exp(-t_1/T_2)$, which remains unchanged throughout any series of experiments wherein the length of the gradient pulse is varied. Under these conditions the signal amplitude is given by

$$\ln\left[\frac{S(\delta)}{S(0)}\right] = -\gamma_I^2 D\delta^2 g^2 \left(\frac{t_1}{2} - \frac{\delta}{3}\right) \exp\left(-\frac{t_1}{T_2}\right), \tag{208}$$

where D is the diffusion constant, g is the magnitude of the gradient, and δ is its duration.[244] The field is assumed to be homogeneous in the absence of the gradient.

A large gradient, which establishes a correspondingly large variation in precession rates across the sample, makes the experiment more sensitive to slow motions. Hence arises the now familiar multiplicity of quanta, and with it the modified equation[245,246]

$$\ln\left[\frac{S(\delta)}{S(0)}\right] = -n^2\gamma_I^2 D\delta^2 g^2 \left(\frac{t_1}{2} - \frac{\delta}{3}\right) \exp\left[-\frac{t_1}{T_2(n)}\right]. \tag{209}$$

The multiple-quantum version of the diffusion experiment, shown in Fig. 59*b*, simply uses a $\pi/2$–$\tau/2$–π–$\tau/2$–$\pi/2$ or $\pi/2$–τ–$\pi/2$ preparation sequence instead of a single $\pi/2$ pulse to create the coherence, and adds a final $\pi/2$ mixing pulse to monitor it. Signals from different orders may be separated either by coherence transfer echo filtering (Section IV.D.3) as illustrated, or by phase cycling. The transverse relaxation times $T_2(n)$, in general dependent upon the order of coherence, limit the time available for observation of the diffusive motion.

Some idea of the benefits of multiple-quantum spin echo experiments is conveyed in Table III, which reproduces data pertaining to the diffusion of dichloromethane, a two-spin ^{1}H system, in three different liquid crystalline phases.[245] In this table are compared values of translational diffusion constants for motion parallel and perpendicular to the applied magnetic field (Fig. 60*a*), obtained by monitoring the behavior of single-quantum and double-quantum spin echoes. It is hoped that this kind of information on the spatial anisotropy of diffusion in these systems will provide insight into the dynamics of the ordering process. Here even the modest twofold intensification of the gradient in the double-quantum experiment significantly

TABLE III

Relaxation Times (seconds) and Translational Diffusion Constants (10^{-6} cm^2/s) for Liquid Crystalline Solutions of Dichloromethane.[a]

Solvent	Phase 5 (N)[b]	8CB (N)[c]	8CB (Sm A)[d]
Concentration of CH_2Cl_2 (mol-%)	9.4	4.7	4.7
Temperature (°C)	25.2	32.7	22.9
Dipolar splitting (Hz)	4246	3425	4950
T_1	1.22 + 0.03	1.31 ± 0.04	1.0 ± 0.1
Single-quantum spin echo decays, $n = 1$			
$T_2(1)$	0.65 ± 0.04	0.70 ± 0.03	0.15 ± 0.01
D_z	1.92 ± 0.04	1.92 ± 0.13	1.16 ± 0.07
D_x	1.38 ± 0.16	1.78 ± 0.21	1.20 ± 0.15
Double-quantum spin echo decays, $n = 2$			
T_2	0.354 ± 0.006	0.411 ± 0.007	0.263 ± 0.004
D_z	1.95 ± 0.04	2.04 ± 0.05	1.04 ± 0.04
D_x	1.30 ± 0.07	1.73 ± 0.13	1.34 ± 0.06

[a]From Ref. 245.
[b]Merck Licristal phase 5 (nematic).
[c]4′-octyl-4-cyanobiphenyl (nematic).
[d]8CB smectic A.

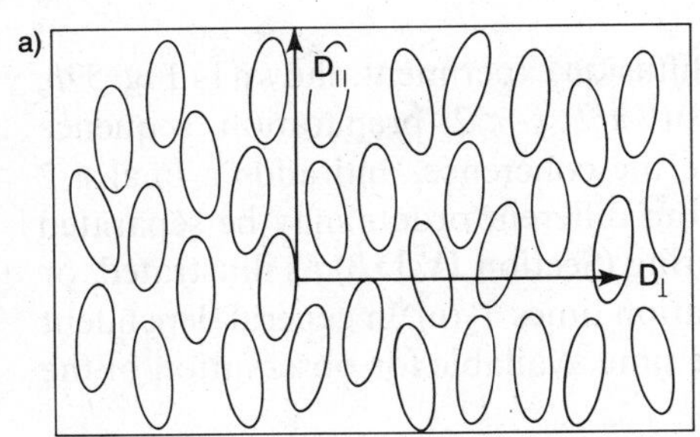

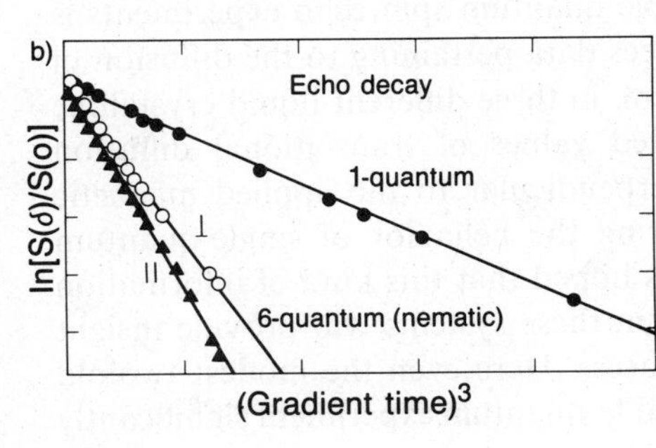

Figure 60. (*a*) Diffusion of molecules in an anisotropic medium in directions parallel and perpendicular to the applied magnetic field. (*b*) Plots of echo amplitude versus the cube of the gradient time for benzene. 6-quantum echoes observed during diffusion in a nematic solvent are compared with single-quantum echoes observed during diffusion in an isotropic phase. The intensification of the gradient in the multiple-quantum experiment is sufficient to reveal a 20% anisotropy in the diffusion rates.

improves the quality of the data, as evidenced by the reduced experimental uncertainty attached to the double-quantum results. The improvement in accuracy permits finer differences between small diffusion constants to be measured.

Even greater enhancements have been observed for 2-butyne and benzene in nematic phases, both six-spin systems.[246] The increased sensitivity of the multiple-quantum diffusion experiment is immediately apparent both in Fig. 60*b*, where six-quantum echoes are used to measure parallel and perpendicular diffusion constants differing by only 20%, and in Fig. 61, where the complete dependence of the echo amplitude for benzene on the order of coherence is illustrated. The intensification is of course most pronounced for the total spin coherence, $n = 6$, which in addition to being most sensitive to the resonance offset is unaffected by the spin–spin couplings. At the same time we should recall that selective excitation methods of the kind discussed in Section IV.F are generally needed to overcome the low signal-to-noise ratios observed in spectra of the highest orders.

F. Relaxation and Correlated Random Fields

To this point we have largely ignored the relaxation processes that restore a perturbed spin system to its original state of equilibrium. Nevertheless, coherences and nonequilibrium population differences inevitably disappear, often (but not always) decaying exponentially with characteristic time constants.[247] In general, every coherence term in the density operator has associated with it an individual transverse relaxation time T_2, while every population term has associated with it a longitudinal, or spin–lattice,

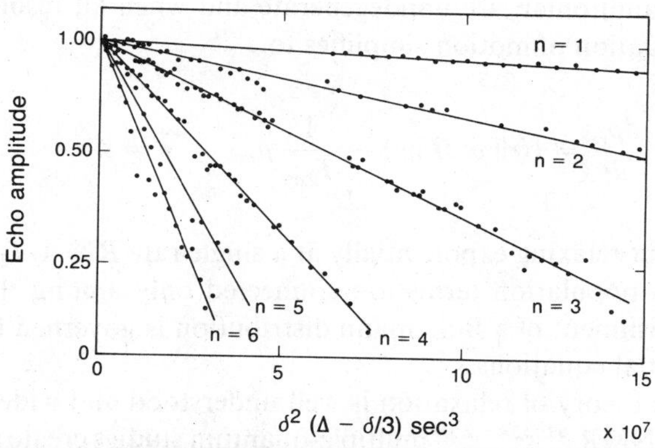

Figure 61. Results of the diffusion experiment for all *n*-quantum orders of benzene dissolved in a liquid crystal. The slopes of the lines vary as n^2. $\Delta \equiv t_1/2$. (Reproduced from Ref. 246 with permission. ©1983 American Institute of Physics.)

relaxation time T_1. Both the transverse and the longitudinal recovery processes derive from random fluctuations of magnetic fields originating either internally, with the spins themselves, or externally, with fluctuations of the lattice. Thus understood, any apparent damping of the oscillatory behavior of the system arising solely from the dynamic influence of a static Hamiltonian is deterministic, and therefore not relaxation in the sense indicated here. Rather, only those phenomena not susceptible to or not subjected to time reversal manipulations are classified as irreversible relaxation. Such processes, typically motions of the spins or lattice that cause the local fields to vary randomly with time, permanently reduce the norm of the density operator, and are accounted for by the addition of specific relaxation terms to the Liouville–von Neumann equation. The damped oscillations of the coherent states of the system then appear as broadenings of the resonances in the frequency domain. Standard techniques such as spin echo,[243,248,249] spin locking,[250-252] and inversion-recovery sequences,[253] appropriately adapted, may be used to obtain multiple-quantum line widths and relaxation times.[65,80]

In the absence of any rapidly oscillating terms, the modified equation of motion

$$\frac{d\rho_{rs}}{dt} = i\langle r|[\rho, H]|s\rangle + \sum_{t,u} R_{rs,tu}(\rho_{tu} - \rho_{tu}^{\mathrm{eq}}), \tag{210}$$

introduced by Redfield, allows each density operator component to return to equilibrium at a rate determined both by the instantaneous values of the other density operator components *of the same order* and by the values of the elements of a time-independent relaxation matrix **R**.[254,255] When all eigenvalues of the Hamiltonian are nondegenerate and when all resonances are resolved, the equation of motion simplifies to

$$\frac{d\rho_{rs}}{dt} = i\langle r|[\rho, H]|s\rangle - \frac{1}{T_{2,rs}}\rho_{rs}, \qquad r \neq s, \tag{211}$$

with each element relaxing exponentially at a single rate $R = 1/T_{2,rs}$. Under these conditions population terms are connected only among themselves, and the reestablishment of a Boltzmann distribution is governed by a set of coupled differential equations.

Although the theory of relaxation is well understood and widely applied in conventional NMR,[61,254-259] multiple-quantum studies create additional opportunities by providing experimental access to the complete relaxation matrix.[65,260–270] The elements of this matrix

$$R_{rs,tu} = \frac{1}{2\hbar^2}\{J_{rtsu}[(M_s - M_u)\omega_0] + J_{rtsu}[(M_r - M_t)\omega_0]$$

$$- \delta_{su}\sum_v J_{vtvr}[(M_v - M_t)\omega_0]$$

$$-\delta_{rt}\sum_v J_{vsvu}[(M_v - M_u)\omega_0]\} \tag{212}$$

depend upon the quantities

$$J_{rstu}(\omega) = \int_{-\infty}^{\infty} \overline{\langle r|H_{\mathrm{rel}}(\tau)|s\rangle\langle t|H_{\mathrm{rel}}(\tau + \tau')|u\rangle} \exp(-i\omega\tau')\, d\tau'$$

$$\equiv \int_{-\infty}^{\infty} G(\tau, \tau') \exp(-i\omega\tau')\, d\tau', \tag{213}$$

which give the spectral densities of the ensemble averaged correlation function of the time-dependent Hamiltonian responsible for the relaxation. For the same reasons that n-quantum coherences respond in different ways to the various static internal Hamiltonians, they are also affected in different ways by the various relaxation mechanisms. For example, zero-quantum coherences are invariant to any time-dependent Hamiltonians linear in I_z, whereas all other coherences perceive the effect as increasing linearly with n. Fluctuating local fields of this sort may arise in a number of ways, perhaps through a modulation of the anisotropic Zeeman and chemical shift interactions by rotational diffusion or through weak coupling interactions with external dipoles found on paramagnetic impurities and nuclei of other molecules.[256] Also apparent from the basic properties of coherence is the invariance of the *total* spin coherence to any fluctuations of a bilinear Hamiltonian—for instance, an intramolecular dipolar or quadrupolar mechanism. Moreover, since a multiple-quantum coherence in a coupled system requires the participation of more than one spin, its relaxation is sensitive to correlated local field differences at all the nuclear sites concerned, and therefore provides a measure of the degree to which motion in the system is correlated.

This notion of correlated fluctuations is depicted schematically in Fig. 62. To illustrate the phenomenon with a specific example, we consider the effects of paramagnetic relaxation agents on the two J-coupled ^{1}H nuclei in 2,3-dibromothiophene. The high-resolution zero-, single-, and double-quantum spectra of this molecule are shown in Fig. 63 together with spectra

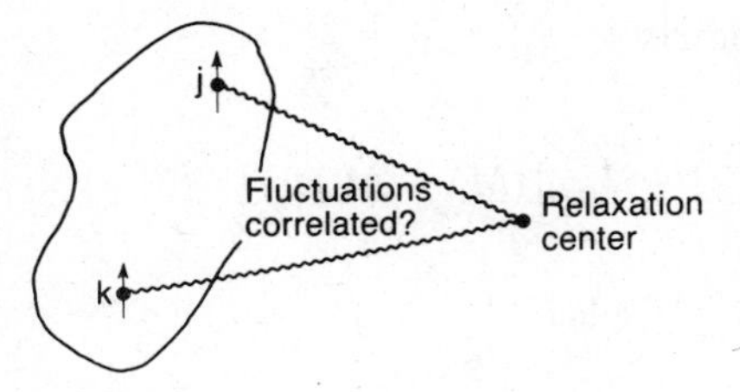

Figure 62. Multiple-quantum relaxation and correlated motions. Spin relaxation results from fluctuations in the local fields perceived by the nuclei. When the fluctuations affect more than one position, as they do when the coherence involves more than one spin, the relaxation is sensitive to correlations in the fields at the different sites.

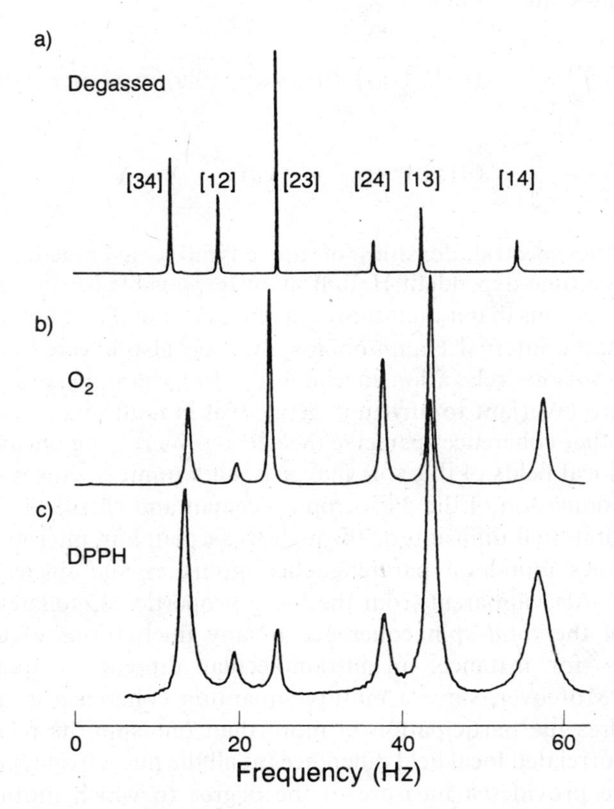

Figure 63. Multiple-quantum ^1H spectra of 2,3-dibromothiophene obtained in the presence of different relaxation agents. (*a*) Degassed solution in hexafluorobenzene. (*b*) Solution saturated with O_2. (*c*) Solution containing 0.015*M* DPPH. In each set [2, 3] is the zero-quantum resonance and [1, 4] is the double-quantum resonance; all others are single-quantum. The line widths provide an indication of the degree to which the fluctuating field of the paramagnetic species is correlated at the two nuclear positions in 2,3-dibromothiophene. (Reproduced from Ref. 62 with permission. © 1978 Taylor and Francis, Ltd.)

obtained in the presence of small concentrations of oxygen and 1,1-diphenyl-2-picryl-hydrazyl (DPPH).[62] The line widths of the zero-quantum and double-quantum resonances differ according to the variations of the electronic dipolar fields felt by *both* nuclei, an effect that depends upon the size of the paramagnetic molecule. The larger molecule, DPPH, with its more diffuse electronic distribution is expected to produce fields of similar magnitude at each site; whereas the smaller molecule, oxygen, which can approach the nuclear system more closely, is more likely to relax one of the spins selectively. Indeed, analysis of the line widths reveals that the perceived local fields of DPPH are correlated to a higher degree than those of oxygen when measured by a two-site correlation coefficient C_{jk} defined through the relationship

$$\overline{B_{j\alpha}(\tau)B_{k\beta}(\tau + \tau')} = C_{jk}\sqrt{\overline{B_{j\alpha}^2} \times \overline{B_{k\alpha}^2}} \exp(-\tau'/\tau_c), \quad \alpha, \beta = \text{x, y, z}. \tag{214}$$

Here τ_c is the correlation time and $B_{j\alpha}$ and $B_{k\alpha}$ are Cartesian components of the local field of the electron at each of the two nuclei. Other experiments and calculations pertaining to cross correlations in local fields have also been reported for a two-spin heteronuclear system, in which frequency-selective excitation was employed to generate coherence between specific pairs of levels;[95] and for neighboring methyl groups executing threefold motions about their symmetry axes.[271]

When a molecule with symmetry is relaxed by a fluctuating local field, the extent of cross correlation influences the pathways available, often opening up routes otherwise forbidden. Unlike the situation existing under normal excitation by radiofrequency, states belonging to *different* irreducible representations of the molecular symmetry group can be connected by a fluctuating dipolar Hamiltonian, provided that the fluctuations are incompletely correlated.[271,272] The three hydrogen nuclei constituting the methyl group in acetonitrile, CH_3CN, offer an illustrative example. Here the eight eigenstates are the symmetry-adapted linear combinations

$$|A_{3/2}\rangle = |+++\rangle$$

$$|A_{1/2}\rangle = \frac{1}{\sqrt{3}}(|-++\rangle + |+-+\rangle + |++-\rangle)$$

$$|A_{-1/2}\rangle = \frac{1}{\sqrt{3}}(|+--\rangle + |-+-\rangle + |--+\rangle)$$

$$|A_{-3/2}\rangle = |---\rangle$$

$$|E^a_{1/2}\rangle = \frac{1}{\sqrt{3}}(|++-\rangle + \varepsilon|+-+\rangle + \varepsilon^*|-++\rangle)$$

$$|E^a_{-1/2}\rangle = \frac{1}{\sqrt{3}}(|--+\rangle + \varepsilon|-+-\rangle + \varepsilon^*|+--\rangle)$$

$$|E^b_{1/2}\rangle = \frac{1}{\sqrt{3}}(|++-\rangle + \varepsilon^*|+-+\rangle + \varepsilon|-++\rangle)$$

$$|E^b_{-1/2}\rangle = \frac{1}{\sqrt{3}}(|--+\rangle + \varepsilon^*|-+-\rangle + \varepsilon|+--\rangle),$$

where $\varepsilon = \exp[i2\pi/3]$. These combinations transform individually under the symmetry operations of C_3 as the designated irreducible representations. With the three nuclei and the one electron S arranged as shown in Fig. 64, the electron–nucleus dipole–dipole interaction

$$H_{\text{rel}}(\tau) = \sum_{j=1,2,3} \hbar^2 \gamma_I \gamma_S \left[\frac{\mathbf{I}_j \cdot \mathbf{S}}{r_j^3} - 3 \frac{(\mathbf{I}_j \cdot \mathbf{r}_j)(\mathbf{S} \cdot \mathbf{r}_j)}{r_j^5} \right] \tag{215}$$

may be rewritten using the symmetry-adapted operators

$$\mathbf{I}_\mu = \frac{1}{\sqrt{3}}(\mathbf{I}_1 + \lambda \mathbf{I}_2 + \lambda^* \mathbf{I}_3) \tag{216a}$$

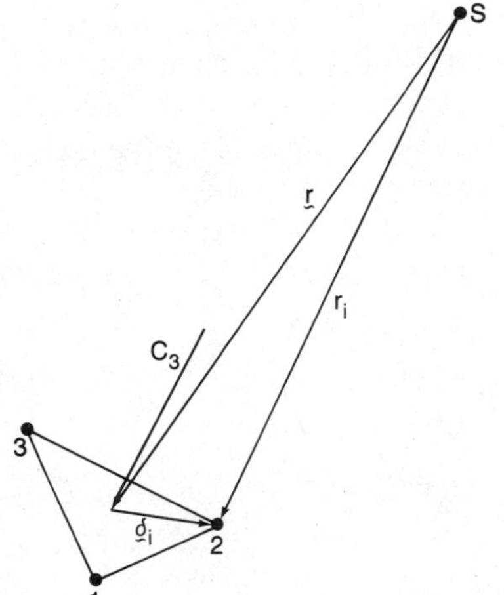

Figure 64. Geometry of interacting methyl group and electron S. Subscripts pertain to each of the three methyl positions. (Reproduced from Ref. 272 with permission. © 1980 American Institute of Physics.)

and

$$\boldsymbol{\delta}_{\bar{\mu}} = \frac{1}{\sqrt{3}}(\boldsymbol{\delta}_1 + \lambda^*\boldsymbol{\delta}_2 + \lambda\boldsymbol{\delta}_3), \tag{216b}$$

where $\mu = A, E_a$, or E_b for $\lambda = 1, \varepsilon$, or ε^*. We then have

$$\begin{aligned} H_{\text{rel}}(\tau) = {} & \sqrt{3}\hbar^2\gamma_I\gamma_S\left[\frac{\mathbf{I}_A\cdot\mathbf{S}}{r^3} - \frac{3(\mathbf{I}_A\cdot\mathbf{r})(\mathbf{S}\cdot\mathbf{r})}{r^5}\right] \\ & - \frac{3\hbar^2\gamma_I\gamma_S}{r^5}\sum_{\mu=E_a,E_b}[(\mathbf{I}_\mu\cdot\boldsymbol{\delta}_{\bar{\mu}})(\mathbf{S}\cdot\mathbf{r}) + (\mathbf{I}_\mu\cdot\mathbf{r})(\mathbf{S}\cdot\boldsymbol{\delta}_{\bar{\mu}}) \\ & + (\mathbf{I}_\mu\cdot\mathbf{S})(\mathbf{r}\cdot\boldsymbol{\delta}_{\bar{\mu}})] + \frac{15\hbar^2\gamma_I\gamma_S}{r^7}\sum_{\mu=E_a,E_b}(\mathbf{I}_\mu\cdot\mathbf{r})(\mathbf{S}\cdot\mathbf{r})(\mathbf{r}\cdot\boldsymbol{\delta}_\mu) \end{aligned} \tag{217}$$

for dilute concentrations of electrons ($\delta/r \ll 1$).[273] Thus decomposed, the Hamiltonian connects states of the same symmetry through its first term, which contains only A operators, and states of different symmetry through its second and third terms, which contain E_a and E_b operators. Whether the symmetry is actually broken, though, depends on the extent of cross-correlation. The decay rates

$$R_{3/2,-3/2} = \frac{1}{\hbar^2}\left[\frac{3}{2}J_A(0) + \frac{1}{2}J_A(\omega_{0I}) + J_E(\omega_{0I})\right] \tag{218a}$$

$$R_{3/2,-1/2} = R_{1/2,-3/2} = \frac{1}{\hbar^2}\left[\frac{2}{3}J_A(0) + \frac{5}{6}J_A(\omega_{0I}) + \frac{1}{3}J_{E_a}(0) + \frac{2}{3}J_{E_a}(\omega_{0I})\right] \tag{218b}$$

$$R_{3/2,1/2} = R_{-1/2,-3/2} = \frac{1}{\hbar^2}\left[\frac{1}{6}J_A(0) + \frac{5}{6}J_A(\omega_{0I}) + \frac{1}{3}J_{E_a}(0) + \frac{2}{3}J_{E_a}(\omega_{0I})\right] \tag{218c}$$

are determined by spectral densities of symmetry A and E_a, which in turn are related to the Fourier transforms of the one-site autocorrelation function G_{jj} and the two-site cross-correlation function G_{jk} through

$$J_A(0) = 2G_{jj}(0)(1 + 2\xi) \tag{219a}$$

$$J_A(\omega_{0I}) = 2G_{jj}(\omega_{0I})(1 + 2\xi) \tag{219b}$$

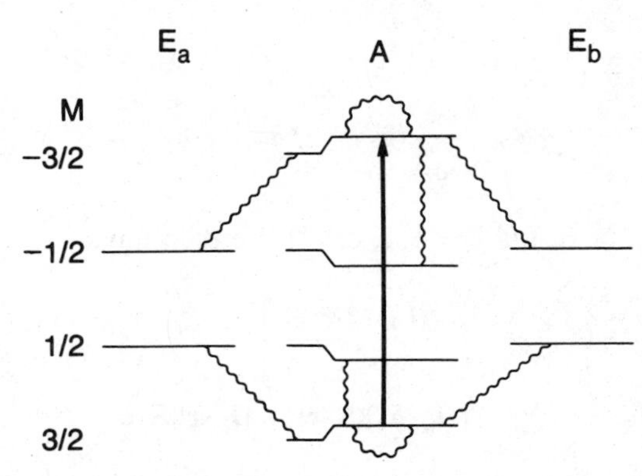

Figure 65. Possible relaxation pathways for the triple-quantum transition in oriented acetonitrile. Spectral densities with A symmetry connect states belonging to the same irreducible representation. Incompletely correlated fluctuations break the symmetry, and lead to relaxation between A and E states. (From Ref. 271.)

$$J_{E_a}(0) = J_{E_b}(0) = 2G_{jj}(0)\,(1 - \xi) \tag{219c}$$

$$J_{E_a}(\omega_{0I}) = J_{E_b}(\omega_{0I}) = 2G_{jj}(\omega_{0I})\,(1 - \xi), \tag{219d}$$

where G_{jj} and G_{jk} are defined in Eq. (213) and $\xi = G_{jk}/G_{jj}$. Hence for completely correlated fluctuations ($\xi = 1$), through which each nucleus experiences the same field from the electron, only the symmetry-conserving pathways allowed by the A spectral density functions are available; while for completely uncorrelated fluctuations ($\xi = 0$), by which each nucleus is relaxed independently, intersymmetry relaxation is allowed (Fig. 65). In the general case, the rates and pathways depend upon the correlation time of the fluctuations, the Larmor frequencies of the nucleus and electron, and the ratio of the cross-correlation function to the autocorrelation function.[272]

Acknowledgments

The preparation of this review was facilitated by the cooperation of Dr. A. Bax, Dr. G. Bodenhausen, Dr. D. M. Doddrell, Dr. G. Drobny, Dr. R. R. Ernst, Dr. R. Freeman, Dr. T. Hashi, Dr. R. Kaptein, Dr. C. P. Slichter, and Dr. D. P. Weitekamp, who consented to the reproduction of their published figures and who kindly provided us with high-quality copies of the drawings. We thank Dr. M. Mehring for a critical reading of the manuscript prior to publication, and Dr. G. J. Ray of Amoco Corporation for communicating to us the results of a computerized literature search. We are also grateful to several other individuals for help in preparing the manuscript: Ralph Dennis of the Technical Information Department of the

Lawrence Berkeley Laboratory coordinated all phases of the illustration work. Phyllis Munowitz and Jean Baum provided timely assistance with proofreading and illustration, and Theresa Allyn typed the final version of the manuscript under pressure of a deadline. This work was supported by the Director, Office of Energy Research, Office of Basic Energy Sciences, Materials Science Division of the U.S. Department of Energy, under Contract DE-AC03-76SF00098.

References

1. S. Sinton and A. Pines, *Chem. Phys. Lett.* **76**, 263 (1980).
2. W. A. Anderson, *Phys. Rev.* **104**, 850 (1956).
3. J. I. Kaplan and S. Meiboom, *Phys. Rev.* **106**, 499 (1957).
4. S. Yatsiv, *Phys. Rev.* **113**, 1522 (1959).
5. W. A. Anderson, R. Freeman, and C. A. Reilly, *J. Chem. Phys.* **39**, 1518 (1963).
6. K. A. McLauchlan and D. H. Whiffen, *Proc. Chem. Soc.*, 144 (1962).
7. A. D. Cohen and K. A. McLauchlan, *Disc. Faraday Soc.* **34**, 132 (1962).
8. A. D. Cohen and D. H. Whiffen, *Mol. Phys.* **7**, 449 (1964).
9. J. I. Musher, *J. Chem. Phys.* **40**, 983 (1964).
10. M. L. Martin, G. J. Martin, and R. Couffignol, *J. Chem. Phys.* **49**, 1985 (1968).
11. K. M. Worvill, *J. Magn. Reson.* **18**, 217 (1975).
12. P. Bucci, M. Martinelli, and S. Santucci, *J. Chem. Phys.* **52**, 4041 (1970).
13. P. Bucci, M. Martinelli, and S. Santucci, *J. Chem. Phys.* **53**, 4524 (1970).
14. J. Biemond, J. A. B. Lohman, and C. MacLean, *J. Magn. Reson.* **16**, 402 (1974).
15. H. Wennerström, N. O. Persson, and B. Lindman, *J. Magn. Reson.* **13**, 348 (1974).
16. G. Lindblom, H. Wennerström, and B. Lindman, *J. Magn. Reson.* **23**, 177 (1976).
17. R. B. Creel, E. D. von Meerwall, and R. G. Barnes, *Chem. Phys. Lett.* **49**, 501 (1977).
18. R. E. McDonald and T. K. McNab, *Phys. Rev. Lett.* **32**, 1133 (1974).
19. D. Dubbers, K. Dörr, H. Ackermann, F. Fujara, H. Grupp, M. Grupp, P. Heitjans, A. Körblein, and H. J. Stöckmann, *Z. Physik A* **282**, 243 (1977).
20. R. R. Ernst and W. A. Anderson, *Rev. Sci. Instr.* **37**, 93 (1966).
21. U. Haeberlen, *High Resolution NMR in Solids: Selective Averaging, Adv. Magn. Reson.*, Supplement 1, Academic, New York, 1976.
22. M. Mehring, *Principles of High Resolution NMR in Solids*, 2nd ed. Springer, Berlin, 1983.
23. J. Jeener, Ampere International Summer School, Basko Polje, Yugoslavia, unpublished (1971).
24. W. P. Aue, E. Bartholdi, and R. R. Ernst, *J. Chem. Phys.* **64**, 2229 (1976).
25. H. Hatanaka, T. Terao, and T. Hashi, *J. Phys. Soc. Jpn.* **39**, 835 (1975).
26. H. Hatanaka and T. Hashi, *J. Phys. Soc. Jpn.* **39**, 1139 (1975).
27. G. Bodenhausen, *Progr. NMR Spectrosc.* **14**, 137 (1981).
28. D. P. Weitekamp, *Adv. Magn. Reson.* **11**, 111 (1983).
29. G. Drobny, *Ann. Rev. Phys. Chem.* **26**, 451 (1985).
30. A. Abragam, *The Principles of Nuclear Magnetism*, Oxford University Press, Oxford and London, 1963.
31. C. Slichter, *Principles of Magnetic Resonance*, 2nd ed., Springer, Berlin, 1978.

32. J. H. van Vleck, *Phys. Rev.* **74**, 1168 (1948).
33. F. J. Dyson, *Phys. Rev.* **75**, 486 (1949).
34. R. C. Tolman, *The Principles of Statistical Mechanics*, Oxford University Press, London, 1938.
35. J. von Neumann, *Mathematical Foundations of Quantum Mechanics*, Princeton University Press, Princeton, NJ, 1955.
36. I. I. Rabi, N. F. Ramsey, and J. Schwinger, *Rev. Mod. Phys.* **26**, 169 (1954).
37. H. Weyl, *The Theory of Groups and Quantum Mechanics*, Dover, New York, 1931. (Translated from second revised German edition by H. P. Robertson.)
38. P. A. M. Dirac, *The Principles of Quantum Mechanics*, 4th ed., Oxford University Press, Oxford, 1958.
39. J. Schwinger, *Phys. Rev.* **82**, 914 (1951).
40. J. V. Lepore, *Phys. Rev.* **119**, 821 (1960).
41. M. H. Cohen and F. Reif, *Sol. State Phys.* **5**, 321 (1957).
42. R. M. Wilcox, *J. Math. Phys.* **8**, 962 (1967).
43. U. Haeberlen and J. S. Waugh, *Phys. Rev.* **175**, 453 (1968).
44. W. Magnus, *Commun. Pure Appl. Math* **7**, 649 (1954).
45. J. S. Waugh, L. M. Huber, and U. Haeberlen, *Phys. Rev. Lett.* **20**, 180 (1968); P. Mansfield, *J. Phys. C* (*Sol. State Phys.*) **4**, 1444 (1971); P. Mansfield, M. J. Orchard, D. C. Stalker, and K. H. B. Richards, *Phys. Rev. B* **7**, 90 (1973); W. K. Rhim, D. D. Elleman, and R. W. Vaughan, *J. Chem. Phys.* **58**, 1772 (1973); **59**, 3740 (1973).
46. F. Bloch, *Phys. Rev.* **111**, 841 (1958).
47. L. R. Sarles and R. M. Cotts, *Phys. Rev.* **111**, 853 (1958).
48. M. Mehring, A. Pines, W. K. Rhim, and J. S. Waugh, *J. Chem. Phys.* **54**, 3239 (1971).
49. U. Fano, *Rev. Mod. Phys.* **29**, 74 (1957).
50. R. P. Feynmann, F. L. Vernon, and R. W. Hellwarth, *J. Appl. Phys.* **28**, 49 (1957).
51. S. Vega and A. Pines, *J. Chem. Phys.* **66**, 5624 (1977).
52. S. Vega, *J. Chem. Phys.* **68**, 5518 (1978).
53. A. Wokaun and R. R. Ernst, *J. Chem. Phys.* **67**, 1752 (1977).
54. O. W. Sørensen, G. W. Eich, M. H. Levitt, G. Bodenhausen, and R. R. Ernst, *Progr.* NMR Spectrosc. **16**, 163 (1983).
55. K. J. Packer and K. M. Wright, *Mol. Phys.* **50**, 797 (1983).
56. F. J. M. van de Ven and C. W. Hilbers, *J. Magn. Reson.* **54**, 512 (1983).
57. C. N. Banwell and H. Primas, *Mol. Phys.* **6**, 225 (1962).
58. N. C. Pyper, *Mol. Phys.* **21**, 1 (1971); N. C. Pyper, *Mol. Phys.* **22**, 433 (1971); B. C. Sanctuary, *J. Chem. Phys.* **64**, 4352 (1976).
59. G. Drobny, A. Pines, S. Sinton, D. P. Weitekamp, and D. Wemmer, *Faraday Symp. Chem. Soc.* **13**, 49 (1979).
60. A. D. Bain, *J. Magn. Reson.* **37**, 209 (1980); A. D. Bain, *J. Magn. Reson.* **39**, 335 (1980); A. D. Bain, J. Bornais, and S. Brownstein, *Can. J. Chem.* **59**, 723 (1981).
61. R. A. Hoffman, *Adv. Magn. Reson.* **4**, 87 (1970).
62. A. Wokaun and R. R. Ernst, *Mol. Phys.* **36**, 317 (1978).
63. A. Wokaun and R. R. Ernst, *Chem. Phys. Lett.* **52**, 407 (1977).

64. G. Pouzard, S. Sukumar, and L. D. Hall, *J. Am. Chem. Soc.* **103**, 4209 (1981).
65. G. Bodenhausen, R. L. Vold, and R. R. Vold, *J. Magn. Reson.* **37**, 93 (1980).
66. J. B. Murdoch, W. S. Warren, D. P. Weitekamp, and A. Pines, *J. Magn. Reson.* **60**, 205 (1984).
67. Y.-S. Yen and A. Pines, *J. Chem. Phys.* **78**, 3579 (1983).
68. R. H. Schneider and H. Schmiedel, *Phys. Lett.* **30A**, 298 (1969).
69. W. K. Rhim, A. Pines, and J. S. Waugh, *Phys. Rev. Lett.* **25**, 218 (1970).
70. W.-K. Rhim, A. Pines, and J. S. Waugh, *Phys. Rev. B* **3**, 684 (1971).
71. A. Pines, W.-K. Rhim, and J. S. Waugh, *J. Magn. Reson.* **6**, 457 (1972).
72. G. Drobny, *Chem. Phys. Lett.* **109**, 132 (1984).
73. K. Nagayama, P. Bachman, K. Wüthrich, and R. R. Ernst, *J. Magn. Reson.* **31**, 133 (1978).
74. E. Y. C. Lu and L. E. Wood, *Phys. Lett.* **44A**, 68 (1973).
75. R. G. Brewer and E. L. Hahn, *Phys. Rev. A* **11**, 1641 (1975).
76. M. M. T. Loy, *IBM J. Res. Dev.* **23**, 504 (1979).
77. D. P. Weitekamp, K. Duppen, and D. A. Wiersma, *Phys. Rev. A* **27**, 3089 (1983).
78. S. Vega, T. W. Shattuck, and A. Pines, *Phys. Rev. Lett.* **37**, 43 (1976).
79. D. Wemmer, Ph.D. Thesis, University of California, Berkeley, 1978 (published as Lawrence Berkeley Lab. Rep. LBL-8042).
80. H. Hatanaka and T. Hashi, *Phys. Lett.* **67A**, 183 (1978).
81. H. Hatanaka, T. Ozawa, and T. Hashi, *J. Phys. Soc. Jpn.* **42**, 2069 (1977).
82. D. G. Gold and E. L. Hahn, *Phys. Rev. A* **16**, 324 (1977).
83. R. C. Hewitt, S. Meiboom, and L. C. Snyder, *J. Chem. Phys.* **58**, 5089 (1973); L. C. Snyder and S. Meiboom, *J. Chem. Phys.* **58**, 5096 (1973).
84. J. W. Emsley, J. C. Lindon, and J. M. Tabony, *J. Chem. Soc. Faraday Trans.* **269**, 10 (1973).
85. A. Pines, D. J. Ruben, S. Vega, and M. Mehring, *Phys. Rev. Lett.* **36**, 110 (1976).
86. A. Pines, S. Vega, and M. Mehring, *Phys. Rev. B* **18**, 112 (1978).
87. D. Suwelack, M. Mehring, and A. Pines, *Phys. Rev. B* **19**, 238 (1979).
88. S. A. Werner, R. Colella, A. W. Overhauser, and C. F. Eagen, *Phys. Rev. Lett.* **35**, 1053 (1975).
89. M. E. Stoll, A. J. Vega, and R. W. Vaughan, *Phys. Rev. A* **16**, 1521 (1977).
90. M. E. Stoll, E. K. Wolff, and M. Mehring, *Phys. Rev. A* **17**, 1561 (1978).
91. E. K. Wolff and M. Mehring, *Phys. Lett.* **70A**, 125 (1979).
92. M. Mehring, E. K. Wolff, and M. E. Stoll, *J. Magn. Reson.* **37**, 475 (1980).
93. M. E. Stoll, *Phil. Trans. R. Soc. London A* **299**, 565 (1981).
94. M. Mehring, *Bull. Magn. Reson.* **3**, 83 (1983).
95. M. E. Stoll, A. J. Vega, and R. W. Vaughan, *J. Chem. Phys.* **67**, 2029 (1977).
96. M. Polak and R. W. Vaughan, *Phys. Rev. Lett.* **39**, 1677 (1977).
97. M. Polak and R. W. Vaughan, *J. Chem. Phys.* **69**, 3232 (1978).
98. M. Polak, A. J. Highe, and R. W. Vaughan, *J. Magn. Reson.* **37**, 357 (1980).
99. P. M. Henrichs and L. J. Schwartz, *J. Magn. Reson.* **28**, 477 (1977).
100. S. B. Ahmad and K. J. Packer, *Mol. Phys.* **37**, 47, 59 (1979).

101. P. M. Henrichs and L. J. Schwartz, *J. Chem. Phys.* **69**, 622 (1978); *J. Chem. Phys.* **70**, 2586 (1979).
102. H. Hatanaka and T. Hashi, *Phys. Rev. B* **21**, 2677 (1980).
103. J. Stepisnik, *Physica B + C (Amsterdam)*, **100B + C**, 245 (1980).
104. R. Kaiser, *J. Magn. Reson.* **40**, 439 (1980).
105. G. L. Hoatson and K. J. Packer, *Mol. Phys.* **40**, 1153 (1980).
106. H. Hatanaka and C. S. Yannoni, *J. Magn. Reson.* **42**, 330 (1981).
107. H. Hatanaka and T. Hashi, *Phys. Rev. B* **27**, 4095 (1983).
108. Y. Zur and S. Vega, *Chem. Phys. Lett.* **80**, 381 (1981).
109. S. Vega and Y. Naor, *J. Chem. Phys.* **75**, 75 (1981).
110. A. Pines, D. Wemmer, J. Tang, and S. Sinton, *Bull. Am. Phys. Soc.* **23**, 21 (1978).
111. L. Müller, *J. Am. Chem. Soc.* **101**, 4481 (1979).
112. A. D. Bain, *J. Magn. Reson.* **56**, 418 (1984).
113. G. Bodenhausen, H. Kogler, and R. R. Ernst, *J. Magn. Reson.* **58**, 370 (1984).
114. L. Braunschweiler, G. Bodenhausen, and R. R. Ernst, *Mol. Phys.* **48**, 535 (1983).
115. S. Schäublin, A. Höhener, and R. R. Ernst, *J. Magn. Reson.* **13**, 196 (1974).
116. G. Bodenhausen and R. Freeman, *J. Magn. Reson.* **36**, 221 (1979).
117. O. W. Sørensen, M. H. Levitt, and R. R. Ernst, *J. Magn. Reson.* **55**, 104 (1983).
118. M. H. Levitt, *Progr. NMR Spectrosc.*, **18**, 61 (1986).
119. M. H. Levitt and R. Freeman, *J. Magn. Reson.* **33**, 473 (1979).
120. T. M. Barbara, R. Tycko, and D. P. Weitekamp, *J. Magn. Reson.* **62**, 54 (1985).
121. D. P. Weitekamp, J. R. Garbow, and A. Pines, *J. Magn. Reson.* **46**, 529 (1982).
122. A. A. Maudsley, A. Wokaun, and R. R. Ernst, *Chem. Phys. Lett.* **55**, 9 (1978).
123. A. Bax, P. G. de Jong, A. F. Mehlkopf, and J. Smidt, *Chem. Phys. Lett.* **69**, 567 (1980).
124. Y. S. Yen and D. P. Weitekamp, *J. Magn. Reson.* **47**, 476 (1982).
125. D. P. Weitekamp, J. R. Garbow, and A. Pines, *J. Chem. Phys.* **77**, 2870 (1982).
126. A. Kumar, *J. Magn. Reson.* **30**, 227 (1978); M. A. Thomas and A. Kumar, *J. Magn. Reson.* **47**, 535 (1982).
127. J. B. Murdoch, Ph.D. Thesis, University of California, Berkeley, 1982 (published as Lawrence Berkeley Lab. Rep. LBL-15254).
128. W. S. Warren, Ph.D. Thesis, University of California, Berkeley, 1980 (published as Lawrence Berkeley Lab. Rep. LBL-11885).
129. I. J. Lowe and R. E. Norberg, *Phys. Rev.* **107**, 46 (1957).
130. W. S. Warren, D. P. Weitekamp, and A. Pines, *J. Chem. Phys.* **73**, 2084 (1980).
131. J. Baum, M. Munowitz, A. N. Garroway, and A. Pines, *J. Chem. Phys.* **83**, 2015 (1985).
132. W. S. Warren, S. Sinton, D. P. Weitekamp, and A. Pines, *Phys. Rev. Lett.* **43**, 1791 (1979).
133. W. S. Warren and A. Pines, *J. Chem. Phys.* **74**, 2808 (1981).
134. G. Drobny, A. Pines, S. Sinton, W. S. Warren, and D. P. Weitekamp, *Phil. Trans. R. Soc. London A* **299**, 585 (1981).
135. W. S. Warren, D. P. Weitekamp, and A. Pines, *J. Magn. Reson.* **40**, 581 (1980).
136. W. S. Warren, J. B. Murdoch, and A. Pines, *J. Magn. Reson.* **60**, 236 (1984).
137. W. S. Warren and A. Pines, *Chem. Phys. Lett.* **88**, 441 (1982).
138. A. Bax, R. Freeman, and S. Kempsell, *J. Am. Chem. Soc.* **102**, 4849 (1980).

139. G. Bodenhausen and C. M. Dobson, *J. Magn. Reson.* **44**, 212 (1981).
140. U. Piantini, O. W. Sørensen, and R. R. Ernst, *J. Am. Chem. Soc.* **104**, 6800 (1982).
141. A. J. Shaka and R. Freeman, *J. Magn. Reson.* **51**, 169 (1983).
142. A. Bax, R. Freeman, and S. P. Kempsell, *J. Magn. Reson.* **41**, 349 (1980).
143. A. Bax and R. Freeman, *J. Magn. Reson.* **41**, 507 (1980).
144. A. Bax, R. Freeman, and T. A. Frenkiel, *J. Am. Chem. Soc.* **103**, 2102 (1981).
145. A. Bax, R. Freeman, T. A. Frenkiel, and M. H. Levitt, *J. Magn. Reson.* **43**, 478 (1981).
146. D. L. Turner, *Mol. Phys.* **44**, 1051 (1981).
147. R. Richarz, W. Ammann, and T. Wirthlin, *J. Magn. Reson.* **45**, 270 (1981).
148. O. W. Sørensen, R. Freeman, T. Frenkiel, T. H. Mareci, and R. Schuck, *J. Magn. Reson.* **46**, 180 (1982).
149. T. H. Mareci and R. Freeman, *J. Magn. Reson.* **48**, 158 (1982).
150. D. L. Turner, *J. Magn. Reson.* **53**, 259 (1983).
151. M. H. Levitt and R. R. Ernst, *Mol. Phys.* **50**, 1109 (1983).
152. E. M. Menger, S. Vega, and R. G. Griffin, *J. Magn. Reson.* **56**, 338 (1984).
153. P. J. Hore, E. R. P. Zuiderweg, K. Nicolay, K. Dijkstra, and R. Kaptein, *J. Am. Chem. Soc.* **104**, 4286 (1982).
154. P. J. Hore, R. M. Scheek, A. Volbeda, R. Kaptein, and J. H. van Boom, *J. Magn. Reson.* **50**, 328 (1982).
155. P. J. Hore, R. M. Scheek, and R. Kaptein, *J. Magn. Reson.* **52**, 339 (1983).
156. H. Kessler, H. Oschkinat, O. W. Sørensen, H. Kogler, and R. R. Ernst, *J. Magn. Reson.* **55**, 329 (1983).
157. M. Rance, O. W. Sørensen, G. Bodenhausen, G. Wagner, R. R. Ernst, and K. Wüthrich, *Biochem. Biophys. Res. Commun.* **117**, 479 (1983).
158. M. Ikura and K. Hikichi, *J. Am. Chem. Soc.* **106**, 4275 (1984).
159. M. Rance, O. W. Sørensen, W. Leupin, H. Kogler, K. Wüthrich, and R. R. Ernst, *J. Magn. Reson.* **61**, 67 (1985).
160. F. J. M. van de Ven, C. A. G. Haasnoot, and C. W. Hilbers, *J. Magn. Reson.* **61**, 181 (1985).
161. U. B. Sørensen, H. J. Jakobsen, and O. W. Sørensen, *J. Magn. Reson.* **61**, 382 (1985).
162. M. H. Levitt and R. R. Ernst, *Chem. Phys. Lett.* **100**, 119 (1983).
163. M. H. Levitt and R. R. Ernst, *J. Chem. Phys.* **83**, 3297 (1985).
164. A. Bax and R. Freeman, *J. Magn.* Reson. **44**, 542 (1981).
165. K. Nagayama, A. Kumar, K. Wüthrich, and R. R. Ernst, *J. Magn. Reson.* **40**, 321 (1980).
166. G. Wagner and K. Wüthrich, *J. Mol. Biol.* **155**, 347 (1982).
167. D. Bright, I. E. Maxwell, and J. de Boer, *J. Chem. Soc. Perkin Trans.* **2**, 2101 (1973).
168. P.-K. Wang, C. P. Slichter, and J. H. Sinfelt, *Phys. Rev. Lett.* **53**, 82 (1984).
169. J. C. Knights, in *The Physics of Hydrogenated Amorphous Silicon I*, J. D. Toannopoulos and G. Lucovsky, Eds., Springer, New York, 1984.
170. J. Baum, K. Gleason, J. Reimer, A. Garroway, and A. Pines, *Phys. Rev. Lett.* **56**, 1377 (1986).
171. S. Emid, *Physica* **128B**, 79 (1985).
172. A. A. Maudsley and R. R. Ernst, *Chem. Phys. Lett.* **50**, 368 (1977).
173. A. Minoretti, W. P. Aue, M. Reinhold, and R. R. Ernst, *J. Magn. Reson.* **40**, 175 (1980).
174. G. A. Morris and R. Freeman, *J. Am. Chem. Soc.* **101**, 760 (1979).

175. D. P. Burum and R. R. Ernst, *J. Magn. Reson.* **39**, 163 (1980).
176. M. R. Bendall, D. M. Doddrell, and D. T. Pegg, *J. Am. Chem. Soc.* **103**, 4603 (1981).
177. H. J. Jakobsen, O. W. Sørensen, W. S. Brey, and P. Kanyha, *J. Magn. Reson.* **48**, 328 (1982).
178. D. T. Pegg, D. M. Doddrell, and M. R. Bendall, *J. Chem. Phys.* **77**, 2745 (1982).
179. D. M. Doddrell, D. T. Pegg, and M. R. Bendall, *J. Magn. Reson.* **48**, 323 (1982).
180. D. T. Pegg and M. R. Bendall, *J. Magn. Reson.* **55**, 114 (1983).
181. O. W. Sørensen and R. R. Ernst, *J. Magn. Reson.* **51**, 477 (1983).
182. M. H. Levitt, O. W. Sørensen, and R. R. Ernst, *Chem. Phys. Lett.* **94**, 541 (1983).
183. D. T. Pegg and M. R. Bendall, *J. Magn. Reson.* **60**, 347 (1984).
184. J. M. Bulsing, W. M. Brooks, J. Field, and D. M. Doddrell, *J. Magn. Reson.* **56**, 167 (1984).
185. J. M. Bulsing, W. M. Brooks, J. Field, and D. M. Doddrell, *Chem. Phys. Lett.* **104**, 229 (1984).
186. J. M. Bulsing and D. M. Doddrell, *J. Magn. Reson.* **61**, 197 (1985).
187. H. Bildsøe, S. Dønstrup, H. J. Jakobsen, and O. W. Sørensen, *J. Magn. Reson.* **53**, 154 (1983).
188. O. W. Sørensen, S. Dønstrup, H. Bildsøe, and H. J. Jakobsen, *J. Magn. Reson.* **55**, 347 (1983).
189. A. A. Maudsley, L. Müller, and R. R. Ernst, *J. Magn. Reson.* **28**, 463 (1977).
190. G. Bodenhausen and R. Freeman, *J. Magn. Reson.* **28**, 471 (1977).
191. G. Bodenhausen and R. Freeman, *J. Am. Chem. Soc.* **100**, 320 (1978).
192. G. Bodenhausen and D. J. Ruben, *Chem. Phys. Lett.* **69**, 185 (1980).
193. A. G. Redfield, *Chem. Phys. Lett.* **96**, 537 (1983).
194. A. Bax, R. H. Griffey, and B. L. Hawkins, *J. Magn. Reson.* **55**, 301 (1983).
195. A. Bax, R. H. Griffey, and B. L. Hawkins, *J. Am. Chem. Soc.* **105**, 7188 (1983).
196. D. H. Live, D. G. Davis, W. C. Agosta, and D. Cowburn, *J. Am. Chem. Soc.* **106**, 6104 (1984).
197. M. Gochin, D. P. Weitekamp, and A. Pines, *J. Magn. Reson.* **63**, 431 (1985).
198. S. R. Hartmann and E. L. Hahn, *Phys. Rev.* **128**, 2042 (1962).
199. F. M. Lurie and C. P. Slichter, *Phys. Rev. A* **133**, 1108 (1964).
200. A. Pines, M. Gibby, and J. S. Waugh, *J. Chem. Phys.* **59**, 569 (1973).
201. R. D. Bertrand, W. B. Moniz, A. N. Garroway, and G. C. Chingas, *J. Am. Chem. Soc.* **100**, 5227 (1978).
202. R. D. Bertrand, W. B. Moniz, A. N. Garroway, and G. C. Chingas, *J. Magn. Reson.* **32**, 465 (1978).
203. L. Müller and R. R. Ernst, *Mol. Phys.* **38**, 963 (1979).
204. P. Brunner, M. Reinhold, and R. R. Ernst, *J. Chem. Phys.* **73**, 1086 (1980).
205. M. Reinhold, P. Brunner, and R. R. Ernst, *J. Chem. Phys.* **74**, 184 (1981).
206. S. Vega, T. W. Shattuck, and A. Pines, *Phys. Rev. A* **22**, 638 (1980).
207. S. Vega, *Phys. Rev. A* **23**, 3152 (1981).
208. H. Kogler, O. W. Sørensen, G. Bodenhausen, and R. R. Ernst, *J. Magn. Reson.* **55**, 157 (1983).
209. J. R. Garbow, D. P. Weitekamp, and A. Pines, *Chem. Phys. Lett.* **93**, 504 (1982).
210. C. J. Turner, *Progr. NMR Spectrosc.* **16**, 311 (1984) and references therein.

211. D. P. Weitekamp, J. R. Garbow, J. B. Murdoch, and A. Pines, *J. Am. Chem. Soc.* **103**, 3578 (1981).
212. J. R. Garbow, D. P. Weitekamp, and A. Pines, *J. Chem. Phys.* **79**, 5301 (1983).
213. L. Braunschweiler and R. R. Ernst, *J. Magn. Reson.* **53**, 521 (1983).
214. N. Chandrakumar and S. Subramanian, *J. Magn. Reson.* **62**, 346 (1985).
215. J. W. Emsley, J. Feeney, and L. H. Sutcliffe, *High Resolution Nuclear Magnetic Resonance Spectroscopy*, Pergamon, Oxford, 1965.
216. P. L. Corio, *Structure of High Resolution NMR Spectra*, Academic, New York, 1966.
217. A. Saupe and G. Englert, *Phys. Rev. Lett.* **11**, 462 (1963).
218. G. Englert and A. Saupe, *Z. Naturforsch.* **19a**, 172 (1964).
219. P. Diehl and C. L. Khetrapal, *NMR Studies of Molecules Oriented in the Nematic Phase of Liquid Crystals*, Springer, Berlin, 1969.
220. J. W. Emsley and J. C. Lindon, *NMR Spectroscopy Using Liquid Crystal Solvents*, Pergamon, Oxford, 1975.
221. S. W. Sinton, D. B. Zax, J. B. Murdoch, and A. Pines, *Mol. Phys.* **53**, 333 (1984).
222. M. Tinkham, *Group Theory and Quantum Mechanics*, McGraw-Hill, New York, 1964.
223. W. S. Warren and A. Pines, *J. Am. Chem. Soc.* **103**, 1613 (1981).
224. J. S. Waugh, *Proc. Nat. Acad. Sci. U.S.A.* **73**, 1394 (1976).
225. R. K. Hester, J. L. Ackerman, B. L. Neff, and J. S. Waugh, *Phys. Rev. Lett.* **36**, 1081 (1976).
226. E. F. Rybaczewski, B. L. Neff, J. S. Waugh, and J. S. Sherfinski,*J. Chem. Phys.* **67**, 1231 (1977).
227. M. E. Stoll, A. J. Vega, and R. W. Vaughan, *J. Chem. Phys.* **65**, 4093 (1976).
228. M. Linder, A. Höhener, and R. R. Ernst, *J. Chem. Phys.* **73**, 4959 (1980).
229. M. G. Munowitz, R. G. Griffin, G. Bodenhausen, and T. H. Huang, *J. Am. Chem. Soc.* **103**, 2529 (1981).
230. M. Munowitz and R. G. Griffin, *J. Chem. Phys.* **76**, 2848 (1982); **77**, 2217 (1982).
231. M. Munowitz, W. P. Aue, and R. G. Griffin, *J. Chem. Phys.* **77**, 1686 (1982).
232. M. G. Munowitz and R. G. Griffin, *J. Chem. Phys.* **78**, 613 (1983).
233. T. Terao, H. Miura, and A. Saika, *J. Chem. Phys.* **75**, 1573 (1981).
234. P. C. Lauterbur, *Nature* (*London*) **242**, 190 (1973).
235. L. Kaufman, L. E. Crooks, and A. R. Margulis, Eds., *Nuclear Magnetic Resonance in Medicine*, Igaku-Shoin, New York, 1981.
236. P. Mansfield and P. C. Morris, *NMR Imaging in Biomedicine*, *Adv. Magn. Reson.*, Supplement 2, Academic, New York, 1982.
237. P. A. Bottomley, *Rev. Sci. Instr.* **53**, 1319 (1982).
238. E. R. Andrew, *Accts. Chem. Res.* **16**, 114 (1983).
239. P. Mansfield, P. K. Grannell, A. N. Garroway, and D. C. Stalker, *Proc. 1st Specialized Colloque Ampere*, J. W. Hennel, Ed., Krakow, 1973, p. 16.
240. P. Mansfield and P. K. Grannell, *J. Phys. C* **6**, 1442 (1973).
241. R. A. Wind and C. S. Yannoni, *J. Magn. Reson.* **36**, 269 (1979).
242. A. N. Garroway, J. Baum, M. G. Munowitz, and A. Pines, *J. Magn. Reson.* **60**, 337 (1984).
243. H. Y. Carr and E. M. Purcell, *Phys. Rev.* **94**, 630 (1954).
244. E. O. Stejskal and J. E. Tanner, *J. Chem. Phys.* **42**, 288 (1965).

245. J. F. Martin, L. S. Selwyn, R. R. Vold, and R. L. Vold, *J. Chem. Phys.* **76**, 2632 (1982).
246. D. Zax and A. Pines, *J. Chem. Phys.* **78**, 6333 (1983).
247. F. Bloch, *Phys. Rev.* **70**, 460 (1946).
248. E. L. Hahn, *Phys. Rev.* **80**, 580 (1950).
249. S. Meiboom and D. Gill, *Rev. Sci. Instr.* **29**, 688 (1958).
250. A. G. Redfield, *Phys. Rev.* **98**, 1787 (1955).
251. I. Solomon, *Compt. Rend.* **248**, 92 (1959).
252. I. Solomon, *Compt. Rend.* **249**, 1631 (1959).
253. R. L. Vold, J. S. Waugh, M. P. Klein, and D. E. Phelps, *J. Chem. Phys.* **48**, 3831 (1968).
254. A. G. Redfield, *IBM J. Res. Dev.* **1**, 19 (1957).
255. A. G. Redfield, *Adv. Magn. Reson.* **1**, 1 (1965).
256. N. Bloembergen, E. M. Purcell, and R. V. Pound, *Phys. Rev.* **73**, 679 (1948).
257. R. K. Wangsness and F. Bloch, *Phys. Rev.* **89**, 728 (1953).
258. F. Bloch, *Phys. Rev.* **102**, 104 (1956).
259. R. L. Vold and R. R. Vold, *Progr. NMR Spectrosc.* **12**, 79 (1978).
260. H. Bildsøe, J. P. Jakobsen, and K. Schaumburg, *J. Magn. Reson.* **23**, 137 (1975).
261. J. P. Jakobsen, H. K. Bildsøe, and K. Schaumburg, *J. Magn. Reson.* **23**, 153 (1975).
262. J. P. Jakobsen and K. Schaumburg, *J. Magn. Reson.* **24**, 173 (1976).
263. R. R. Vold and R. L. Vold, *J. Chem. Phys.* **66**, 4018 (1977).
264. G. Bodensausen, N. M. Szeverenyi, R. L. Vold, and R. R. Vold, *J. Am. Chem. Soc.* **100**, 6265 (1978).
265. R. Poupko, R. L. Vold, and R. R. Vold, *J. Magn. Reson.* **34**, 67 (1979).
266. R. L. Vold, R. R. Vold, R. Poupko, and G. Bodenhausen, *J. Magn. Reson.* **38**, 141 (1980).
267. R. R. Vold, R. L. Vold, N. M. Szeverenyi, *J. Phys. Chem.* **85**, 1934 (1981).
268. D. Jaffe, R. R. Vold, and R. L. Vold, *J. Magn. Reson.* **46**, 475 (1982).
269. D. Jaffe, R. L. Vold, and R. R. Vold, *J. Magn. Reson.* **46**, 496 (1982).
270. D. Jaffe, R. L. Vold, and R. R. Vold, *J. Chem. Phys.* **78**, 4852 (1983).
271. J. Tang, Ph.D. Thesis, University of California, Berkeley, 1981 (published as Lawrence Berkeley Lab. Rep. LBL-13605).
272. J. Tang and A. Pines, *J. Chem. Phys.* **72**, 3290 (1980).
273. S. Clough and T. Hobson, *J. Phys. C.* **7**, 3387 (1974).

THE RESPONSE OF POLYMER MOLECULES IN A FLOW

MYUNG S. JHON AND G. SEKHON

Department of Chemical Engineering
Carnegie-Mellon University
Pittsburgh, Pennsylvania 15213

R. ARMSTRONG
Sandia National Laboratories
Livermore, California 194550-0096

CONTENTS

I. INTRODUCTION

When polymer molecules or suspended particles are immersed in a Newtonian solvent (this system is a polymer solution or suspension), dynamic couplings between the two result. In other words, the polymer molecules disturb the Newtonian solvent, and at the same time the solvent changes the polymer conformation. This apparent dynamic coupling at the microscopic level is experimentally observable. The overall fluid behavior may no longer be Newtonian (it becomes viscoelastic if the polymer is flexible), and the polymer molecules can change drastically from their equilibrium values. Therefore, in studying the dynamics of polymeric solutions, it is possible to adopt two different, complementary viewpoints: (1) the study of the overall rheological properties of fluids, or (2) the study of the response of the polymer molecules.

Case (1) falls into the category of rheology, with either a continuum or a molecular approach. The major effort of the continuum mechanicians' approach has been an attempt to determine the best set of "rheological equations of state"—the functional relationship between force (stress) and deformation (strain).[1–5] Molecular theoretical approaches provide a partial justification of the rheological equations of state conjectured by continuum mechanicians. Although this viewpoint is extremely useful in solving practical engineering problems, such as polymer processing, we will not review this subject. We will deal mainly with case (2), investigating the response of the polymer molecules when the Newtonian solvent is in motion. Polymer physicists have thoroughly investigated the polymer configuration in quiescent flow or weak flow fields, and the subject is well understood at present (see, for example, Yamakawa[3] and Bird et al[5]).

This chapter will investigate the dynamic behavior of polymer molecules in solution when exposed to an arbitrary flow field and confined in some finite region. Since this subject is extremely complex, rigorous theoretical developments are lacking at present. Therefore, we will impose simplifications as necessary without losing the general meaning.

In Section II we will develop a description of the dynamics of a single chain in solution when exposed to an arbitrary flow field and confined within some finite region. We express this equation in either Langevin form or Fokker–Planck form. The most salient feature of the equation developed in this section is, in contrast with previous theories, the recognition of the important physical variables. These are the undisturbed velocity profile $\mathbf{v}_0$, the hydrodynamic interaction $\mathbf{T}_{ij}$, and the connector forces $\mathbf{F}^{(c)}$ between the beads. These variables are studied in Section III.A.

In the absence of hydrodynamic interactions, and also for the case of simple (or homogeneous) flow, the theory is well developed, at least for the infinitely dilute solution. Interesting problems on this subject may be found in the excellent text by Bird et al.[5] Section III will deal with an arbitrary deterministic flow field (including nonhomogeneous flow), subtle hydrodynamic interactions due to confined geometry, and some aspects of nonlinear theory not treated in Bird et al. These generalizations are relevant to many areas of engineering that deal with the porous media, including separation technology, size exclusion chromatography, and enhanced oil recovery from porous rocks. The following two topics will be discussed to illustrate certain physical concepts.

The first topic is polymer flow through a micropore. Near the pore entrance the flow is almost elongational. For the strong flow case, the linear spring model (Rouse chain) may give a mathematical singularity. To avoid the mathematical artifact arising from the linear force law, we will introduce nonlinear spring forces. Introducing the nonlinear spring results in complicated mathematics, known as the nonlinear Langevin equation, which we will examine carefully in Section III.B.1.

Polymer migration phenomena is the second topic. When polymer molecules are exposed to a flow field in a confined geometry, they may not follow the bulk flow, due to nonhomogeneity in bulk flow and due to hydrodynamic interaction. Sometimes the polymer molecule may be behind or ahead of the bulk flow, or sometimes it may cross the streamline. The Langevin equation will be used to calculate the trajectory of the center of mass of the polymer molecule. If the polymer molecule crosses the streamline, it will cause nonuniformity in the concentration profile. This will be studied in Section III.B.2 using the Fokker–Planck equation.

In Section IV we will study the polymer conformation in turbulent flow where the fluid velocity is no longer deterministic but random. The fluctuating velocity in turbulent flow can only be defined statistically; therefore we will formulate an effective probability equation for the polymer chain, which can describe the system with random velocity fields. This will be given in Section IV. A. From this equation we will show that the effective interaction between any two points in turbulent flow is reduced relative to the bare interaction. The effective interaction (which we refer to as renormalized interaction in Section IV.B) will be used in computing the polymer conformation in turbulent flow. If we add a minute amount of high-molecular-weight polymer to a turbulent fluid flowing through a pipe, we can achieve a reduction in drag.[6] Therefore the polymer can be used in reducing energy costs for pumping fluids through pipelines. In the last section we will explain the turbulent drag reduction mechanism from the change in polymer conformation by correlating the polymer conformation with the frictional resistance.

II. SINGLE-CHAIN DYNAMICS

This section will examine the motion of a single-chain polymer molecule immersed in a Newtonian solvent. The Newtonian solvent is exposed to a finite strain rate in an arbitrary flow field, which can be laminar or turbulent, and confined in some finite region in space, as shown in Fig. 1.

We will first examine the fluid dynamics in the absence of polymers. Later we will study how the presence of polymers modifies the fluid flow, and how the fluid flow influences polymer dynamics. From these results we will derive the dynamic equations for the single chain in either the Langevin or the Fokker–Planck form. These equations will be compared with classic results in this area.

A. Fluid Dynamics in the Absence of a Polymer

The fundamental equations in fluid dynamics are derived from the principles of conservation of mass and momentum.[7] The conservation of mass leads to the continuity equation

$$\frac{\partial}{\partial t}\rho + \nabla\cdot(\rho\mathbf{v}(\mathbf{r},t)) = 0 \tag{2.1}$$

Here $\mathbf{v}(\mathbf{r}, t)$ is the velocity at position $\mathbf{r}$ and time t, and ρ is the fluid density. Also, Cauchy's momentum equation can be obtained from the momentum conservation law,

$$\rho\left(\frac{\partial}{\partial t} + \mathbf{v}\cdot\nabla\right)\mathbf{v} = \nabla\cdot\mathbf{T}(\mathbf{r},t) + \mathbf{F}_{\text{ext}}. \tag{2.2}$$

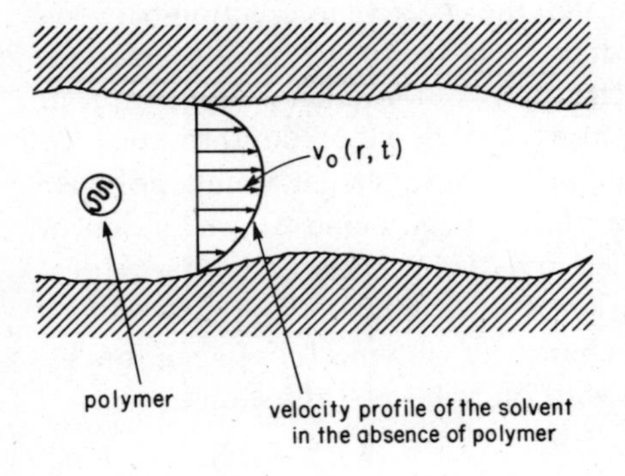

Figure 1. A single polymer chain immersed in a confined geometry with arbitrary flow.

Here $\mathbf{T}$ is shown as the stress tensor, $\nabla \cdot \mathbf{T}$ originates from surface forces, and $\mathbf{F}_{\text{ext}}$ is the body force (gravity and/or electromagnetic force).

Although Eqs. (2.1) and (2.2) are exact, they cannot be useful as they stand. We have 13 unknown variables (ρ, three components of $\mathbf{v}$, and nine components of $\mathbf{T}$), but only four equations. Therefore we must develop nine additional equations to have a complete description of the fluid flow.

The additional nine equations can be obtained from constitutive relationships (or rheological equations of state). The functional relationship between the deviatoric stress tensor $\boldsymbol{\tau}$ and the rate-of-strain tensor $\dot{\boldsymbol{\gamma}}$ can be written in the following form:

$$\mathbf{T} = -P\boldsymbol{\delta} + \boldsymbol{\tau}. \tag{2.3}$$

Here P is the isotropic pressure. The Newtonian fluids have the simplest constitutive relationship, that is $\boldsymbol{\tau}$ is linearly proportional to $\dot{\boldsymbol{\gamma}}$ with proportionality constant η, and

$$\boldsymbol{\tau} = \eta\dot{\boldsymbol{\gamma}} \qquad \text{Newtonian fluids,} \tag{2.4}$$

where

$$\dot{\boldsymbol{\gamma}} = (\nabla\mathbf{v}) + (\nabla\mathbf{v})^{\mathrm{T}}. \tag{2.5}$$

Here $\nabla\mathbf{v}$ is the velocity gradient tensor and T in Eq.(2.5) stands for transpose.

The proportionality constant η in Eq. (2.4) is known as viscosity. Generally a fluid that consists of small molecules obeys the Newtonian fluid constitutive relationship. However, polymer solutions cannot be characterized by a simple linear relationship between the stress tensor and the rate-of-strain tensor[2,8]. Constructing the best form of the rheological equations of state has been the major problem in polymer rheology. Two complementary approaches have been adopted in the past: continuum approach and molecular approach.

Substituting Eqs. (2.3) to (2.5) into Eq. (2.2) results in the equation of motion for the Newtonian fluid,

$$\rho\left[\frac{\partial}{\partial t} + \mathbf{v}\cdot\nabla\right]\mathbf{v} - \eta\nabla^2\mathbf{v} + \nabla P = \mathbf{F}_{\text{ext}}. \tag{2.6}$$

Here we have assumed an isothermal condition, and as a consequence the viscosity is independent of position. Eq. (2.6) is known as the Navier–Stokes equation.

Eqs. (2.1) and (2.6) give a complete description of the fluid flow problem for the Newtonian fluid, since there are four unknown variables (P and $\mathbf{v}$) and we have four equations to solve these variables. To simplify the analysis, we assume that the fluid is incompressible. Under this assumption the continuity equation [Eq. (2.1)] becomes simplified,

$$\nabla \cdot \mathbf{v} = 0. \tag{2.7}$$

We can, in principle, solve the velocity profile from Eqs. (2.6) and (2.7) for the given boundary condition. However, there are very few cases in which $\mathbf{v}$ can be obtained analytically, and we usually have to make approximations to handle Eqs. (2.6) and (2.7). Two dimensionless numbers, the Reynolds number Re and the Strouhal number Sr, are introduced as criteria for various approximations in fluid mechanics (see Ref. 9, chap. 11).

The following two approximations will be used in this chapter.

1. $\mathbf{Re} \ll \mathbf{1}$ *and* $\mathbf{ReSr}^{-1} \ll \mathbf{1}$ (*creeping flow approximation*). In this case the Navier–Stokes equation [Eq. (2.6)] can be approximated as

$$-\eta \nabla^2 \mathbf{v} + \nabla P = \mathbf{F}_{\text{ext}}. \tag{2.8}$$

2. $\mathbf{Re} \ll \mathbf{1}$ *and* $\mathbf{ReSr}^{-1} \simeq \mathbf{1}$ (*linearized Navier–Stokes equation*). Under this approximation, Eq. (2.6) can be approximated as

$$\rho \frac{\partial}{\partial t} \mathbf{v} - \eta \nabla^2 \mathbf{v} + \nabla P = \mathbf{F}_{\text{ext}}. \tag{2.9}$$

Polymer scientists usually use these two approximations. By taking the divergence of either Eq. (2.8) or Eq. (2.9) and using Eq. (2.7), we can uniquely determine the pressure as

$$\nabla^2 P = \nabla \cdot \mathbf{F}_{\text{ext}}. \tag{2.10}$$

Therefore, Eqs. (2.8) to (2.10) imply that $\mathbf{v}$ can be easily and uniquely determined if $\mathbf{F}_{\text{ext}}$ is given.

However, for polymer dynamics studied in turbulent flow ($\text{Re} \gg 1$), these two approximations obviously fail. In this case we must include the nonlinear term $\mathbf{v} \cdot \nabla \mathbf{v}$. Until now no one has attempted to include the $\mathbf{v} . \nabla \mathbf{v}$ term in the study of polymer motion in turbulent flow fields.

B. Fluid Dynamics in the Presence of a Polymer

When a polymer is immersed in a Newtonian solvent as shown in Fig. 1, the polymer molecule disturbs the Newtonian solvent and at the same time the solvent changes the polymer conformation. At the microscopic level there is a dynamic coupling between the motion of the polymer and that of the solvent. This coupling produces remarkable effects which are experimentally measurable—the overall fluid behavior is no longer Newtonian in nature and the polymer can be drastically deformed from its equilibrium configuration. The dynamic coupling force, which is an important input in the equation of motion, will be formulated from the *n*-bead model for the polymer (see Fig. 2) as follows. Bead *i* exerts a force $\sigma_i(t)$ (unknown at the present, but to be determined later) on the fluid at time *t*. Given that bead *i* is at position $\mathbf{R}_i(t)$, the force $\sigma_i(t)$ represents a point source $\mathbf{r} = \mathbf{R}_i(t)$. Hence the collection of all polymer beads yields a force density,

$$\sum_{i=1}^{n} \delta(\mathbf{r} - \mathbf{R}_i(t))\sigma_i(t),$$

acting on the fluid at $\mathbf{r}$ and *t*. Here $\delta(\mathbf{r})$ is the three-dimensional delta function.

Therefore, upon introduction of this force, the fluid flow can be characterized by the following linearized Navier–Stokes equation:

$$\rho \frac{\partial}{\partial t}\mathbf{v} - \eta \nabla^2 \mathbf{v} + \nabla P(\mathbf{r}, t) = \sum_{i=1}^{n} \delta(\mathbf{r} - \mathbf{R}_i(t))\sigma_i(t) + \mathbf{F}_{\text{ext}}$$

$$= \sum_{i=1}^{n} \phi_i(\mathbf{r}, t)\sigma_i(t) + \mathbf{F}_{\text{ext}}. \qquad (2.11)$$

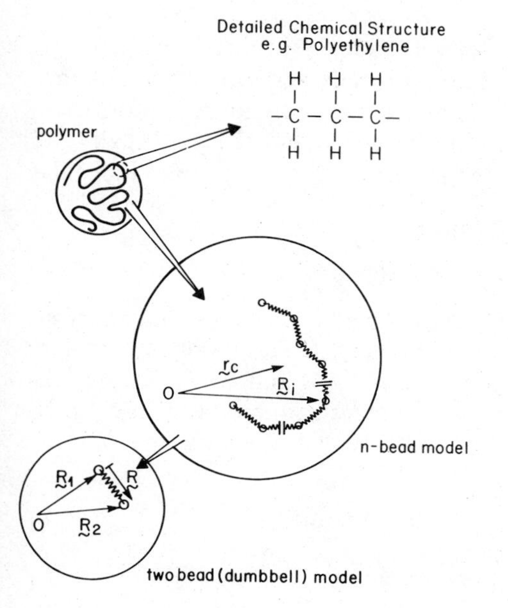

Figure 2. Models for a polymer. In this and following figures, a wavy underscore indicates vectors.

Here we have introduced the definition of the operator

$$\phi_i(\mathbf{r}, t) \equiv \delta(\mathbf{r} - \mathbf{R}_i(t)),$$

which transforms between bead i and fluid $\mathbf{r}$ space.

We will now construct the dynamic equation for the ith bead. According to Newton's second law, the equation of motion for the ith bead becomes

$$m_i \frac{d^2}{dt^2} \mathbf{R}_i(t) = \text{total forces acting on } i\text{th bead.}$$

Here m_i is the mass of the ith bead. In the absence of forces due to the solvent, the beads experience connector forces, excluded volume forces, and the force due to the external field. The presence of the solvent will produce a force equal to $-\sigma_i(t)$ on the polymer bead at $\mathbf{R}_i(t)$ from Newton's third law. Therefore by summing all the forces acting on the ith bead we obtain the dynamic equation for the ith bead,

$$m_i \frac{d^2}{dt^2} \mathbf{R}_i(t) = -\sigma_i(t) + \mathbf{f}_i(t). \tag{2.12}$$

Here $\mathbf{f}_i(t)$ is equal to a sum of forces,

$$\mathbf{f}_i(t) = \mathbf{f}_i^{\text{ext}}(t) + \mathbf{f}_i^* + \mathbf{f}_i^{\text{other}} + \mathbf{F}_i^{(c)}, \tag{2.13}$$

stemming from the external field (ext), random Brownian encounters (*), direct skeletal connections (c) to the nearest beads, and other sources (other) due to long-range interaction (coordinate along the chain axis) between the beads, such as excluded volume effects. Eqs. (2.7) and (2.11) through (2.13) represent a set of coupled equations for the polymer and the Newtonian fluid.

C. Single-Chain Equation

In Section II.B we have derived explicit coupled equations for the polymer and the fluid to study the system depicted in Fig. 1. To simplify our analysis we will discard the inertia term $m_i(d^2/dt^2)\mathbf{R}_i$, $\mathbf{f}_i^{\text{ext}}$, and $\mathbf{f}_i^{\text{other}}$. We will also assume that the simplest model for the connector force (known as linear spring or Rouse chain) is given as

$$-\mathbf{F}_i^{(c)} = \sum_j \frac{3k_B T}{l^2} A_{ij} \mathbf{R}_j(t) \equiv \mathbf{\Delta}_i \mathbf{R}(t) \tag{2.14}$$

where $A_{ij} = 2\delta_{i,j} - \delta_{i-1,j} - \delta_{i+1,j}$ is an element of the Rouse matrix, l is the mean distance between the beads, k_B is Boltzmann's constant, and T is an absolute temperature.† With these simplifying assumptions, Eqs. (2.12) and (2.13) can be written as

$$\text{Here, } H \equiv 3k_BT/l^2 \quad \mathbf{\Delta}_i\mathbf{R}_i(t) = H(2\mathbf{R}_i - \mathbf{R}_{i-1} - R_{i+1}) = -\sigma_i + \mathbf{f}_i^*. \quad (2.15)$$

Eqs. (2.7), (2.11), and (2.15) represent a set of coupled equations for the polymer and the fluid that still requires a hydrodynamic boundary condition to determine the unknown force $\sigma_i(t)$. A boundary condition requiring no slip at the bead surface is mathematically the simplest one and implies that

$$\dot{\mathbf{R}}_i(t) = \mathbf{v}(\mathbf{R}_i(t), t) = \int d^3r\phi_i(\mathbf{r})\mathbf{v}(\mathbf{r}, t) \equiv \phi_i(\bar{\mathbf{r}})\mathbf{v}(\bar{\mathbf{r}}, t). \quad (2.16)$$

Here we use the summation convention, that is, the repeated barred variables imply integration over all barred variables.

Now we are able to calculate the velocity of the fluid in the presence of the polymer molecule. We can formally solve $\mathbf{v}$ in terms of the Green's function $\mathbf{G}$ of a pure fluid (this will be explained Section III.A) from Eqs. (2.10) and (2.11),

$$\mathbf{v}(\mathbf{r}, t) = \mathbf{v}_0(\mathbf{r}, t) + \mathbf{G}(\mathbf{r} - \bar{\mathbf{r}}; t - \bar{t})\cdot \sum_{j=1}^{n} \phi_i(\bar{\mathbf{r}})\sigma_i(\bar{t}). \quad (2.17)$$

The calculation of $\mathbf{v}_0$ and $\mathbf{G}$ requires the solution of complicated fluid mechanical calculations and will also be discussed in Section III.A.

From Eqs. (2.15) through (2.17) we obtain the explicit expression for single-chain dynamics for the ith bead,

$$\begin{aligned}\dot{\mathbf{R}}_i(t) &= \phi_i(\bar{\mathbf{r}})\mathbf{v}_0(\bar{\mathbf{r}}, t) + \phi_i(\bar{\mathbf{r}})\mathbf{G}(\bar{\mathbf{r}} - \bar{\bar{\mathbf{r}}}; t - \bar{t})\cdot \sum_{j=1}^{n} \phi_j(\bar{\bar{\mathbf{r}}})\{-\mathbf{\Delta}_j\mathbf{R}_j(\bar{t}) + \mathbf{f}_j^*(\bar{t})\} \\ &= \mathbf{v}_0(\mathbf{R}_i, t) + \sum_j \mathbf{T}_{ij}\cdot(-\mathbf{\Delta}_j\mathbf{R}_j(\bar{t}) + \mathbf{f}_j^*) \end{aligned} \quad (2.18)$$

with

$$\mathbf{T}_{ij}(t - t') = \phi_i(\bar{\mathbf{r}})\mathbf{G}(\bar{\mathbf{r}} - \bar{\bar{\mathbf{r}}}; \bar{t} - t')\phi_j(\bar{\bar{\mathbf{r}}}) = \mathbf{G}(\mathbf{R}_i(t) - \mathbf{R}_j(t'); t - t'). \quad (2.19)$$

†The explicit form of Eq. (2.14) is written as $-\mathbf{F}_i^{(c)} = (3k_BT/l^2)[2\mathbf{R}_i(t) - \mathbf{R}_{i-1}(t) - \mathbf{R}_{i+1}(t)]$. Note that $\delta_{i,j}$ is the Kronecker delta.

The procedure in deriving Eq. (2.19) is virtually identical to the method developed by Freed and Edwards[10-12], except for a modification due to the confined geometry.[13]

The physical meaning of Eqs. (2.18) and (2.19) can be given as follows. The velocity of the ith bead is the summation of $\mathbf{v}_0(\mathbf{R}_i, t)$, the undisturbed velocity of the fluid at $\mathbf{R}_i$, and the velocity disturbances created by the beads. The term involving $\mathbf{T}_{ij}$ can arise from two different physical origins: the Stokes law type of drag force on the same bead i for $i = j$ and the Oseen type of hydrodynamic interaction from the different beads j if $j \neq i$.

The details are clarified in Section III.A on Green's function. Eq. (2.18) is known as a chain Langevin equation. The alternative form is to express the dynamics of the polymer chain in terms of a bead's probability function ψ, which will be given below. There are many methods[14,15] of deriving the Fokker–Planck equation. We will follow the method of Akcasu[16] to derive the equation for ψ,

$$\begin{aligned}\frac{\partial \psi(\{\mathbf{R}_k\}, t)}{\partial t} = &- \sum_i \frac{\partial}{\partial \mathbf{R}_i} \cdot \mathbf{v}_0(\mathbf{R}_i, t)\psi(\{\mathbf{R}_k\}, t) \\ &+ \sum_{ij} \frac{\partial}{\partial \mathbf{R}_i} \cdot \mathbf{T}_{ij}(t - \bar{t})\cdot(-\Delta_j \mathbf{R}_j(\bar{t})\psi(\{\mathbf{R}_k\}, \bar{t})) \\ &+ k_B T \sum_{ij} \frac{\partial}{\partial \mathbf{R}_i} \cdot \mathbf{T}_{ij}(t - \bar{t})\cdot \frac{\partial}{\partial \mathbf{R}_j} \psi(\{\mathbf{R}_k\}, \bar{t}).\end{aligned} \tag{2.20}$$

Here the probability function $\psi(\{\mathbf{R}_i\})\ t$ is interpreted as the probability of finding the beads at $\mathbf{R}_i$, $(i = 1, 2, \ldots, n)$, at time t.

Eq. (2.20) is virtually an exact equation for the probability function of a single chain in confined geometry as shown in Fig. 1. The important consideration is the recognition of the confined geometrical effect through $\mathbf{T}_{ij}$. The functional form of $\mathbf{T}_{ij}$ for unbound geometry is given in Altenberger.[17]

If we use the creeping flow approximation instead of linearized hydrodynamics, then $\mathbf{T}_{ij}$ becomes $\mathbf{G}_{ij}\delta(t - t')$. For the case of an unbound geometry, $\mathbf{G}_{ij}$ is related to the Stokes force term for $i = j$, whereas $\mathbf{G}_{ij}$ is the Oseen tensor for $i \neq j$, and

$$\mathbf{G}_{ij}^{(c)} = \zeta^{-1}\delta_{ij}\boldsymbol{\delta} + \mathbf{K}(\mathbf{R}_i - \mathbf{R}_j)(1 - \delta_{ij}). \tag{2.21}$$

Here $\mathbf{G}_{ij}^{(c)}$ is fluid propagator between the ith and jth beads for the unbound geometry, $\boldsymbol{\delta}$ is a unit tensor, $\zeta = 6\pi\eta a$ (a being the bead diameter) is the usual bead friction, and $\mathbf{K}(r)$ is normally chosen as the Oseen tensor evaluated for the point source, that is, $\mathbf{K}(r) = 1/r(\delta + \mathbf{rr}/r^2)$ with $|\mathbf{r}| = r$. Note that the point source assumption yields an instability in the numerical analysis that will be briefly discussed later.

With these simplifying approximations, Eq. (2.20) reduces to the conventional Kirkwood–Riseman equation,[18]

$$\frac{\partial \psi}{\partial t} = -\sum_i \frac{\partial}{\partial \mathbf{R}_i} \cdot [\mathbf{v}_0(\mathbf{R}_i, t)\psi] + \sum_{i,j} \frac{\partial}{\partial \mathbf{R}_i} \cdot \mathbf{G}_{ij}^{(c)} \cdot (-\Delta_j \mathbf{R}_j)\psi$$

$$+ k_B T \sum_{i,j} \frac{\partial}{\partial \mathbf{R}_i} \cdot \mathbf{G}_{ij}^{(c)} \cdot \frac{\partial}{\partial \mathbf{R}_j} \psi. \tag{2.22}$$

In an approach opposite to ours, Zwanzig[19] derived the chain Langevin equation [similar to Eq. (2.18)] from the Kirkwood–Riseman equation. This approach is well presented in Freed.[12] During the last few decades the Kirkwood–Riseman equation [Eq. (2.22)] has played an important role in the study of polymer dynamics. To avoid mathematical complexity, one often chooses the equilibrium averaged value for $\mathbf{G}_{ij}^{(c)}$. (This is known as preaveraging.) However, it was found numerically that the preaveraged form of the diffusion tensor, $\langle \mathbf{G}_{ij}^{(c)} \rangle$, is not positive definite.[20,21] It is expected that this nonpositive definite property arises from a mathematical artifact of the point particle assumption in the derivation of $\mathbf{K}$ given in Eq. (2.21). Several authors[22–24] have attempted to remove the singularities by introducing the finite size effect as in the derivation of hydrodynamic interaction.

D. Chain Langevin Equation for the Dumbbell

So far we have derived an explicit equation of motion for the single chain, both in Langevin equation form [Eq. (2.18)] and in Fokker–Planck equation form [Eq. (2.20)] for a confined geometry with an arbitrary flow field. In deriving these equations we have used the n-bead model. In practice the n-bead model involves extensive bookkeeping problems. To avoid this complexity, we will study the two-bead case, often called the dumbbell model, illustrated in Figs. 2 and 3.

For the dumbbell model, the Langevin equation [Eq. (2.18)] can be explicitly written in the form

$$\dot{\mathbf{R}}_1 = \mathbf{v}_0(\mathbf{R}_1, t) + \mathbf{T}_{11}(t-\bar{t}) \cdot (\mathbf{F}_1^{(c)}(\bar{t}) + \mathbf{f}_1^*(\bar{t})) + \mathbf{T}_{12}(t-\bar{t}) \cdot (\mathbf{F}_2^{(c)}(\bar{t}) + \mathbf{f}_2^*(\bar{t})) \tag{2.23a}$$

and

$$\dot{\mathbf{R}}_2 = \mathbf{v}_0(\mathbf{R}_2, t) + \mathbf{T}_{21}(t-\bar{t}) \cdot (\mathbf{F}_1^{(c)}(\bar{t}) + \mathbf{f}_1^*(\bar{t}) + \mathbf{T}_{22}(t-\bar{t}) \cdot (\mathbf{F}_2^{(c)}(\bar{t}) + \mathbf{f}_2^*(\bar{t})) \tag{2.23b}$$

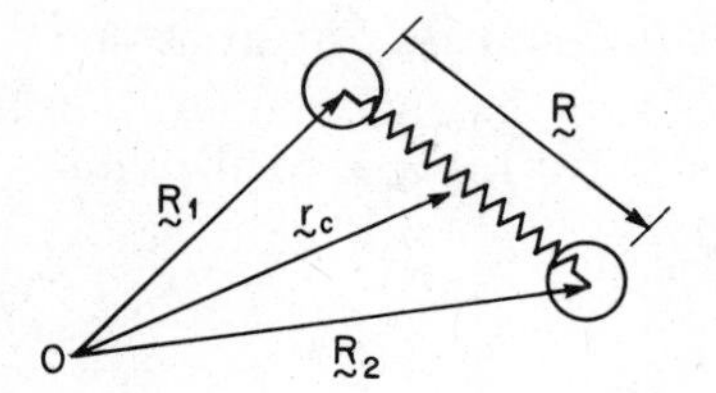

Figure 3. Center-of-mass coordinates for the dumbbell.

It is often convenient to express the motion of the dumbbell in center-of-mass coordinates, as illustrated in Fig. 3. In this coordinate system, $\mathbf{R}$ and $\mathbf{r}_c$ are given by

$$\begin{aligned} \mathbf{R} &\equiv \mathbf{R}_2 - \mathbf{R}_1 \\ \mathbf{r}_c &\equiv \tfrac{1}{2}(\mathbf{R}_1 + \mathbf{R}_2). \end{aligned} \tag{2.24}$$

The center-of-mass coordinate scheme shown in Fig. 3 is a natural coordinate system in the study of the motion of the dumbbell. The study of $\mathbf{R}$ will provide information on the polymer deformation, an important property in determining the overall rheological response of the fluids. The study of $\mathbf{r}_c$ will provide useful information on the motion of the polymer's center of mass. (It is useful to think of the polymer as an equivalent sphere.)

The calculation of $\mathbf{T}_{ij}$ requires fluid mechanical calculations. However, to simplify our analysis we will neglect the hydrodynamic interaction between the beads and set $\mathbf{T}_{ij}(t - t') \doteq \zeta^{-1}\delta_{ij}\boldsymbol{\delta}\delta(t - t')$. The random force $\mathbf{f}^*$ can be written alternatively as $\mathbf{f}_i^* = -k_B T(\partial/\partial\mathbf{R}_i)\ln\psi$.[5]

With these simplifications Eq. (2.23) can be written as

$$\dot{\mathbf{R}}_1 = \mathbf{v}_0(\mathbf{R}_1, t) + \zeta^{-1}\left[\mathbf{F}_1^{(c)} - k_B T \frac{\partial}{\partial \mathbf{R}_1}\ln\psi\right] \tag{2.25a}$$

and

$$\dot{\mathbf{R}}_2 = \mathbf{v}_0(\mathbf{R}_2, t) + \zeta^{-1}\left[\mathbf{F}_2^{(c)} - k_B T \frac{\partial}{\partial \mathbf{R}_2}\ln\psi\right], \tag{2.25b}$$

or, in terms of center-of-mass coordinates, as

$$\dot{\mathbf{R}} = \dot{\mathbf{R}}_2 - \dot{\mathbf{R}}_1 = [\mathbf{v}_0(\mathbf{R}_2, t) - \mathbf{v}_0(\mathbf{R}_1, t)] - \frac{2k_B T}{\zeta}\frac{\partial}{\partial \mathbf{R}}\ln\psi - \frac{2}{\zeta}\mathbf{F}^{(c)} \tag{2.26a}$$

and

$$\dot{\mathbf{r}}_c = \frac{1}{2}(\dot{\mathbf{R}}_1 + \dot{\mathbf{R}}_2) = \frac{1}{2}[\mathbf{v}_0(\mathbf{R}_1, t) + \mathbf{v}_0(\mathbf{R}_2, t)] + \frac{k_B T}{2\zeta}\frac{\partial}{\partial \mathbf{r}_c}\ln\psi. \tag{2.26b}$$

Note that $\mathbf{F}_1 = -\mathbf{F}_2 = \mathbf{F}^{(c)}$. If ψ is independent of $\mathbf{r}_c$, which is the case for the homogeneous flow (see Section III.A), Eqs. (2.25) and (2.26) reduce to Eqs. (10.2-6) to (10.2-9) of Bird et al.[5] Table I summarizes the single-chain dynamics equation.

TABLE I. Summary of Single-Chain Dynamics

Physical situation	Langevin type equation	Fokker–Planck type equation
Confined geometry Arbitrary flow n-bead model	Eq. (2.18), see Ref. 17.	Modified Kirkwood–Riseman equation, Eq. (2.20)
Unbounded geometry Arbitrary flow n-bead model	Rouse equation (Ref. 25, ignore hydrodynamic interaction). Zimm equation (Ref. 26, include hydrodynamic equation). *Note:* Eq. (2.22) can give instability due to point source assumption.	Kirkwood-Riseman equation, see Ref. 18. Eq. (2.22), see Ref. 19.
Confined geometry Arbitrary flow Two-bead model Ignoring hydrodynamic interaction	Eqs. (2.24) and (2.25). Assumed $\mathbf{f}^* = -k_B T(\partial/\partial \mathbf{R}_i)\ln\psi$.	Trivial
Unbounded geometry Homogeneous flow Two-bead model Ignoring hydrodynamic interaction	Eqs. (10.2-6) to (10.2-9) of Ref. 5.	Eq. (10.2-13) of Ref. 5.

III. POLYMER RESPONSE IN DETERMINISTIC FLOW

In this section we will give some case studies of the polymer response in simple confined geometry and flow fields. Note that most of the engineering applications of polymer solutions involve flowing situations and situations bounded by some kind of wall. For the case of an unbounded simple flow, known as a homogeneous flow, the response of the polymer is well documented in Bird et al.[5] Our purpose in this section is to further examine the existing studies presented in Bird et al.[5]

A. Study of $\mathbf{v}_0$, $\mathbf{T}_{ij}$, and $\mathbf{F}_i^{(c)}$

Before we get into details, we will study three important physical variables: $\mathbf{v}_0$, $\mathbf{T}_{ij}$, and $\mathbf{F}_i^{(c)}$. The first two involve fluid mechanical calculations and are dependent on confined geometrical shapes through boundary conditions; $\mathbf{F}_i^{(c)}$, however, depends solely on the polymer properties.

1. Undisturbed Fluid Velocity $\mathbf{v}_0(\mathbf{r}, t)$

In the absence of a polymer, the velocity profile can, in principle, be computed from the Navier–Stokes equation and the continuity equation given by Eqs. (2.6) and (2.7). Several exact solutions of $\mathbf{v}_0$ will be illustrated.[2,7,9,27]

a. Couette Flow (Drag Flow). This type of flow is generated by the motion of boundaries. No pressure gradient is generally imposed on the system. Let us examine the flow generated between infinite parallel planes, one of which moves relative to the other, as shown in Fig. 4*A*. In this flow geometry, Eqs. (2.6) and (2.7) with a no-slip boundary condition will give the velocity field

$$\mathbf{v}_0 = \frac{Uy}{B}\,\boldsymbol{\delta}_x. \tag{3.1}$$

Here B is the distance between two plates, U is the relative velocity, and $\boldsymbol{\delta}_x$ is the unit vector in the x direction. Similar types of drag flow in different

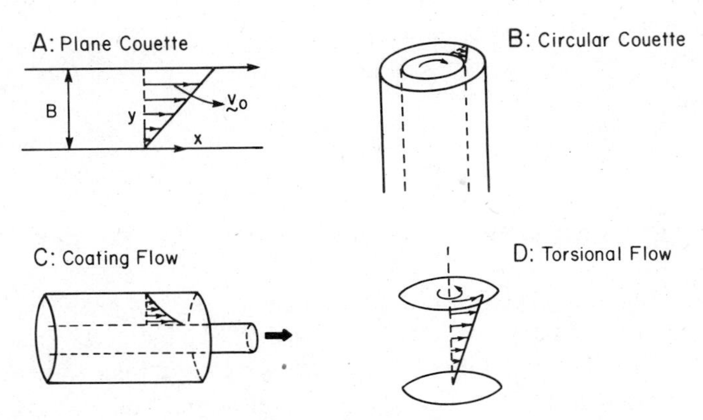

Figure 4. Drag flow.

geometries are the circular Couette flow (Fig. 4*B*), the coating flow (Fig. 4*C*), and the torsional flow (Fig. 4*D*). It is easy to handle the Navier–Stokes equation in these simple flows. The exact velocity profile is available in the standard texts on fluid mechanics.[7,9] These constitute the typical drag flow geometry which polymer engineers often study.

b. Pressure-Driven Flow. Let us consider the steady *laminar* flow of a constant-property fluid through two infinite parallel planes with imposed pressure gradient, as shown in Fig. 5*A*. In this situation we can easily calculate $\mathbf{v}_0$ from the Navier–Stokes equation,

$$\mathbf{v}_0 = \frac{(P_1 - P_2)B^2}{8\eta L}\left[1 - \left(\frac{2y}{B}\right)^2\right]\delta_x. \tag{3.2}$$

Here L is the length of the pipe, and P_1 and P_2 imply the pressure at the inlet and the outlet (assume $P_1 > P_2$). Even though the parallel-plate problem is the simplest, the most frequently encountered flow problems in engineering involve circular (cylindrical) geometry.

For the case of flow in a pipe (or tube) with a circular cross section as shown in Fig. 5*B*, we can easily calculate $\mathbf{v}_0$ from the Navier–Stokes equation

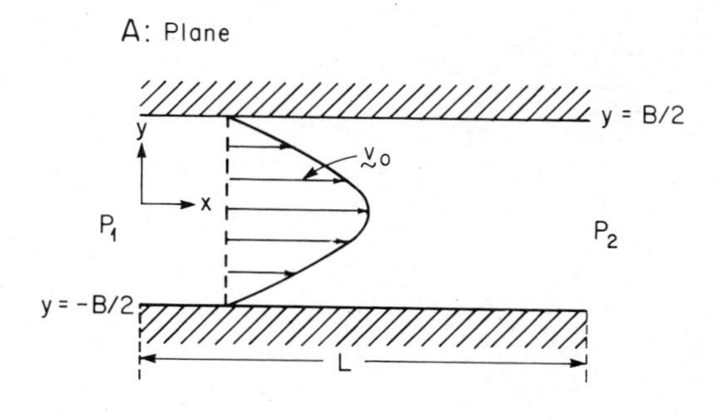

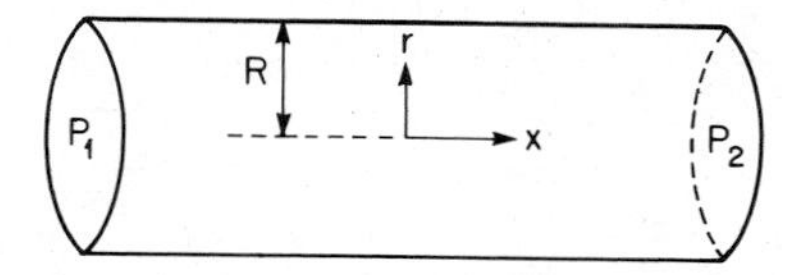

Figure 5. Pressure-driven flow.

in cylindrical coordinates,

$$\mathbf{v}_0 = v_{\max}\left[1 - \left(\frac{r}{R}\right)^2\right]. \tag{3.3}$$

Note that $v_{\max}$ is the maximum velocity, or the velocity at the centerline,

$$v_{\max} = \frac{(P_1 - P_2)R^2}{4\eta L}. \tag{3.4}$$

Here R is the radius of the tube.

c. *Extensional Flow.*† Extensional flow can be achieved by a thought experiment with four rollers, as shown in Fig. 6*A*. At the centerline (the origin in a two-dimensional plot) the velocity profile is easily obtained. We assumed that the pressure is uniform.

$$\mathbf{v}_0 = \dot{\varepsilon} x \delta - \dot{\varepsilon} y \delta_x. \tag{3.5}$$

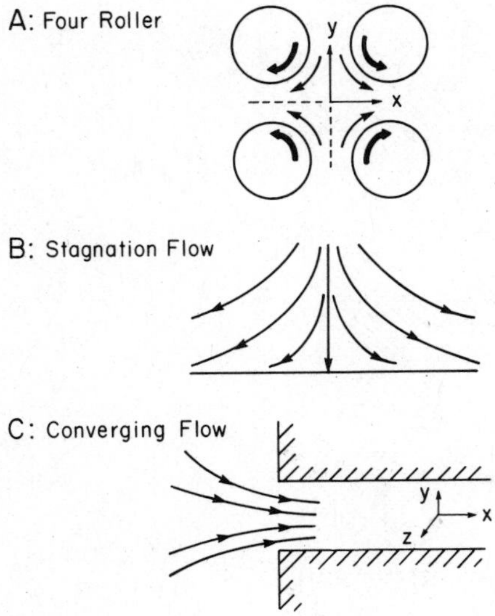

A: Four Roller

B: Stagnation Flow

C: Converging Flow

Figure 6. Extensional flow.

†See Refs. 2, 27, and 28.

This flow is known as planar elongational flow. Here $\dot{\varepsilon}$ is the elongational rate. This type of flow is the same as the plane stagnation flow (Hiemenz flow) shown in Fig. 6*B*.

Another example to be studied later is the flow converging into a small channel, as shown in Fig. 6*C*. At the centerline of the pore entrance $\mathbf{v}_0$ can be approximately written as

$$\mathbf{v}_0 = \dot{\varepsilon}x\boldsymbol{\delta}_x - \tfrac{1}{2}\dot{\varepsilon}y\boldsymbol{\delta}_y - \tfrac{1}{2}\dot{\varepsilon}z\boldsymbol{\delta}_z. \tag{3.6}$$

This type of flow is called uniaxial extensional or uniaxial elongational flow. So far we have shown simple flow geometries, which were often used in the past to study the motion of polymers in imposed bulk flows. For a general flow field, $\mathbf{v}_0(\mathbf{r})$ can be expressed in the form

$$\begin{aligned}\mathbf{v}_0(\mathbf{r}) &= \mathbf{v}_0(\mathbf{r}')|_{\mathbf{r}'=0} + \mathbf{r}\cdot\nabla'\mathbf{v}_0(\mathbf{r}')|_{\mathbf{r}'=0} + \frac{1}{2!}(\mathbf{rr}):\nabla'\nabla'\mathbf{v}_0(\mathbf{r}')|_{\mathbf{r}=0} + \cdots \\ &= \exp(\mathbf{r}\cdot\nabla')\mathbf{v}_0(\mathbf{r}')|_{\mathbf{r}'=0}.\end{aligned} \tag{3.7}$$

From this expression we can classify the flow field as follows.[5] The flow is called homogeneous if all derivatives vanish except the first. Otherwise it is called nonhomogeneous flow. For example, the plane Couette flow and the extensional flow are homogeneous, whereas the pressure-driven flow is nonhomogeneous. The homogeneous flow can be written as

$$\mathbf{v}_0(\mathbf{r}) = \mathbf{v}_0(\mathbf{r}')|_{\mathbf{r}'=0} + \mathbf{r}\cdot\nabla'\mathbf{v}_0(\mathbf{r}')|_{\mathbf{r}'=0} = \mathbf{v}_0(\mathbf{r}')|_{\mathbf{r}'=0} + \mathbf{K}\cdot\mathbf{r} \tag{3.8}$$

where $\mathbf{K} = (\nabla\mathbf{v}_0)^{\mathrm{T}}$ is the velocity gradient tensor.

The rate-of-strain tensor (measure of deformation) is often defined† as $\dot{\boldsymbol{\gamma}}$ and the vorticity tensor (measure of rotation) as $\boldsymbol{\Omega}$ in rheology,[2]

$$\dot{\boldsymbol{\gamma}} = \mathbf{K} + \mathbf{K}^{\mathrm{T}} \tag{3.9a}$$

and

$$\boldsymbol{\Omega} = \mathbf{K}^{\mathrm{T}} - \mathbf{K}. \tag{3.9b}$$

For example, $\boldsymbol{\Omega} = 0$ for the extensional flow, whereas $\boldsymbol{\Omega} \neq 0$ for the plane Couette flow.

†We use $\mathbf{S}$ in place of $\dot{\boldsymbol{\gamma}}$ in the Appendix when reviewing turbulence.

2. *Hydrodynamic Interaction* $\mathbf{T}_{ij}$

As stated in Section II, the single most important physical parameter entered in the probability equation is the $\mathbf{T}_{ij}$ term. Let us first study the term for $i = j$. As we discussed in the Kirkwood–Riseman equation [Eq. (2.22)], $\mathbf{T}_{ii}$ in unbounded geometry becomes the isotropic diffusion tensor, $\mathbf{T}_{ii} = \zeta^{-1}\boldsymbol{\delta}$. Here $\zeta = 6\pi\eta a$. However, when a bead is in the proximity of a wall, as shown in Fig. 7, diffusion is inhibited in directions both parallel and perpendicular to the wall. The perpendicular component, however, drops off much faster than the parallel one.[29] We will therefore assume that the parallel component remains unchanged from free diffusion. The perpendicular part of the mobility was calculated by Cox and Brenner,[30]

$$(\mathbf{T}_{ii})_{zz} = \frac{1}{\zeta} f_i(H),$$

where $H = (z_i - a)/a$. Here $z_i - a$ is the distance of the ith bead from the wall, and a is the bead radius. The term $f_i(H)$ has the following asymptotic form:

$$f_i(H) \sim \begin{cases} H, & \text{as } H \to 0 \\ \dfrac{H - \frac{1}{8}}{H + 1}, & \text{as } H \to \infty. \end{cases}$$

Therefore, for simplicity, we may adopt the combined asymptotic form

$$(\mathbf{T}_{ii})_{zz} = \frac{1}{\zeta}\frac{H}{H + 1}.$$

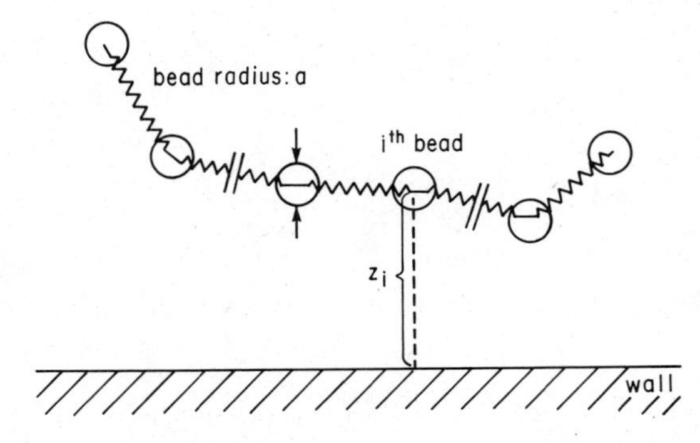

Figure 7. Hydrodynamic interaction between ith bead and wall.

Thus $\mathbf{T}_{ii}$ is not isotropic due to bead–wall interaction, and the approximate form can be written in the following tensor form:

$$\mathbf{T}_{ii} = \frac{1}{\zeta}\begin{pmatrix} 1 & 0 & 0 \\ 0 & 1 & 0 \\ 0 & 0 & \dfrac{H}{H+1} \end{pmatrix} = \frac{1}{\zeta}\begin{pmatrix} 1 & 0 & 0 \\ 0 & 1 & 0 \\ 0 & 0 & \dfrac{z_i/a - 1}{z_i/a} \end{pmatrix}. \tag{3.10}$$

To explain the hydrodynamic interaction term $\mathbf{T}_{ij(i \neq j)}$, we will first solve Eqs. (2.9) and (2.10) for unbound geometry.[12] To solve for the velocity disturbance in an unbound geometry, it is convenient to cast these equations into the Fourier transforms in space and time. By taking Fourier transformation,

$$A(\mathbf{k}, \omega) \equiv \int d^3r \int_{-\infty}^{\infty} dt\, e^{+i\mathbf{k}\cdot\mathbf{r} - i\omega t} A(\mathbf{r}, t),$$

of Eqs. (2.9) and (2.10), we obtain

$$i\omega\rho\mathbf{v}(\mathbf{k}, \omega) + \eta k^2 \mathbf{v}(\mathbf{k}, \omega) - i\mathbf{k}P(\mathbf{k}, \omega) = \mathbf{F}_{\text{ext}}(\mathbf{k}, \omega) \tag{3.11}$$

and

$$-k^2 P(\mathbf{k}, \omega) = -i\mathbf{k}\cdot\mathbf{F}_{\text{ext}}(\mathbf{k}, \omega). \tag{3.12}$$

By eliminating $P(\mathbf{k}, \omega)$ from Eqs. (3.11) and (3.12), we obtain an explicit equation for $\mathbf{v}(\mathbf{k}, \omega)$,

$$\mathbf{v}(\mathbf{k}, \omega) = \mathbf{G}^{\text{u}}(\mathbf{k}, \omega)\cdot\mathbf{F}_{\text{ext}}(\mathbf{k}, \omega). \tag{3.13}$$

Here $\mathbf{G}^{\text{u}}$ is the fluid propagator, or Green's function, for unbound geometry, and the explicit form for $\mathbf{G}^{\text{u}}$ becomes

$$\mathbf{G}^{\text{u}}(\mathbf{k}, \omega) = \frac{\boldsymbol{\delta} - \mathbf{kk}/k^2}{i\omega\rho + \eta k^2}. \tag{3.14}$$

We can express Eqs. (3.13) and (3.14) in real space form by taking the inverse Fourier transform,[17]

$$\mathbf{v}(\mathbf{r}, t) = \int d^3\bar{r} \int_{-\infty}^{\infty} d\bar{t}\, \mathbf{G}^{\text{u}}(\mathbf{r} - \bar{\mathbf{r}}; t - \bar{t})\cdot\mathbf{F}_{\text{ext}}(\bar{\mathbf{r}}, \bar{t}) \tag{3.15}$$

and

$$\mathbf{G}^{u}(\mathbf{R}; s) = \int \frac{d^3k}{(2\pi)^3} \int \frac{d\omega}{2\pi} e^{-i\mathbf{k}\cdot\mathbf{R}+i\omega s} \frac{\boldsymbol{\delta} - \mathbf{kk}/k^2}{i\omega\rho + \eta k^2}$$

$$= \frac{1}{8\pi^2\rho}\left[(\boldsymbol{\delta} + \hat{R}\hat{R})\frac{\sqrt{\pi}}{2}\left(\frac{\rho}{\eta s}\right)^{3/2} \exp\left(-\frac{\rho R^2}{4\eta s}\right)\right.$$

$$\left. - (\boldsymbol{\delta} - 3\hat{R}\hat{R})\pi \frac{\partial^2}{\partial R^2}\left\{R^{-1} \operatorname{erf}\left(\frac{\rho R^2}{2\eta s}\right)^{1/2}\right\}\right]. \tag{3.16}$$

Here $\mathbf{R} \equiv \mathbf{r} - \mathbf{r}'$, $s \equiv t - t'$, $\hat{R} \equiv \mathbf{R}/|\mathbf{R}|$, and

$$\operatorname{erf} x = \frac{2}{\sqrt{\pi}} \int_0^x dz\, e^{-z^2}$$

is an error function.

This result for $\mathbf{G}$ was obtained from the linearized Navier–Stokes equation [Eq. (2.9)]. However, if we use the creeping flow approximation [Eq. (2.8)], $\mathbf{G}$ can be obtained by setting $\omega = 0$ in Eq. (3.11),

$$\mathbf{G}^{(c)}(\mathbf{k}) = \lim_{\omega \to 0} \mathbf{G}^{u}(\mathbf{k}, \omega) = \frac{\boldsymbol{\delta} - \mathbf{kk}/k^2}{\eta k^2}, \tag{3.17}$$

or, in real space form,

$$\mathbf{G}^{(c)}(\mathbf{r} - \mathbf{r}') = \int_{-\infty}^{\infty} d(t - t')\, \mathbf{G}^{u}(\mathbf{r} - \mathbf{r}'; t - t')$$

$$= \frac{1}{8\pi\eta} \frac{1}{|\mathbf{r} - \mathbf{r}'|} \left\{\boldsymbol{\delta} + \frac{(\mathbf{r} - \mathbf{r}')(\mathbf{r} - \mathbf{r}')}{(\mathbf{r} - \mathbf{r}')^2}\right\}. \tag{3.18}$$

Note that $\mathbf{G}^{(c)}$ is the same as the Oseen tensor for the point source denoted by $\mathbf{K}$ in Eq. (2.21). To explain the physical meaning of $\mathbf{G}$, we will study instantaneous and local force density, that is, $\mathbf{F}_{\text{ext}} = \mathbf{F}\delta(\mathbf{r} - \mathbf{r}_0)\delta(t - t_0)$. In this situation, Eq. (3.15) can be written as

$$\mathbf{v}(\mathbf{r}, t) = \mathbf{G}^{u}(\mathbf{r} - \mathbf{r}_0; t - t_0)\cdot\mathbf{F}. \tag{3.19}$$

Eq. (3.19) can be interpreted as follows. If we disturb the point $\mathbf{r}_0$ at time t_0 with **F**, then this force is transmitted through the Newtonian media by waves, as shown in Fig. 8*A*, and produces the velocity disturbance at position **r** and $t(\mathbf{v}(\mathbf{r}, t))$. Therefore **G** can be interpreted as a propagator, and can be characterized solely from the property of fluid media. Also, in mathematical language, **G** is a Green's function since it is related with a solution of a nonhomogeneous differential equation. The propagator $\mathbf{G}^{u}$ calculated from the linearized Navier–Stokes equation has time dependence. This is expected, due to a certain time delay for a signal at $\mathbf{r}_0$, to reach **r**. However, the propagator $\mathbf{G}^{(c)}$ obtained from the creeping flow approximation is the low-frequency limit of $\mathbf{G}^{u}$ and is independent of time.

Now we will study the fluid propagator in the presence of boundaries. To obtain **G** for the arbitrary confined geometry shown in Fig. 8*B*, we must sum all of the contributions due to the waves reflected from the boundaries. **G** can be formally obtained (creeping flow approximation) as

$$\mathbf{G}(\mathbf{r}, \mathbf{r}') = \frac{1}{\eta}\left\{S(\mathbf{r}, \mathbf{r}')\boldsymbol{\delta} - \int d^3\bar{r}\, S(\mathbf{r}, \bar{\mathbf{r}})\boldsymbol{\nabla}\bar{\boldsymbol{\nabla}} S(\bar{\mathbf{r}}, \mathbf{r}')\right\}. \tag{3.20}$$

A: Unbound Geometry

v (r,t)
F
r₀
r

B: Arbitrarily Shaped Confined Geometry

r
r'

C: Flat Plate

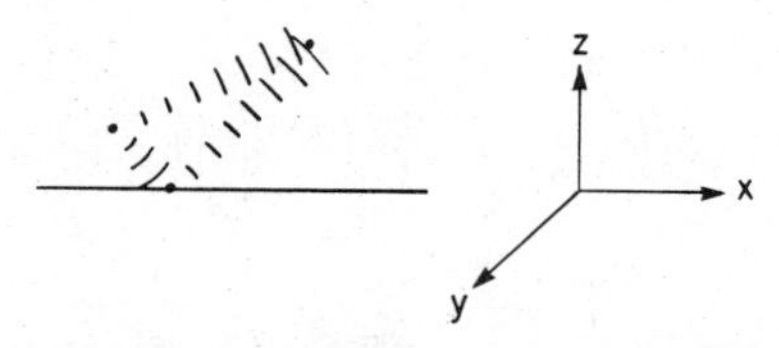

Figure 8. Physical meaning of fluid propagator. A force **F** propagates through the waves.

Here S can be calculated from

$$\nabla^2 S(\mathbf{r};\mathbf{r}') = \delta(\mathbf{r}-\mathbf{r}') \tag{3.21}$$

with proper boundary conditions (no slip at wall). Therefore $\mathbf{G}$ depends on the geometry of the system through the boundary conditions, and, in principle, we can calculate $\mathbf{G}$ for a given geometry from Eqs. (3.20) and (3.21).

As an illustration, we will choose the simple geometry shown in Fig. 8C. Due to the reflected waves, $\mathbf{G}$ is spatially anisotropic. Note that $\mathbf{G}$ is isotropic for the unbound geometry. Therefore we can write Eq. (3.19) as

$$\begin{pmatrix} \mathbf{v}_{\|} \\ v_{\perp} \end{pmatrix} = \begin{pmatrix} \mathbf{G}_{\|,\|}, & \mathbf{G}_{\|,\perp} \\ \mathbf{G}_{\perp,\|}, & G_{\perp,\perp} \end{pmatrix} \cdot \begin{pmatrix} \mathbf{F}_{\|} \\ F_{\perp} \end{pmatrix}, \tag{3.22}$$

or, explicitly in component form,

$$\begin{aligned} \mathbf{v}_{\|} &= \mathbf{G}_{\|,\|}\cdot\mathbf{F}_{\|} + \mathbf{G}_{\|,\perp}F_{\perp} \\ v_{\perp} &= \mathbf{G}_{\perp,\|}\cdot\mathbf{F}_{\|} + G_{\perp,\perp}F_{\perp}. \end{aligned} \tag{3.23}$$

Here $\mathbf{v}_{\|}$ and $\mathbf{F}_{\|}$ are the components of velocity and force parallel to the wall surface (x–y plane) and are two-dimensional vectors. On the other hand, $v_{\perp}$ and $F_{\perp}$ are the components of velocity and force perpendicular to the surface of the wall (z direction). Now let us give a physical meaning to $\mathbf{G}$'s, explaining $\mathbf{G}_{\|,\|}$ only. If $\mathbf{F}$ has only parallel components (that is, $F_{\perp} = 0$), then $\mathbf{v}_{\|} = \mathbf{G}_{\|,\|}\cdot\mathbf{F}_{\|}$ from Eq. (3.23). Therefore, $\mathbf{G}_{\|,\|}$ can be interpreted as the fluid propagator which transmits the parallel component of the force to the parallel component of velocity disturbances. $\mathbf{G}_{\perp,\|}$, $\mathbf{G}_{\|,\perp}$, and $G_{\perp,\perp}$ can be similarly explained. Fig. 9 provides their physical meanings pictorially.

The explicit form of $\mathbf{G}$'s is obtained from Eqs. (3.20) and (3.21). As an example, we shall give an expression for $G_{\perp,\perp}$ (scalar.)[13] (Note that the symmetry is broken only in the z direction, that is, $\mathbf{G}$ is a function of $x - x'$, $y - y'$, z, and z'.)

$$\begin{aligned} G_{\perp,\perp}(\boldsymbol{\rho}-\boldsymbol{\rho}';z,z') &= \frac{1}{8\pi\eta}\left\{\frac{1}{[(\boldsymbol{\rho}-\boldsymbol{\rho}')^2+(z-z')^2]^{1/2}} - \frac{1}{[(\boldsymbol{\rho}-\boldsymbol{\rho}')^2+(z+z')^2]^{1/2}}\right\} \\ &+ \frac{1}{8\pi\eta}\left\{\frac{(z-z')^2}{[(\boldsymbol{\rho}-\boldsymbol{\rho}')^2+(z-z')^2]^{3/2}} - \frac{(z+z')^2}{[(\boldsymbol{\rho}-\boldsymbol{\rho}')^2+(z+z')^2]^{3/2}}\right\}. \end{aligned} \tag{3.24}$$

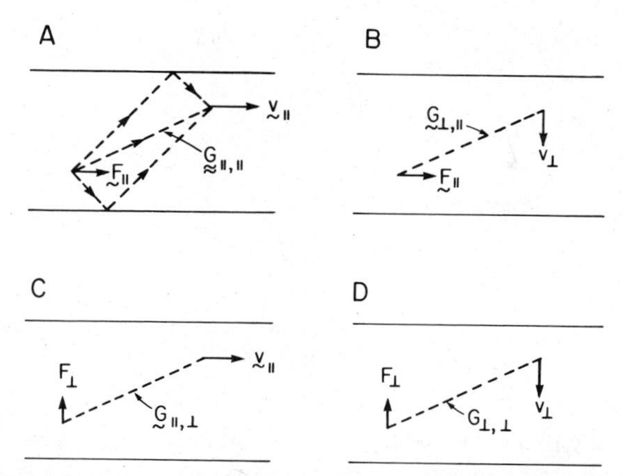

Figure 9. Physical meaning of **G** for two flat plates. (G is a tensor.)

Here $\boldsymbol{\rho} - \boldsymbol{\rho}' \equiv (x - x')\boldsymbol{\delta}_x + (y - y')\boldsymbol{\delta}_y$.

As expected, **G** becomes the Oseen tensor if the wall effect can be ignored, that is, z and z' are located far away from the wall. If the bead size can be ignored (point source assumption), then we can obtain $\mathbf{T}_{ij}$ from Eq. (2.19). **G** in Eq. (3.24) does not have time dependence, since we have imposed creeping flow approximation. From Eq. (3.24), $\mathbf{T}_{ij}$ for the dumbbell model can be expressed in terms of center-of-mass coordinates [Eq. (2.24)],

$$
\begin{aligned}
(\mathbf{T}_{12}), (\boldsymbol{\rho} - \boldsymbol{\rho}'; Z, z_c) \\
&= \frac{1}{8\pi\eta}\left\{\frac{1}{R} - \frac{1}{[(\boldsymbol{\rho} - \boldsymbol{\rho}')^2 + (2z_c)^2]}\right\} \\
&\quad + \frac{1}{8\pi\eta}\left\{\frac{Z^2}{R^{3/2}} - \frac{(2z_c)^2}{[(\boldsymbol{\rho} - \boldsymbol{\rho}')^2 + (2z_c)^2]^{3/2}}\right\}. \qquad (3.25)
\end{aligned}
$$

Here $\boldsymbol{\rho} - \boldsymbol{\rho}'$, z, and z' are given in Fig. 10. Further, $Z = z' - z$ and $z_c = \frac{1}{2}(z + z')$.

So far we have discussed mathematical details of the hydrodynamic interaction. We will now give a simple and intuitive explanation of hydrodynamic interaction. Let us say that you are swimming in a quiet ocean; the fluid exerts a force on you. This drag force is a Stokes force. Every time you exert a force on the fluid (say at $\mathbf{r}$ and t), then a propagation of waves will occur, producing velocity disturbance at $\mathbf{r}'$ and t'. (Here $t' > t$ due to causality.) This propagation is related to the Oseen tensor. Now suppose that

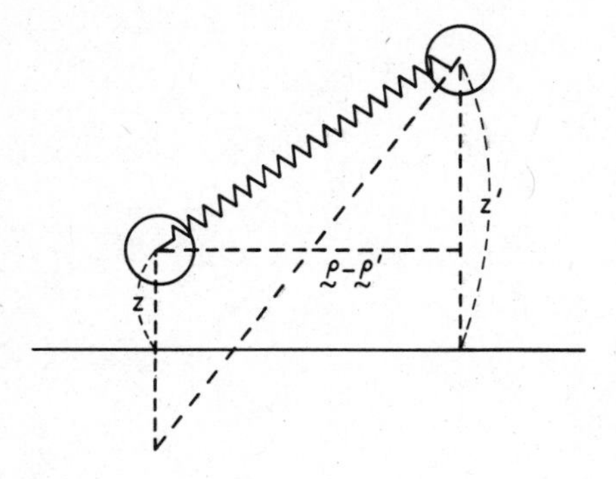

Figure 10. Pictorial representation of the coordinate system for a dumbbell in the presence of one flat wall.

you are swimming in a small swimming pool instead of the ocean. In this case there will be reflected waves from the boundaries. Therefore the Stokes law and the Oseen tensor will be modified due to boundary effects. This is a naive physical picture of hydrodynamic interaction in a confined geometry. If we apply it to the polymer solution, you are the bead and the water is the solvent. Since the bead is small, we can consider it a point source.

3. *Connector Force* $\mathbf{F}_i^{(c)}$†

The connector force is a property of the polymer only. For an n-bead chain we usually use the linear spring model (known as the Rouse chain). We will study various models of the connector force and choose a two-bead case (dumbbell model) to simplify our illustration. The simplest connector force is known as the Hookean spring. This is the same as the Rouse chain for the n-bead situation,

$$\mathbf{F}^{(c)} = H\mathbf{R}. \tag{3.26}$$

As Sections III.B.1 and IV will show, the Hookean spring is unstable under strong flow. To avoid this type of catastrophic situation, nonlinear springs are often introduced. The most commonly used nonlinear spring is known as the Warner model or FENE (finitely extendable nonlinear elastic) model. The force law for FENE is given as

$$\mathbf{F}^{(c)} = \frac{H}{1 - (R/R_0)^2}\mathbf{R}, \qquad \text{for } R < R_0. \tag{3.27}$$

Although the FENE model is useful for studying the polymer response in the deterministic flow, we find that it is very difficult to use for modeling

†See Ref. 5.

the polymer conformation in the turbulent flow, as will be shown in Section IV. Therefore we propose an alternative nonlinear spring force, called anharmonic spring. The force law for the anharmonic spring is

$$\mathbf{F}^{(c)} = H(1 + \varepsilon R^2)\mathbf{R}. \tag{3.28}$$

Here ε is the measure of nonlinearity. Note that both the FENE and the anharmonic spring are two constant models. We will introduce the connector potential $\phi^{(c)}$ in Section IV. The relationship between $\mathbf{F}^{(c)}$ and $\phi^{(c)}$ is

$$\mathbf{F}^{(c)}(R) = \nabla\phi^{(c)}(R) \qquad \text{or} \qquad \phi^{(c)} = \int_0^{\mathbf{R}} \mathbf{F}^{(c)}(R')\, d\mathbf{R}'. \tag{3.29}$$

B. Case Studies

So far we have examined all parts of the equations for the probability function of polymer beads. This section will present a few examples of polymer motion in the deterministic flow with confined flow geometry. Although many interesting physical phenomena are associated with polymer conformation in deterministic flow, we will discuss only a few specific cases which we have investigated during the past few years.

1. Polymer Dynamics in Strong Flow

Consider the flow of a solution of flexible polymers into a small pore, as depicted in Fig. 11, and study the polymer conformation.[31] It is intuitively obvious that the polymer molecule will form an equilibrium configuration due to Brownian motion far from the pore entrance. However, at or near the pore entrance, the polymer molecule will tend to elongate along the axis of the pore. Note that the flow near the pore mouth is approximately extensional, as explained in the previous section. This problem could be relevant to many areas of engineering, including separation technology, size exclusion chromatography, and membrane science. Let us qualitatively study this experiment. The extensional rate $\dot{\varepsilon}$ near the pore mouth can be

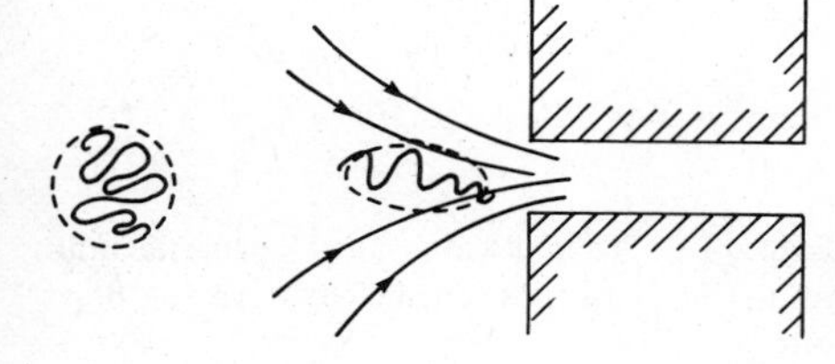

Figure 11. Flexible polymer chain entering a microspore. Near the pore entrance the polymer chain is greatly deformed.

given as $\dot{\varepsilon} \sim Q/R^3$, where Q is the flow rate and R is the pore radius. The following two flow regimes will be studied: (1) weak flow situation $(\dot{\varepsilon}\lambda_H \ll 1)$ —virtually no deformation, that is, the polymer is almost in equilibrium—and (2) strong flow region $(\dot{\varepsilon}\lambda_H > 1)$—large deformation of polymer chains. Here λ_H is the longest relaxation time of a polymer.

In case (1) the polymer chain behaves similarly to a rigid sphere. The polymer deformation can be studied from the Hookean spring law. Case (2) is more interesting, especially when $\lambda \equiv \sqrt{\langle R^2 \rangle_{eq}}/R > 1$, here $\sqrt{\langle R^2 \rangle_{eq}}$ is the radius of gyration of the polymer in quiescent flow. In this case the polymer conformation cannot be studied from the Hookean spring law. Since the linear spring causes instability, we must introduce the nonlinear spring model.

When $\lambda > 1$, the polymer will not pass through the pores unless it deforms drastically. Daoudi and Brochard[32] and deGennes[31] applied a scaling theory and obtained a remarkable result. They show that the reflection coefficient σ, the fraction of polymers rejected by the pore, does not depend on the polymer or pore size. They speculated that σ versus Q may be a first-order phase transition, that is, σ changes from 1 to 0 very sharply as Q increases past a critical value Q_c which is solely determined by the solvent viscosity. However, Long and Anderson[33,34] performed very careful experiments with linear monodisperse polystyrenes at very low concentrations in CCl_4/CH_3OH mixed solvent and showed that the experimental results do not give even qualitative agreement with theoretical predictions. Their results show no sharp transition in σ as Q increases. σ decays slowly to zero as Q becomes larger. We believe that the previous theory gives qualitatively incorrect results due to the use of a linear spring. Theories dealing with a strong flow situation are available in Hinch[35] among others, and may be useful in studying the above problem. A comparison between theory and experiment is briefly sketched in Fig. 12. Karis and Jhon[36] studied the case of rigid rods entering the pore in conjunction with magnetic processing

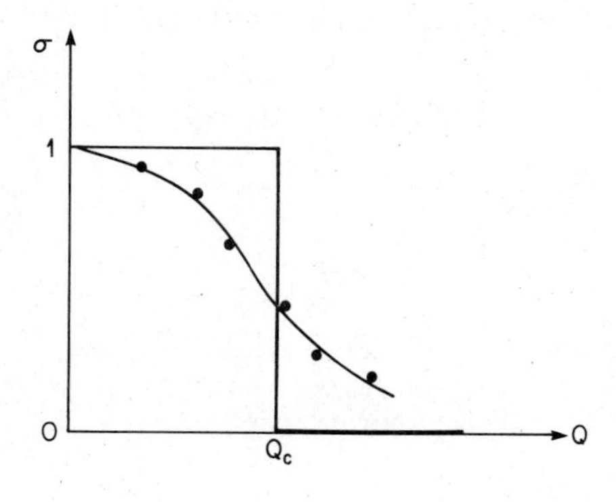

Figure 12. Theoretical and experimental results for reflection coefficient σ versus flow rate Q.

technology. We will develop the nonlinear Langevin theory to describe the polymer conformation in a strong flow situation. To simplify the analysis, we present the following thought experiment.

Let us study the polymer conformation located at the center of four rotating rollers, as shown in Fig. 13. At the centerline the polymer molecule will experience a planar elongational flow (explained in Section III.A.1). From Eqs. (3.5) and (3.8) we can easily calculate

$$\mathbf{v}_0(\mathbf{R}_2, t) - \mathbf{v}_0(\mathbf{R}_1, t) = \mathbf{K}\cdot(\mathbf{R}_2 - \mathbf{R}_1), \tag{3.30a}$$

and $\mathbf{K}$ becomes

$$\mathbf{K} = \dot{\varepsilon}\delta_x\delta_x - \dot{\varepsilon}\delta_y\delta_y. \tag{3.30b}$$

By treating a polymer as a dumbbell, and also by adapting the simplest equation of motion for dumbbells (free draining limit) given by Eq. (2.26a), we obtain

$$\dot{\mathbf{R}} = \mathbf{K}\cdot\mathbf{R} - \frac{2}{\zeta}\mathbf{F}^{(c)} + \mathbf{f}_{\mathbf{R}}^{*}. \tag{3.31}$$

Here

$$\mathbf{f}_{\mathbf{R}}^{*} = -\frac{2k_{\mathrm{B}}T}{\zeta}\frac{\partial}{\partial\mathbf{R}}\ln\psi$$

is the Brownian random force.

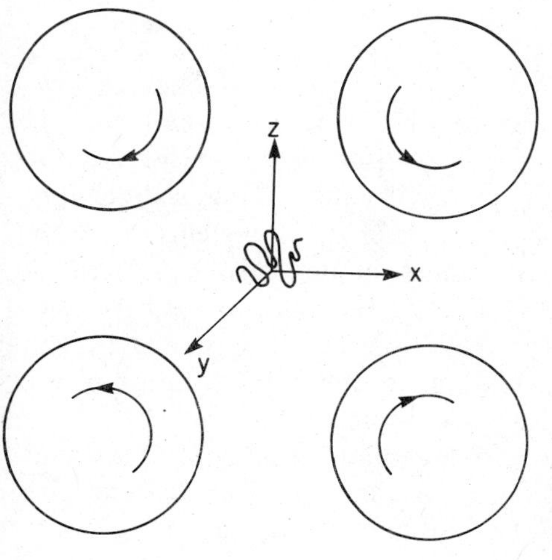

Figure 13. Thought experiment. If the polymer is situated between the centerlines of the four rollers, what is its conformation?

To illustrate the essence of the stability problem for Eq. (3.31), we will first study the x-component equation with linear spring [see Eq. (3.26)],

$$\dot{X} = \left(\dot{\varepsilon} - \frac{2}{\zeta} H\right) X + (\mathbf{f}_{\mathbf{R}}^{*})_x. \tag{3.32}$$

We will assume $(\mathbf{f}_{\mathbf{R}}^{*})_x = 0$, addressing only the deterministic part in Eq. (3.32). Then $X(t)$ can be easily calculated,

$$X(t) = X(t = 0) \exp\left[\left(\dot{\varepsilon} - \frac{2}{\zeta} H\right) t\right]. \tag{3.33}$$

The limiting behavior of $X(t)$ is worthy of note:

$$\lim_{t \to \infty} X(t) = \begin{cases} 0, & \dot{\varepsilon} < \dfrac{2}{\zeta} H \quad \text{weak flow} \\ \infty, & \dot{\varepsilon} > \dfrac{2}{\zeta} H \quad \text{strong flow.} \end{cases} \tag{3.34}$$

Therefore the linear spring is stable in a weak flow and unstable for the strong flow. To avoid this catastrophic situation, we introduce the anharmonic spring given by Eq. (3.28). Then the x-component equation becomes

$$\dot{X} = \left(\dot{\varepsilon} - \frac{2}{\zeta} H\right) X - \frac{2}{\zeta} H\varepsilon X^3 + (\mathbf{f}_{\mathbf{R}}^{*})_x. \tag{3.35}$$

Here we set $Y = Z = 0$, that is, $R = X$, to simplify the analysis. The nonlinear term $-(2/\zeta)H\varepsilon X^3$ is absolutely necessary to guarantee mathematical stability and thus to provide a physically realistic picture. However, the addition of this term yields a nonlinear Langevin equation that is difficult to handle. Fortunately many physical and nonphysical systems (such as chemical reaction, nonlinear optics, tunnel diodes, Malthus–Verhulst problem, and magnetic systems) deal with similar nonlinear equations,[37] and techniques to resolve them have been developed in the past. Rather than providing detailed mathematical calculations to obtain the solution of Eq. (3.35), we will draw an analogy with the magnetic phase transition.[38]

For $\dot{\varepsilon} \cong 0$, the polymer molecule is in a random configuration (similar to a paramagnetic system where spins are all randomly oriented and give zero magnetization). For the strong flow situation, however, the polymer molecule is elongated along the x axis. This situation is analogous to a ferromagnetic system where spins are lined up in a preferred direction with nonzero magnetization. To make the analogy quantitative, we cast Eq. (3.35) into the following form (assume $\mathbf{f}_{\mathbf{R}}^{*} = 0$ for simplicity):

$$\dot{X} = -\frac{\partial V}{\partial X} \tag{3.36a}$$

with

$$V = -\frac{1}{2}\left(\dot{\varepsilon} - \frac{2}{\zeta} H\right) X^2 + \frac{H\varepsilon}{2\zeta} X^4. \tag{3.36b}$$

X is analogous to the magnetization M, and V is analogous to the Helmholtz free energy F. According to Landau or the mean field theory,[38] the free energy as a function of magnetization can be written as

$$F = a(T - T_c)M^2 + bM^4, \qquad a, b \text{ positive,} \tag{3.37}$$

and is shown in Fig. 14.

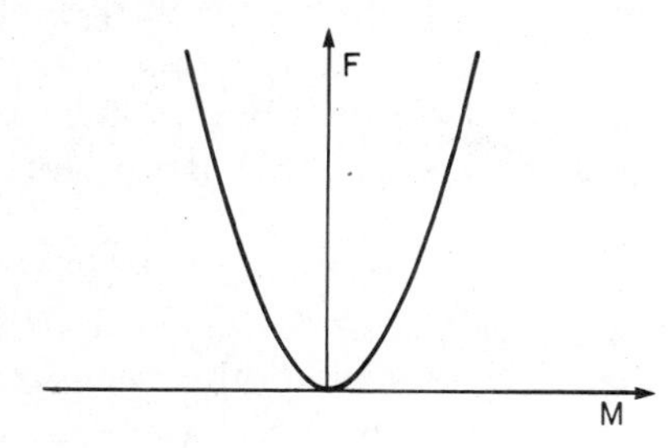

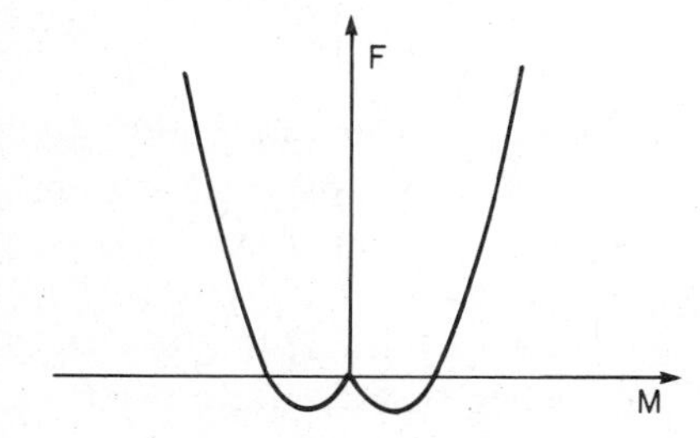

Figure 14. Plot of free energy F versus magnetization M. T_c stands for critical temperature.

Figure14A illustrates the situation for $T > T_c$. In this case $M = 0$ gives minimum free energy and is the stable fixed point. For $T < T_c$, $M = \pm M_0$ gives minimum free energy while $M = 0$ is an unstable fixed point, as shown in Figure 14B. Completing the analogy, we obtain the following for the polymeric system:

$$\text{for } \dot{\varepsilon} < \frac{2}{\zeta} H, \qquad X = 0 \quad \text{is a stable fixed point,}$$

$$\text{for } \dot{\varepsilon} > \frac{2}{\zeta} H, \qquad X = 0 \quad \text{is an unstable fixed point,} \tag{3.38}$$

and

$$X = \pm \frac{\sqrt{\zeta\dot{\varepsilon}/2H - 1}}{\varepsilon} \quad \text{are stable fixed points.}$$

From this analogy we speculate that σ versus Q exhibits a second-order phase transition rather than the first-order phase transition predicted by previous theories.[31,32]

So far we have assumed $\mathbf{f}_R^* = 0$ to qualitatively explain the result given in Fig. 12. We can calculate ψ by including $\mathbf{f}_R^*$ in the above analysis. The standard procedure of obtaining ψ is well documented in Suzuki.[39] Here we present only the qualitative results. Figure 15 illustrates the time evolution of the probability function $\psi(X, t)$. The dashed lines imply the trajectory $X(t)$ in the absence of Brownian motion, that is, $\mathbf{f}_R^* = 0$. The limiting value for $X(t)$ as time approaches infinity is the stable fixed point given by Eq. (3.38). The spreading of the wave packet with time requires complex calculations and will not be reported here.

Note that for the weak flow the Kirkwood–Riseman equation (Rouse chain or linear spring) gives a physically realistic picture, as shown in Fig. 15A. However, for the strong flow case it is essential to include the nonlinear spring model. Consequently we must study the nonlinear Langevin equation. While everyone seems to have a favorite method of tackling this equation, we use Suzuki's method[39] to derive the result given in Fig. 15B.

2. *Polymer Migration in Newtonian Fluids*

The purpose of this section is to illustrate the importance of the bulk flow term $\mathbf{v}_0$ and of the hydrodynamic interaction term $\mathbf{T}_{ij}$ in polymer response to a flow field with confined geometry. We shall study the following thought experiments given in Fig. 16. If the polymer molecules are placed in a flow field $\mathbf{v}_0$ with confined geometry, do they move with the fluid? If not, what is the origin of the forces that cause the polymer's deviation in trajectory from

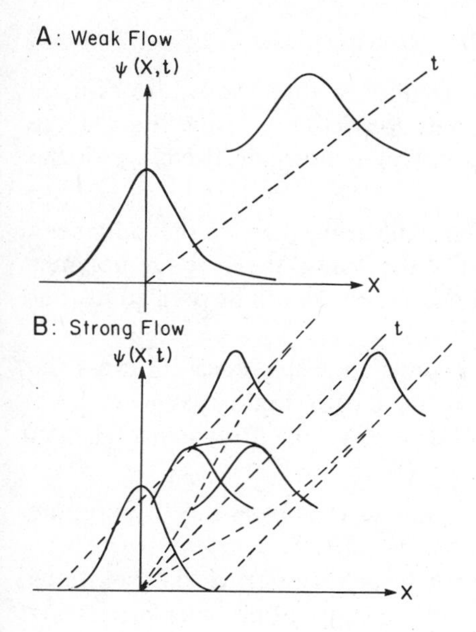

Figure 15. Time evolution of the probability function.

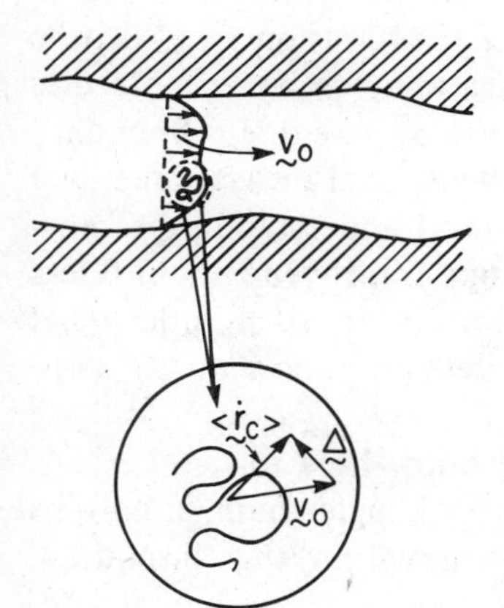

Figure 16. The essence of the problem is to study the motion of the polymer in a confined geometry with given flow. $\mathbf{v}_0$ stands for fluid velocity in the absence of the polymer molecule.

the main stream? The migration of a polymer molecule can be characterized by computing $\mathbf{\Delta}$, defined by

$$\mathbf{\Delta} = \langle \dot{\mathbf{r}}_c \rangle - \mathbf{v}_0(\mathbf{r}_c, t), \tag{3.39}$$

where $\mathbf{r}_c$ denotes the center of mass of the polymer and $\dot{\mathbf{r}}_c$ the velocity of the center of mass. Here angular brackets imply an average over all internal coordinates (for dumbbells, it is R) of the polymer.

This thought experiment results in the following classes:

1. $\boldsymbol{\Delta} = 0$, that is, $\langle \dot{\mathbf{r}}_c \rangle = \mathbf{v}_0(\mathbf{r}_c, t)$. This implies that the center-of-mass velocity of the polymer is the same as the velocity of the fluid at the center-of-mass position. Hence the polymer molecule travels with the fluid.
2. $\boldsymbol{\Delta} \neq 0$, that is, $\langle \dot{\mathbf{r}}_c \rangle \neq \mathbf{v}_0(\mathbf{r}_c, t)$. Here the trajectory of the polymer's center of mass is different from that of the fluid. Therefore the polymer molecule does not travel with the fluid. This class can be divided further into the following:
   a. $\boldsymbol{\Delta}$ is parallel to $\mathbf{v}_0(\mathbf{r}_c, t)$. The polymer molecule does not cross the streamline but simply is behind or ahead of the mainstream.
   b. $\boldsymbol{\Delta}$ has a component perpendicular to $\mathbf{v}_0(\mathbf{r}_c, t)$. Polymer migration occurs across the streamline.

Thus to understand the migration phenomena it is essential to compute $\langle \dot{\mathbf{r}}_c \rangle$ from the single-chain equation given in Section II.

The migration of polymer molecules (or suspended particles) in solutions and melts is an important phenomenon in the study of polymer processing or of anything that involves flow. For example, migration during injection or extrusion processes causes a nonuniform product. Migration may also be useful in studying lubrication problems. The migration of polymer molecules toward or away from a solid surface may result in apparent slip boundary conditions at the wall and thereby may affect rheological measurements of polymeric liquids. It is very likely that an enhanced understanding of how the polymer molecules move would lead to design improvements in filters for polymer solutions. Migration is also important in such biomedical engineering transport problems as red blood cell migration in the non-Newtonian serum of a blood vessel.

To understand the above problem we must know how to handle concentration effects. This section will address only single-chain situations applicable to an infinitely dilute solution. Even within this limitation, however, the theory has not yet been fully developed.

Previous investigators[40–44] have used either the chain Langevin equation [similar to Eq. (2.18)] or the Kirkwood–Riseman equation [Eq. (2.22)] in the free draining limit to describe the polymer migration caused by a nonhomogeneous flow field. Shafer, Laiken, and Zimm[40,41] showed that observed migration of DNA molecules in circular Couette flow results directly from the flow's curved nature. Aubert et al.[42,43] found some form of migration in all flows, curved or uncurved; however, in parallel flows they found no migration perpendicular to the direction of flow. Rather, the polymer merely lagged or preceded the flow along a streamline. Further, when they approximated the curved flow as a quadratic, they found that cross-streamline migration occurs in curvilinear flows (such as circular

Couette). Simple geometrical interpretation of the effects of flow on migration for the various geometries is given in Sekhon et al.[45] All of these theories claim that the origin of polymer migration is due to the nonhomogeneous flow.

Sekhon et al.[45] and Lhuiller[46] first recognized that hydrodynamic interaction can play an important role in the study of the polymer migration. They used the Kirkwood–Riseman equation to study the effect of hydrodynamic interaction on the polymer migration. To study the polymer migration, Jhon and Freed[13] later used modified hydrodynamic interaction. They used this method because of the confined geometrical boundaries as explained in Section III.A. Let us examine polymer migration from the chain Langevin equation [Eq. (2.18)]. To calculate the migration velocity $\mathbf{\Delta}$ it is necessary to compute

$$\langle \dot{\mathbf{r}}_c \rangle \equiv \frac{1}{n} \left\langle \sum_{i=1}^{n} \mathbf{R}_i \right\rangle .$$

From Eqs. (2.18) and (3.39) we obtain

$$\mathbf{\Delta} = \mathbf{\Delta}^{\mathrm{F}} + \mathbf{\Delta}^{\mathrm{H}}. \tag{3.40}$$

Here the quantity

$$\mathbf{\Delta}^{\mathrm{F}} = \frac{1}{n} \left\langle \sum_{i=1}^{n} \mathbf{v}_0(\mathbf{R}_i, t) \right\rangle - \mathbf{v}_0(\mathbf{r}_c, t) \tag{3.41a}$$

describes the effect of bulk flow on polymer migration, and

$$\mathbf{\Delta}^{\mathrm{H}} = \frac{1}{n} \left\langle \sum_{i} \mathbf{T}_{ij} \cdot (\mathbf{F}_j^{(c)} + \mathbf{f}_j^*) \right\rangle \tag{3.41b}$$

includes the confined geometrical effect through $\mathbf{T}_{ij}$ and the form of the connector force. In the absence of hydrodynamic interaction ($\mathbf{T}_{ij} = 0$), the migration velocity is determined solely by the bulk flow. We shall illustrate this using the dumbbell model ($n = 2$). In this situation,

$$\begin{aligned}
\mathbf{\Delta} &= \frac{1}{2} \langle (\mathbf{v}_0(\mathbf{R}_1, t) + \mathbf{v}_0(\mathbf{R}_2, t)) \rangle - \mathbf{v}_0(\mathbf{r}_c, t) \\
&= \frac{1}{2} \left\langle \left\{ \exp\left(\frac{\mathbf{R}}{2} \cdot \mathbf{\nabla}' \right) + \exp\left(-\frac{\mathbf{R}}{2} \cdot \mathbf{\nabla}' \right) \right\} \mathbf{v}_0(\mathbf{r}')|_{\mathbf{r}'=\mathbf{r}_c} \right\rangle - \mathbf{v}_0(\mathbf{r}_c, t) \\
&= 2 \left\langle \sinh^2 \left(\frac{\mathbf{R}}{4} \cdot \mathbf{\nabla}' \right) \mathbf{v}_0(\mathbf{r}')|_{\mathbf{r}'=\mathbf{r}_c} \right\rangle = \left\langle \frac{1}{8} (\mathbf{R} \cdot \mathbf{\nabla}')^2 \mathbf{v}_0(\mathbf{r}')|_{\mathbf{r}'=\mathbf{r}_c} + \cdots \right\rangle .
\end{aligned} \tag{3.42}$$

We have used Eqs. (2.24) and (3.7) in deriving Eq. (3.42).

For the two-bead case, $\langle\ \rangle$ can be explicitly written as

$$\langle W(\mathbf{R}, \mathbf{r}_c)\rangle \equiv \int \mathbf{d}^3 R\, W(\mathbf{R}, \mathbf{r}_c)\psi(\mathbf{R}, \mathbf{r}_c; t). \tag{3.43}$$

Here $\psi(\mathbf{R}, \mathbf{r}_c; t)$ is the probability function for the dumbbell, $\Psi(\mathbf{R}_1, \mathbf{R}_2; t)$, expressed in center-of-mass coordinates [Eq. (2.24)]. Theoretically $\psi(\mathbf{R}_1, \mathbf{R}_2; t)$ can be obtained from Eq. (2.20). However, there is no exact solution for ψ at present. Therefore, we must rely on an approximate solution to compute the quantities given above. For example, an equilibrium average could be chosen to estimate $\boldsymbol{\Delta}$ as given by Eq. (3.42). The results are as follows. For homogeneous flow, such as plane Couette flow and extensional flow, $\boldsymbol{\Delta} = 0$. This implies that the polymer molecule travels with the fluid. However, $\boldsymbol{\Delta} \neq 0$ for nonhomogeneous flows. For example, for the pressure-driven flow $\boldsymbol{\Delta}$ is parallel to $\mathbf{v}_0(\mathbf{r}_c, t)$, but negative. This means that the polymer molecule does not cross the streamline, but remains behind it. For circular Couette flow and coating flow $\boldsymbol{\Delta}$ has a nonzero component perpendicular to $\mathbf{v}_0(\mathbf{r}_c, t)$. Therefore the polymer molecule migrates across the streamline.

So far we have *not* considered the hydrodynamic interaction. However, its inclusion will have a subtle effect upon polymer migration. Without hydrodynamic interaction caused by the wall, the polymer molecule does not cross the streamline in the pressure-driven flow situation. The presence of hydrodynamic interaction modifies the polymer motion, enabling the polymer to migrate across the streamline even for pressure-driven flows with rectilinear geometry.

Let us now investigate the effect on polymer migration due to the hydrodynamic interaction. Eq. (3.41b) can be written as (we set $\mathbf{f}_j^* = 0$ for simplicity)

$$\boldsymbol{\Delta}^{\mathrm{H}} = \frac{1}{n}\left\langle \sum_i \mathbf{T}_{ii}\cdot\mathbf{F}_i^{(c)} \right\rangle + \frac{1}{n}\left\langle \sum_{i\neq j}\sum \mathbf{T}_{ij}\cdot\mathbf{F}_j^{(c)} \right\rangle. \tag{3.44}$$

Consider the first term in Eq. (3.44). $\mathbf{T}_{ii}$ originates from bead–wall interactions and is given by Eq. (3.10). Even though Eq. (3.10) is derived for the plate, it can be used for an arbitrary confined geometry because the size of the beads is generally small and the hydrodynamic interaction is significant only a few bead diameter distances from the wall. Thus the curvature effect of the wall can be neglected, and Eq. (3.10) can be used for an arbitrary geometry. Recall that z refers to the distance from the wall. Therefore we can use Eq. (3.10) to calculate the effect of bead–wall interaction on polymer migration by computing

$$\frac{1}{n}\left\langle \sum_i \mathbf{T}_{ii}\cdot\mathbf{F}_i^{(c)} \right\rangle.$$

The calculation appears to be trivial; its result has not yet been reported.

Now we will study the second term in Eq. (3.44). $\mathbf{T}_{ij}$ denotes the modified hydrodynamic interaction due to the boundaries, and requires complicated fluid dynamic calculations. In general $\mathbf{T}_{ij}$ depends on the shape of the geometry through the boundary condition; its analytic form is available only for simple geometry. For the example of a dumbbell in the proximity of a single flat, $\mathbf{T}_{ij}$ is given by Eq. (3.25). From this result we can calculate the effect of hydrodynamic interaction on polymer migration with a single wall by calculating

$$\frac{1}{n}\left\langle \sum_{i \neq j}\sum \mathbf{T}_{ij} \cdot \mathbf{F}_j^{(c)} \right\rangle .$$

The calculation for the Hookean spring model is given in Jhon and Freed.[13] Studies for the various spring force models become relatively easy with the approximate form of the probability function.

So far we have studied the migration of polymer molecules by determining the trajectory of the polymer molecule. The Langevin equation is best suited for describing that trajectory. However, if the polymer molecules migrate across the streamline, there will be a nonhomogeneity in the concentration profile. In this case the Fokker–Planck type equation is more appropriate for determining the concentration profile.

The major stumbling block with this approach is the mathematical difficulty in obtaining the probability function from Eq. (2.20). Since the calculation of the probability function is almost impossible, we must try to obtain an approximate result. The standard procedure is to construct the moment equations and to calculate the first few moments. Here the nth moments are defined as

$$\langle \mathbf{RR} \cdots \mathbf{RR} \rangle \equiv \int d^3R\, \mathbf{RR} \cdots \mathbf{RR}\psi(\mathbf{R}, \mathbf{r}_c, t). \tag{3.45}$$

The zeroth moment evaluated at $\mathbf{r}_c = \mathbf{r}(C(\mathbf{r}, t))$ is the concentration of the polymer molecules, that is,

$$C(\mathbf{r}, t) \equiv \langle 1 \rangle \equiv \int d^3R\psi(\mathbf{R}, \mathbf{r}_c; t)|_{\mathbf{r}_c = \mathbf{r}}.$$

In general the moment equations generated from Eq. (2.20) will form hierarchy equations, as often appears in the kinetic theory of gases.[47] Therefore we must use either truncations or approximations for the higher order moments in order to close the hierarchy equations.

The projection operator method[48,49] is another procedure for calculating the probability equation into the zeroth moment, $C(\mathbf{r}, t)$. For the free draining limit Aubert et al.[43] have derived the concentration equation from another procedure not outlined here.

Another interesting problem involving pores is the study of adsorbed polymer conformation under solvent flow.[50] Although experimental[51] and theoretical[52] studies of adsorbed polymers in quiescent flow are well understood, the current literature[53-58] contains some inconsistent results for experimental observations of the conformation of adsorbed polymers under flow. To measure the adsorbed polymer layer experimentally, researchers often introduce hydrodynamic thickness L_{H}, defined as the equivalent reduction in pore size. This variable accounts for a reduced solvent flow rate in pores due to adsorbed polymer. Dejardiu and Varoqui[53] observed that L_{H} remained constant as the rate of solvent flow increased. On the other hand, Myard and Gramin[54] observed an increase in L_{H} as the flow rate increased above a certain threshold value. Moreover, Cohen and Metzner[55] found that L_{H} decreased as the flow rate increased. Other interesting experimental results are found in Tsang,[56] Lee and Fuller,[57] and Idol.[58]

In the theoretical approach Dimarzio and Rubin[59] calculated L_{H} for a Rouse chain without considering hydrodynamic interaction. They found that L_{H} does not change with the flow rate. Lee and Fuller[57] studied the effect of nonlinear springs and concluded that L_{H} decreases as the shear rate increases. Armstrong and Jhon[50] examined the effects of hydrodynamic interaction and speculate that L_H increases with increasing shear rates. The results we obtained from the study of polymer migration for the nonadsorbed system may give insight into the thickness of the adsorbed polymer layer. For example, the calculations by Jhon and Freed[13] indicate that hydrodynamic interaction between the wall and the polymer molecules causes nonadsorbed polymer molecules to migrate away from the wall. This finding suggests that the increasing thickness of adsorbed polymer with increasing shear rate is due to the hydrodynamic interaction.

IV. POLYMER CONFORMATION IN TURBULENT FLOW

The previous section dealt with the motion of polymers in deterministic flow. In this section we will study the polymer conformation when exposed to a turbulent flow in which $\mathbf{v}_0$ is random or stochastic and can be only statistically defined. It is not our intention to develop any new theory in Newtonian turbulence, but to predict the polymer conformation by utilizing the existing results in turbulence theory. Since there already exists a great deal of literature on turbulence,[60-62] we will give only a brief review of turbulence in the Appendix.

In spite of many theoretical attempts to understand the underlying mechanism of polymer response in turbulence, the theory is still very much in the primitive stage. Existing theories all start from the Kirkwood–Riseman equation [Eq. (2.22)]. As shown in Section II, the Kirkwood–Riseman equation, derived for $\mathrm{Re} \ll 1$, is not correct for describing the polymer dynamics in turbulent flow. It would be desirable to develop a consistent polymer chain equation descriptive for a high Reynolds number, but at present no one has developed such a theory.

A. Effective Probability Equation for *n*-Bead Chain†

We will adapt the Kirkwood–Riseman equation to study the polymer conformation in turbulent flow. Further, to simplify the analysis we will ignore the hydrodynamic interaction. With these simplifying assumptions, the governing Fokker–Planck equation for $\tilde{\psi}$, the probability density for this system, is

$$\frac{\partial}{\partial t}\tilde{\psi} = -\sum_{i=1}^{n} \frac{\partial}{\partial \mathbf{R}_i} \cdot \left[\left(\mathbf{v}(\mathbf{R}_i, t) - \frac{k_B T}{\zeta}\frac{\partial}{\partial \mathbf{R}_i}\ln\tilde{\psi} - \frac{\mathbf{F}^{(c)}}{\zeta}\right)\tilde{\psi}\right]. \tag{4.1}$$

The solution of Eq. (4.1) will, in general, involve functionals of the random fluctuating velocity $\mathbf{v}'(\mathbf{R}_i, t)$. To be completely accurate we should write $\mathbf{v} = \langle \mathbf{v} \rangle_{\mathbf{v}'} + \mathbf{v}'$, with $\langle \mathbf{v} \rangle_{\mathbf{v}'}$ as the average velocity (deterministic). Here $\langle\ \rangle_{\mathbf{v}'}$ denotes the average over an ensemble of velocity realizations which we presume to be stationary. As we shall see later, the only observable quantity from the flow field in which we will be interested is the gradient of this velocity field: $\nabla\mathbf{v} = \langle \nabla\mathbf{v} \rangle_{\mathbf{v}'} + \nabla\mathbf{v}'$. It is a well-established fact[60] that the fluctuation gradient is much larger than the gradient of the average velocity. So it seems justifiable to assign $\langle \mathbf{v} \rangle_{\mathbf{v}'} = 0$ without loss of generality. Even with these assumptions, the explicit solution of $\tilde{\psi}$ is almost impossible to obtain since $\mathbf{v}'(\mathbf{R}_i, t)$ is a random function of $\mathbf{R}_i$. However, we may assume that ψ can be formally obtained in the functional form of $\mathbf{v}'$,

$$\tilde{\psi} = \tilde{\psi}(\{\mathbf{v}'(\mathbf{R}_i, t)\}; \mathbf{R}_i, t).$$

This can be considered the probability function of the n-bead chain for a particular realization of a turbulent velocity field. To make sense out of this turbulent fluctuating velocity field and recast the equation in terms of statistics of this field, we can write a velocity-averaged probability function $\langle \tilde{\psi}(\{\mathbf{v}'(\mathbf{R}_i, t)\}; \mathbf{R}_i, t) \rangle_{\mathbf{v}'} \equiv \psi(\mathbf{R}_i, t)$. Note that ψ is explicitly dependent only on the bead position and time, and dependent on the statistics of the velocity field.

†See Refs. 63 and 64.

This procedure is a general guideline for calculating ψ. However, this is a brute-force kind of approach and not a useful procedure in handling the stochastic differential equation.[15] The standard procedure is to construct an effective, or renormalized, equation of ψ.

We will now seek a cumulant expansion[65] of the term involving the stochastic velocity field, keeping only the second cumulant [which is the second moment since we require $\langle \mathbf{v}'(\mathbf{R}_i, t)\rangle = 0$]. In order to avoid convolutions resulting from the interaction representation, we assume that the fluctuations in the velocity field are much faster than the characteristic response time of the polymer molecule. Restating this assumption mathematically, we say that the Eulerian time scale T_E of the velocity field is much smaller (say, an order of magnitude smaller) than the longest relaxation time of the polymer λ_H. With this assumption the cumulant expansion yields the renormalized differential equation

$$\frac{\partial}{\partial t}\psi(\mathbf{R}_i, t) = \frac{1}{\zeta}\sum_i \frac{\partial}{\partial \mathbf{R}_i}\cdot[\mathbf{F}_i^{(c)}\psi(\{\mathbf{R}_k\}, t)] + \frac{k_B T}{\zeta}\sum_i \frac{\partial^2}{\partial \mathbf{R}_i^2}\psi(\{\mathbf{R}_k\}, t)$$

$$+ \sum_{i,j}\boldsymbol{\Gamma}(\mathbf{R}_i - \mathbf{R}_j):\frac{\partial^2\psi(\{\mathbf{R}_k\}, t)}{\partial \mathbf{R}_i \partial \mathbf{R}_j} \tag{4.2}$$

with

$$\left\{\int_l^\infty ds\, \langle \mathbf{v}'(\mathbf{r}, s)\mathbf{v}'(\mathbf{r}', s)\rangle_{\mathbf{v}'}\right\} = \boldsymbol{\Gamma}(\mathbf{r} - \mathbf{r}'). \tag{4.3}$$

This is an effective equation for the polymer probability density and is the central result of this section. Eq. (4.2) represents the probability of a certain conformation in a turbulent fluid and should tend toward a stationary (invariant) probability density at large time, since the stochastic velocity field is itself stationary under the assumptions already made. In the forthcoming development the solution for $\partial\psi/\partial t = 0$ will therefore be sought. While we have derived the effective probability equation [Eq. (4.2)] from the stochastic probability equation [Eq. (4.1)] by using a cumulant expansion, the same result can be derived from the Langevin equations for the n-bead model.[66] The advantage of the cumulant expansion method is that it may point the way toward loosening the restrictions on the relative time scales of the flow and polymer. It should be noted that previous studies[66,67] assumed $T_E \ll \lambda_H$, although these were performed by using the second moment of ψ [see Eq. (3.45)].

The terms in Eq. (4.2) that result from the connector force and the Brownian forces appear as the first and second terms, respectively, and are unchanged from similar terms arising in the familiar Kirkwood–Riseman equation. The third term, however, arises from the stochastic velocity field and behaves as a diffusion term with a spatially dependent diffusion tensor. The fact that the stochastic velocity field manifests itself as a diffusion term is not surprising. In a manner similar to the derivation of the canonical Fokker–Planck equation, we have exploited the fact that the velocity field has a correlation time that is short compared with the polymer relaxation time. It may appear that we are considering the turbulent and Brownian forces equivalent, but this could hardly be the case since the progenitors of $k_B T$ and Γ are physically quite different. It cannot be emphasized too strongly that the relative magnitudes of the velocity and the Brownian forces (which differ by at least six orders of magnitude) are irrelevant to the validity of Eq. (4.2). Rather, its validity relies on the fact that both random processes have correlation times significantly shorter than the response time of the polymer (the n-bead chain in this case). We have replaced the real turbulent flow field with a stochastic process that is as similar to turbulence as possible, yet still yields solvable results. An alternate view is that the major contribution of the turbulence is to reduce the connector force between beads.

B. Renormalized Interaction†

Here we will study quantitatively how turbulence reduces the connector force between the beads (renormalized interaction). Even though the renormalization process is an approximate one, this concept simplifies a great deal of algebra. We shall study the renormalized interaction from the dumbbell model. For the two-bead model Eq. (4.2) becomes

$$\frac{\partial\psi}{\partial t}=\frac{1}{\zeta}\frac{\partial}{\partial\mathbf{R}_1}\cdot(\mathbf{F}_1^{(c)}\psi)+\frac{1}{\zeta}\frac{\partial}{\partial\mathbf{R}_2}\cdot(\mathbf{F}_2^{(c)}\psi)+\Gamma(0):\frac{\partial}{\partial\mathbf{R}_1}\frac{\partial}{\partial\mathbf{R}_1}\psi+\Gamma(0):\frac{\partial}{\partial\mathbf{R}_2}\frac{\partial}{\partial\mathbf{R}_2}\psi$$
$$+\,2\Gamma(\mathbf{R}_2-\mathbf{R}_1):\frac{\partial}{\partial\mathbf{R}_1}\frac{\partial}{\partial\mathbf{R}_2}\psi+\frac{k_B T}{\zeta}\left[\frac{\partial}{\partial\mathbf{R}_1}\cdot\frac{\partial}{\partial\mathbf{R}_1}+\frac{\partial}{\partial\mathbf{R}_2}\cdot\frac{\partial}{\partial\mathbf{R}_2}\right]\psi. \quad (4.4)$$

We can recast Eq. (4.4) into the center-of-mass coordinates given by Eq. (2.24),

$$\frac{\partial\psi}{\partial t}=\frac{2}{\zeta}\frac{\partial}{\partial\mathbf{R}}\cdot\left(\frac{\partial\phi^{(c)}}{\partial\mathbf{R}}\psi\right)+2(\Gamma(0)-\Gamma(\mathbf{R})):\frac{\partial}{\partial\mathbf{R}}\frac{\partial}{\partial\mathbf{R}}\psi+\frac{2k_B T}{\zeta}\frac{\partial}{\partial\mathbf{R}}\cdot\frac{\partial}{\partial\mathbf{R}}\psi$$
$$+\,\frac{1}{2}(\Gamma(0)+\Gamma(\mathbf{R})):\frac{\partial}{\partial\mathbf{r}_c}\frac{\partial}{\partial\mathbf{r}_c}\psi+\frac{k_B T}{2\zeta}\frac{\partial}{\partial\mathbf{r}_c}\cdot\frac{\partial}{\partial\mathbf{r}_c}\psi. \quad (4.5)$$

†See Ref. 64.

Since Γ and $\phi^{(c)}$ are independent of $\mathbf{r}_c$, we can see that if at some time during the evolution of the system ψ becomes homogeneous (independent of $\mathbf{r}_c$), it will remain homogeneous from then on. If we then assume that ψ is initially homogeneous, we can neglect terms involving derivatives of $\mathbf{r}_c$ in Eq. (4.5),

$$\frac{\partial \psi}{\partial t} = \frac{2}{\zeta} \frac{\partial}{\partial \mathbf{R}} \cdot \left[\left(\frac{\partial \phi^{(c)}}{\partial \mathbf{R}} \right) \psi \right] + 2(\Gamma(0) - \Gamma(\mathbf{R})) : \frac{\partial}{\partial \mathbf{R}} \frac{\partial}{\partial \mathbf{R}} \psi + \frac{2k_B T}{\zeta} \frac{\partial}{\partial \mathbf{R}} \cdot \frac{\partial}{\partial \mathbf{R}} \psi. \tag{4.6}$$

It may seem that the aspect of homogeneity is overemphasized, but this is where developments in this work clearly diverge from those of the past. Previous investigators[68] developed an equation for the moment of a probability density that moved with fluid particles. These types of analyses involve the additional hardship of using convected (Lagrangian) statistics for the turbulent field. Apparently there is no need to look at the molecule in the convected frame in order to get detailed knowledge about its conformation.

It is a well-accepted fact that the size of the smallest eddies appearing in the turbulence is many times larger than the equilibrium conformation of the molecule.[35] Even if the molecule is stretched out to the greatest length that its monomeric conformation will allow, it might only then be on the order of the size of the *smallest* eddies in a high Reynolds number flow. In light of this, we can represent the velocity field well by assuming a truncated Taylor series expansion,

$$\Gamma(\mathbf{R}) = \Gamma(0) + \frac{\mathbf{RR}}{2} : \left[\frac{\partial}{\partial \mathbf{R}'} \frac{\partial}{\partial \mathbf{R}'} \Gamma(\mathbf{R}') \right]_{\mathbf{R}'=0} \tag{4.7}$$

Now using Eq. (4.7) and the assumption of the isotropy of turbulence, we find

$$\frac{\partial \psi}{\partial t} = \frac{2}{\zeta} \frac{\partial}{\partial \mathbf{R}} \cdot \left[\left(\frac{\partial \phi^{(c)}}{\partial \mathbf{R}} \right) \psi \right] + \frac{2k_B T}{\zeta} \frac{\partial}{\partial \mathbf{R}} \cdot \frac{\partial}{\partial \mathbf{R}} \psi + \mathbf{\Delta}(\mathbf{R}) : \frac{\partial}{\partial \mathbf{R}} \frac{\partial}{\partial \mathbf{R}} \psi \tag{4.8}$$

with

$$\mathbf{\Delta}(\mathbf{R}) = -\mathbf{RR} : \left[\frac{\partial}{\partial \mathbf{R}'} \frac{\partial}{\partial \mathbf{R}'} \Gamma(\mathbf{R}') \right]_{\mathbf{R}'=0} = A[2R^2 \boldsymbol{\delta} - \mathbf{RR}] \ .$$

where

$$A = -u'^2 \left[\frac{\partial}{\partial \mathbf{R}'} \cdot \frac{\partial}{\partial \mathbf{R}'} f \right]_{\mathbf{R}'=0} \qquad \text{(positive number)},$$

u' being the rms velocity fluctuation and f the so-called coefficient of the longitudinal velocity correlation.[60] For the steady-state case when $\partial\psi/\partial t = 0$, ψ can be obtained (we denote the steady-state value of ψ by ψ^{ss}),

$$\frac{\partial \ln \psi^{ss}}{\partial \mathbf{R}} = -\frac{2}{\zeta}\left(\boldsymbol{\Delta}(\mathbf{R}) + \frac{2k_B T}{\zeta}\boldsymbol{\delta}\right)^{-1} \cdot \frac{\partial \phi^{(c)}}{\partial \mathbf{R}}. \tag{4.9}$$

$(\boldsymbol{\Delta}(\mathbf{R}) + 2k_B T/\zeta\boldsymbol{\delta})^{-1}$ has the effect of a spatially dependent diffusion tensor arising from turbulent and Brownian forces. It can be easily computed as follows. From the symmetry argument the inverse of $[\boldsymbol{\Delta}(\mathbf{R}) + 2k_B T/\zeta\boldsymbol{\delta}]$ must be of the form $C_1(R)\boldsymbol{\delta} + C_2(R)\mathbf{RR}$, or

$$[C_1(R)\boldsymbol{\delta} + C_2(R)\mathbf{RR}] \cdot \left(\boldsymbol{\Delta}(\mathbf{R}) + \frac{2k_B T}{\zeta}\boldsymbol{\delta}\right) = \boldsymbol{\delta}, \tag{4.10}$$

and we find $C_1(R)$ and $C_2(R)$ by comparing terms,

$$C_1(R) = \frac{1}{2A(R^2 + \alpha)}$$

$$C_2(R) = \frac{1}{2A(R^2 + \alpha)(R^2 + 2\alpha)}.$$

Therefore we obtain

$$\left(\boldsymbol{\Delta}(\mathbf{R}) + \frac{2k_B T}{\zeta}\boldsymbol{\delta}\right)^{-1} = \frac{1}{2A(R^2 + \alpha)(R^2 + 2\alpha)}[(R^2 + 2\alpha)\boldsymbol{\delta} + \mathbf{RR}] \tag{4.11}$$

where $\alpha = k_B T/\zeta A$.

Therefore, from Eqs. (4.9) and (4.11) we obtain

$$\psi^{ss} = C \exp\left\{-\frac{2\alpha}{k_B T}\int_0^R \frac{1}{R'^2 + 2\alpha}\frac{d\phi^{(c)}(R')}{dR'}\, dR'\right\}. \tag{4.12}$$

Note that C is left as a normalization constant. Note also that the term on the right-hand side involves only scalar quantities. It is clear that if $A \to 0\,(\alpha \to \infty)$, we recover the equilibrium solution,

$$\psi^{ss} \to \psi_{eq} = C \exp\left[\frac{-\phi^{(c)}}{k_B T}\right]. \tag{4.13}$$

We can see that the presence of turbulence has caused a change or a "renormalization" in the connector potential, or schematically,

$$\underbrace{\phi^{(c)}}_{\text{bare potential}} \to \underbrace{\phi_R^{(c)} = \int_0^R dR' \frac{d\phi^{(c)}(R')/dR'}{(R'/\sqrt{2\alpha})^2 + 1}}_{\text{renormalized potential}}. \tag{4.14}$$

The negative gradient of this potential is what really counts in the dynamics of the polymer molecule itself,

$$\underbrace{\mathbf{F}^{(c)} = -\hat{R}\frac{d\phi(R)}{dR}}_{\text{bare force}} \to \underbrace{\mathbf{F}_R^{(c)} = \frac{2k_B T}{A\zeta R^2 + 2k_B T}\mathbf{F}^{(c)}}_{\text{renormalized force}} \tag{4.15}$$

where $\hat{R}$ is the unit vector in the $\mathbf{R}$ direction. We see that the magnitude of the force has been reduced everywhere except at $R = 0$. Clearly, turbulence has a screening effect on the force, reducing its magnitude but not its direction.

Let us illustrate the renormalized force (or potential) for the various spring force models given in Section III.

Case 1: *Hookean spring* ($\mathbf{F}^{(c)} = H\mathbf{R}$). From Eq. (4.15), the effective spring force in turbulent flow becomes

$$\mathbf{F}_R^{(c)} = \frac{2k_B TH}{A\zeta R^2 + 2k_B T}\mathbf{R}. \tag{4.16}$$

Figure 17*A* illustrates this situation. Here the dumbbell may stretch forever in the turbulent flow, since $\mathbf{F}_R^{(c)} \to 0$ as $R \to \infty$. Also, this can be seen by studying the probability function ψ and the second moment of ψ, that is, $\langle \mathbf{RR} \rangle \equiv \int d^3R\, \mathbf{RR}\psi(R)$. From Eqs. (4.12) and (4.16) we obtain ψ,

$$\psi^{ss} = C\left[1 + \frac{R^2}{2\alpha}\right]^{-H/\zeta A}, \tag{4.17}$$

where C is a normalization constant. An interesting mathematical detail with regard to this equation is that for sufficiently large A (turbulent strength) ψ will cease to be normalized. In other words, the integral

$$\int_0^\infty R^2\left[1 + \frac{R^2}{2\alpha}\right]^{-H/\zeta A} dR \tag{4.18}$$

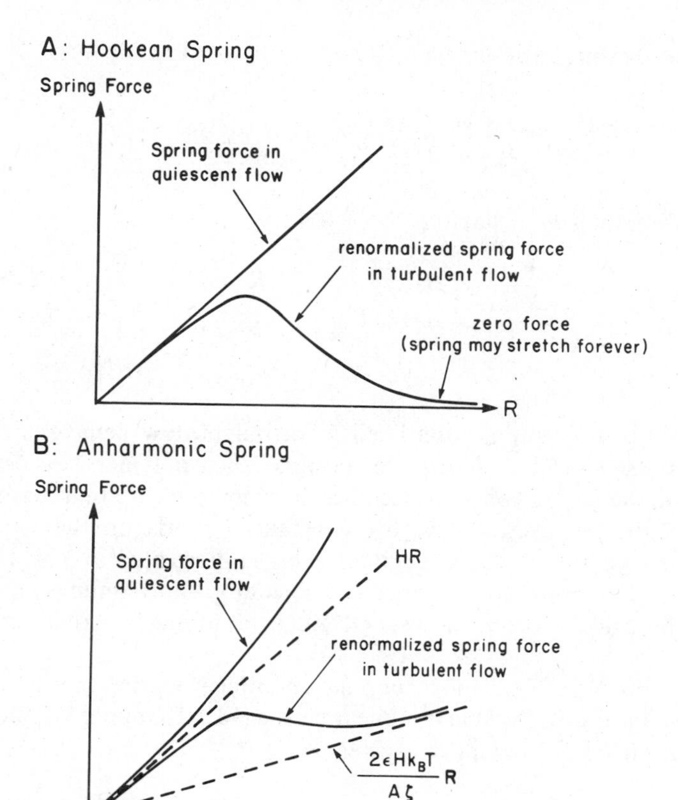

Figure 17. Bare and renormalized spring force.

will not converge for $H/\zeta A < \frac{3}{2}$. By a similar procedure we can find that the second moment of ψ will not converge for $H/\zeta A < \frac{5}{2}$; this result was observed in the analysis done by Lumley.[67] This anomaly has no basis in the physics of the problem since we expect the potential in a real polymer molecule to go to much higher orders in R, though the Hookean potential is a good approximation for small perturbations from equilibrium.

To avoid this catastrophic situation, we will introduce the following nonlinear spring models.

Case 2: Anharmonic spring. The simple nonlinear model is (see Section III),

$$\mathbf{F}^{(c)} = H\mathbf{R}(1 + \varepsilon R^2).$$

Here ε is the measure of nonlinearity. The renormalized spring force

$$\mathbf{F}^{(c)} = \frac{2k_B TH}{A\zeta R^2 + 2k_B T}(1 + \varepsilon R^2)\mathbf{R}$$

has interesting limiting behavior,

$$\mathbf{F}^{(c)} = \begin{cases} H\mathbf{R}, & R \ll 1 \\ \dfrac{2\varepsilon H k_B T}{A\zeta}\mathbf{R}, & R \gg 1. \end{cases} \tag{4.19}$$

Therefore, this anharmonic dumbbell in turbulent flow behaves like two characteristically different harmonic springs. When polymers are near the equilibrium, the spring constant becomes H; when they are in a reasonably extended state, however, the spring constant depends on the Reynolds number through turbulence strength A. This is illustrated in Fig. 17*B*. In this sense this simple model remedies the major shortcomings in linear springs. The calculations of ψ and $\langle R^2 \rangle$ can be found in Armstrong and Jhon.[64]

Case 3: *FENE.* The most popular nonlinear spring model is, we believe, the finitely extendable nonlinear elastic (FENE) model. As shown in Section III, the FENE potential is given as

$$\phi^{(c)} = -\frac{1}{2}HR_0^2 \ln\left[1 - \left(\frac{R}{R_0}\right)^2\right], \qquad R < R_0.$$

This potential has the advantage of never allowing the dumbbell to stretch past R_0. When the potential is fed into Eq. (4.12), we obtain the following result:

$$\psi^{ss} = C\left[\frac{R_0^2 - R^2}{R^2 + 2\alpha}\right]^{H/\zeta A(1 + 2\alpha/R_0^2)} \tag{4.20}$$

Now all integrations are carried out over the internal $(0, R_0)$ and consequently the normalization constant and all the moments of ψ converge. We can easily compute the moments from the above probability function if necessary.

At this point let us comment on the nonlinear spring forces studied here. The study of the polymer conformation in the turbulent flow examined can be applied to practical engineering problems. One of these problems is known as turbulent drag reduction, which will be explained in the next section. In that problem, the important physical quantity is the term associated with the energy dissipation caused by dumbbell molecules, that is, $\langle \mathbf{F}^{(c)}(\mathbf{R}) \cdot \mathbf{F}^{(c)}(\mathbf{R}) \rangle$.

This term is unfortunately singular (infinite) for the FENE model over a physically significant range of H and R_0. This is true regardless of the presence of turbulence and occurs even at thermodynamic equilibrium. This problem is a result of the fact that the FENE spring not only exerts an infinite force at $R = R_0$, but requires an infinite amount of energy to get it there. For this reason the anharmonic model may be best suited for the study of turbulent drag reduction.

The renormalized interaction we studied for the dumbbell model can be generalized as follows. *The interaction between any two points in turbulent flow may be screened and reduced compared to the bare interaction.* Therefore, the polymer conformation for the n-bead model may be studied by solving the following equation:

$$\frac{\partial}{\partial t} \psi(\{\mathbf{R}_k\}; t) = \frac{1}{\zeta} \sum_j \frac{\partial}{\partial \mathbf{R}_j} \cdot (\mathbf{F}_{\mathbf{R}j}^{(c)} \psi(\{\mathbf{R}_k\}; t)) + \frac{1}{\zeta} \sum_j \frac{\partial^2}{\partial \mathbf{R}_j^2} \psi(\{\mathbf{R}_k\}; t) + \sum_{i,j} \Gamma(\mathbf{R}_i - \mathbf{R}_j) : \frac{\partial^2 \psi(\{\mathbf{R}_k\}; t)}{\partial \mathbf{R}_i \partial \mathbf{R}_j} \tag{4.21a}$$

with

$$\mathbf{F}_{\mathbf{R}i}^{(c)} = \frac{2k_B T}{A\zeta(\mathbf{R}_i - \mathbf{R}_{i+1})^2 + 2k_B T} \mathbf{F}_i^{(c)} \tag{4.21b}$$

The solution of this n-bead equation is not available at present, though we know that the result will be qualitatively the same as that of the dumbbell model. So far we have shown that the effective radii of gyration of polymer molecules in a turbulent core become greater than they would be at rest. In other words, the polymer molecules will unravel from their equilibrium conformation when they are exposed to a turbulent flow. So far there is no direct experimental evidence to support the fact that the polymer molecules expand to a spectacular degree when they are exposed to a turbulent flow. However, there is much indirect evidence for the expansion of the polymer molecules in turbulent flow.

If the polymer molecule increases its size in turbulent flow, the long polymer chain may not be stable and hence may degrade into lower molecular weight species. In spite of many experimental observations of polymer degradation in turbulent flow,[69,70] there exists virtually no theory for the degradation due primarily to the lack of information on polymer conformation in turbulent flow. The renormalized interaction idea developed by Armstrong and Jhon[64] combined with a simple chemical reaction model, such as a theory based on the first-passage time,[71] may be an extremely useful tool in the study of the degradation of polymer molecules in turbulent flow. Turbulent drag reduction might constitute more indirect evidence of the expansion of the polymer size, as will be explained in the next section.

C. Turbulent Drag Reduction

Under extreme conditions the addition of a minute amount of polymer to a Newtonian fluid has dramatic effects. The best known of these effects is the reduction of turbulent drag, often by as much as 50%. Ever since the discovery of this unusual flow phenomenon in 1948,[72] researchers have tried to understand it. Despite these attempts, the exact mechanism by which turbulent drag is reduced remains elusive. Although this area has long been the subject of experimental research, we will investigate it theoretically using the results obtained in the previous section.

1. What Is Turbulent Drag Reduction?

To understand what is meant by the term "turbulent drag reduction," consider the following illustration. If we add 10 to 50 ppm of a high-molecular-weight polymer to a turbulent fluid flowing through a pipe, we can achieve a reduction in drag in the range of 50%. By a certain percentage reduction in drag we mean that the energy cost necessary to move the fluid from point to point has been reduced by that percentage.

Polymer additives are used to increase throughput in crude-oil pipelines and are currently in use in the Trans-Alaska Pipeline.[73] Two companies, CONOCO and ARCO, have had commercial success producing polymers for this purpose. Small particles and droplets are also drag reducers but to a less spectacular degree. Consequently, reducing drag by the addition of particles is probably not commercially feasible. It may, however, be an important design criterion in processes concerned with slurry flows (for example, the paper industry). We will only study turbulent drag reduction due to polymer molecules.

Despite these successful applications, research in this area has declined. Part of the reason is the difficulty in getting consistent results. In addition, there is little in the way of theory for design criteria. Much of the blame for this trend must be attributed to the tendency of the polymer to degrade into lower molecular weight species and therefore to lose its ability to reduce drag. This means that additional polymer must be injected periodically throughout the pipe. Such a process is costly and contaminates the liquid with polymer residue at a much higher concentration than the tiny amounts originally necessary to reduce drag. To understand this degradation, we must study drag reduction at the molecular level. However, in the previous section we pointed out that a theory for degradation is currently not available. Therefore we will not address this problem, even though it may be an important factor in studying the turbulent drag reduction.

Not all turbulent flows that contain drag-reducing polymers are reducing drag. Usually there is a particular Reynolds number for a given flow geometry at which the "onset" of drag reduction occurs (as shown in Fig. 21). Below this onset Reynolds number, the friction factor (explained below) is roughly the same as the Newtonian value. Above this onset number, the friction factor diverges from the normal value. The model for drag reduction ought to replicate this well-known fact. Not only does drag reduction occur in very dilute solutions, but the intrinsic drag reduction (the percent drag reduction normalized by the concentration) also declines with increasing concentration.[74]

2. *What Is the Friction Factor? How Does Polymer Influence the f versus Re Correlation?*

In order to study the drag reduction phenomenon quantitatively, it is convenient to introduce the engineering parameter called friction factor f. The definition is available in standard texts on engineering fluid mechanics.[7,9]

Dimensional analysis shows that f is a function only of Reynolds number Re for the smooth, circular pipe or tube. The Reynolds number in this situation is defined as $\mathrm{Re} = 2\rho\langle v_x\rangle R/\eta$. Here $\langle v_x\rangle$ is the average velocity defined as

$$\langle v_x\rangle \equiv \frac{1}{\pi R^2}\int_0^{2\pi}\int_0^R v_x r\,dr\,d\theta = 2\int_0^1 v_x\left(\frac{r}{R}\right)\frac{r}{R}\,d\left(\frac{r}{R}\right).$$

The empirical correlation between f and Re for the Newtonian fluid is shown in Fig. 18*A* (solid line).

Now let us add a minute amount of polymer molecules and examine the f versus Re relationship. Since the viscosity changes only slightly, the Reynolds number of the polymer solution can be approximately the same as that of the Newtonian solution. However, the frictional factor of the polymer solution is quite different from that of the Newtonian fluid. There is virtually no change in the laminar region, but in the turbulent region the frictional factor of the polymer solution is substantially lower than that of the Newtonian fluid (dashed line in Fig. 18*A*).

Now let us study the polymer conformation in turbulent flow. We have computed the radii of gyration of the polymer molecules (treated them as

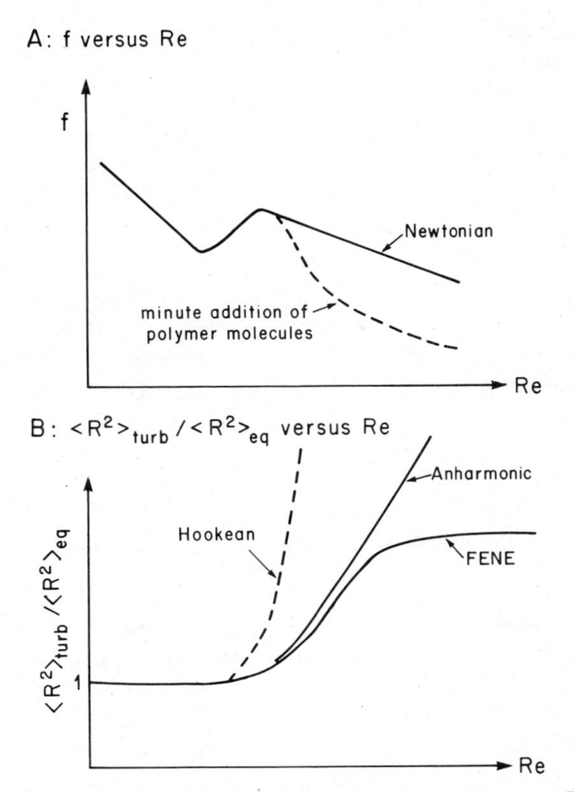

Figure 18. Friction factor and radius of gyration for the polymer versus Reynolds number with or without polymer.

dumbbells) in the turbulent flow $\langle R^2 \rangle_{\text{turb}}$ for the three different spring forces (Hookean, anharmonic, and FENE) discussed in Section IV.B, and plotted $\langle R^2 \rangle_{\text{turb}}/\langle R^2 \rangle_{\text{eq}}$ versus Re in Fig. 18B. Here $\langle R^2 \rangle_{\text{eq}}$ implies the radius of gyration in the quiescent flow. We speculate that the change in the friction factor Δf may be correlated with the change in the radius of gyration $\Delta\langle R^2 \rangle$. Then, based on Fig. 18, we observe that for the Hookean spring, $\Delta\langle R^2 \rangle$ cannot be correlated with Δf. On the other hand, Δf appears to have reasonably good correlations with $\Delta\langle R^2 \rangle$ for either the anharmonic spring or the FENE model. (FENE appears to be better than the anharmonic spring.)

This suggests that either an anharmonic spring or a FENE model can be used in the study of turbulent drag reduction. Armstrong and Jhon[75] showed that Δf is related to the energy dissipation caused by the dumbbell molecule, that is, $\langle \mathbf{F}^{(c)} \cdot \mathbf{F}^{(c)} \rangle$, rather than the radius of gyration $\langle R^2 \rangle_{\text{turb}}$. As shown in the previous section, $\langle \mathbf{F}^{(c)} \cdot \mathbf{F}^{(c)} \rangle$ is unfortunately singular (infinite) for the FENE model. For this reason the anharmonic model is best suited to the needs of the drag reduction theory.

3. *Molecular Theory of Polymer-Induced Turbulent Drag Reduction*

The Appendix briefly reviews the theory of turbulence. Basically we consider turbulence to be the mechanism by which energy is absorbed from the mean shear, by length scales at which there is virtually no viscous dissipation, and is passed down to length scales where viscous dissipation dominates. These two length scales are given by Eqs. (A.9) and (A.10). Now we shall introduce the concepts of viscous sublayer and turbulent core to investigate the origin of drag reduction at a semiquantitative level.[76]

In Fig. 19 we plot $1/\eta_+$ to get an idea of how the spectral range of turbulence varies as a function of the distance from the wall. An interpretation of this plot from what we have discussed in the Appendix is that as long as L_+ is larger than η_+, turbulence or, more precisely, the production of turbulence is possible, that is, as long as the largest eddies are not immediately dissipated by viscosity, turbulence can exist (turbulent core). The area below the dashed line in Fig. 19 represents the region close to the wall where $L_+ < \eta_+$ and where turbulence is not possible; this is called viscous sublayer.

Polymers may be added into the fluid that would elevate the viscosity in the inertial layer but not in the viscous sublayer. Such an addition will translate the line representing η_+ over to the left, as shown in Fig. 20. If an additive does this, it can be seen that the viscous sublayer will increase in experimentally observed thickness, decreasing the total fraction of pipe

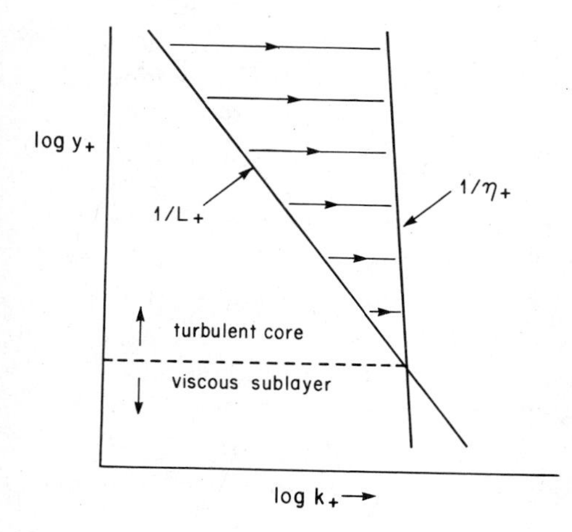

Figure 19. Plot of the nondimensional distance from the wall, y_+ versus L_+ and η_+, the largest and smallest length scales found in the turbulence. Arrows indicate the direction in which kinetic energy flows from production to dissipation.

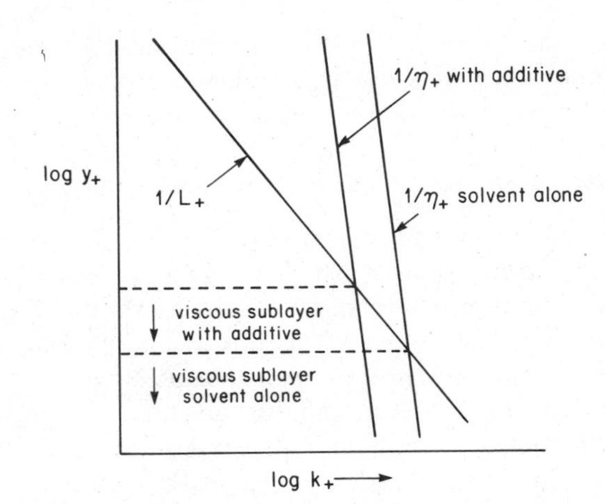

Figure 20. Similar to Fig. 19, but a polymer is present, which shifts the curve for η_+ to the left.

diameter available for the production of turbulence, and thus decreasing the net energy dissipation. This results in a greater mean velocity than would otherwise occur. This increase in mean velocity is sometimes referred to as effective slip,[74] the essence of drag reduction. It is important to recognize that the additive must change the viscosity only in the inertial layer and not in the viscous sublayer, because the parameters y_+ and k_+ are scaled by the solvent viscosity (the viscosity in the viscous sublayer).

We might speculate that the viscosity in the turbulent core is much greater than the viscosity in the laminar sublayer, since the effective radius of gyration in a turbulent flow is much larger than it would be at rest or in a laminar sublayer. Then this simple explanation would appear justified; however, it is not quite correct. Actually the shear viscosity changes little, and the extensional viscosity changes substantially from the foregoing picture.[69]

It has been speculated that the average size of a dissolved polymer molecule increases dramatically when exposed to turbulence, and many have postulated ways in which this increased size affects the turbulence as explained. Our simple model (explained in Section IV.B) shows that a polymer molecule grows by a factor of 10 or more from its equilibrium conformation.[65,77] However, turbulence is probably unaffected by the size of the molecule. Most of what we know about turbulence has been gained through careful scrunity of its kinetic energy budget (explained in the Appendix). So instead of concerning ourselves only with molecular conformation, we also studied polymeric contribution to this budget.

What resulted from this study was a theory based solely on the statistics of a Newtonian turbulence and a few molecular parameters. From these we calculated fluid mechanical frictional factors for turbulent flow through a pipe. The calculated values versus experimental data are shown in Fig. 21. This type of plot is known as von Karman plot. We found that it is absolutely necessary to consider the nonlinear properties of the polymeric solution in order to describe drag reduction accurately. The reason for this is that real polymer solutions behave in vastly different ways, depending on the amount of deformation they are required to endure over some period of time. At low deformation rates, infinitely dilute polymer solutions are not likely to behave differently from their Newtonian solvents. At high deformation rates, however, the nonlinear properties of the polymer come into play and dramatically increase the solution's ability to resist deformation. It is found, for instance, that small amounts of polymer mixed in jet fuel may help prevent the disastrous fires that so often accompany the crash of an airliner. Although the polymer is present in small amounts, there is no overt

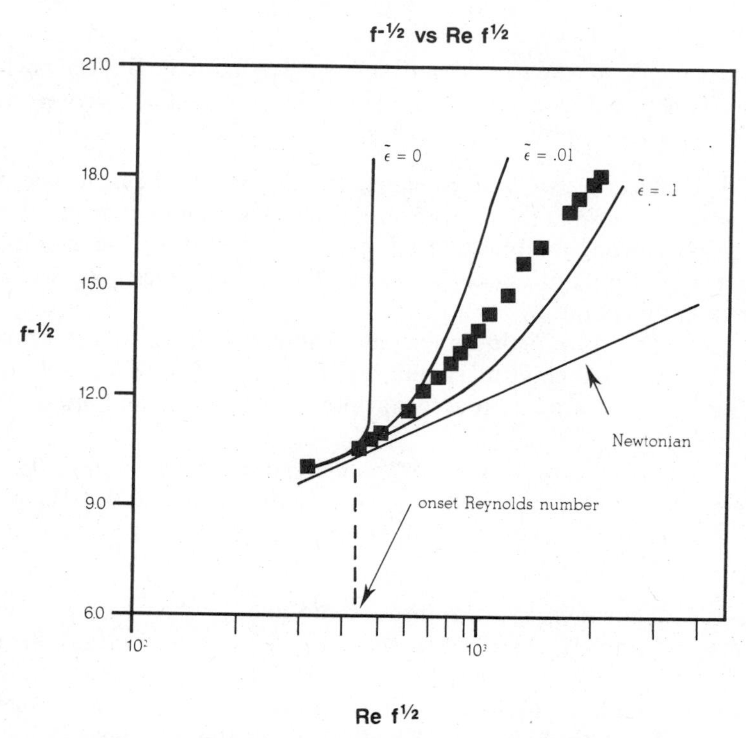

Figure 21. The black squares are data for 10 ppm of polyacrylamide ($M_w = 4.7 \times 10^6$; solvent-water) taken by Virk and Baker.[77] The solid lines are the results for the anharmonic model evaluated at these conditions for several values of the nonlinear parameter ε.

change in the fluid mechanical properties of the jet fuel. However, when fuel is caused to flow at the tremendous rates that might be induced by a crash, the flow resistance is much higher than would normally be expected. This higher resistance manifests itself in less dispersal of the fuel on and about the aircraft, thus reducing the hazards of fire.

If the drag-reducing solution increases flow resistance in the case of fuel dispersal during a plane crash, why does it reduce drag in turbulent fluid flow in a pipe? The answer lies in the fact that the flow resistance of drag-reducing solutions is increased only for certain types of deformation—those involving greater extensional character than shearing character. It turns out that turbulent shearing flows have distinct layers involving different amounts of extensional and shearing character. These different layers are affected by the presence of drag reducers and, in the end, conspire so as to produce a net reduction in drag.

V. CONCLUDING REMARKS

We have reviewed theories describing single-chain dynamics when the polymer is exposed to an arbitrary flow field in a confined geometry. We focused our attention on three major points:

1. The effects of *confined geometry* are important when the size of the polymer is comparable to the pore size. We showed that geometrical effects are manifest through subtle hydrodynamic interactions. The examples of polymer migration and thickness of the adsorbed polymer layer were used to illustrate this point.
2. In strong flows it is important to introduce a *nonlinear connector force* to avoid a mathematical singularity. Techniques for solving a nonlinear Langevin equation are essential for tackling this problem and are presented in this chapter.
3. We have shown an approximate method for calculating polymer conformation in the most difficult *flow fields*, including turbulence. We demonstrated how molecular information gives insight to engineering design criteria.

We limited ourselves to these three points, and the theory we examined is limited to an infinitely dilute system. Even with these restrictions, the exact theory does not exist. It is possible to formally develop an exact theory, as reported in Section II, but only with the use of approximations. Sometimes approximations neglect essential physical ingredients when the system becomes extremely complicated. For example, to describe drag reduction, we used the Kirkwood–Riseman equation (which is not correct for $\mathrm{Re} \gg 1$).

We also assumed that there is no degradation.

Most of the practical problems involve solutions of finite concentration, and the procedures we explained may not be directly applicable for nondilute solutions. In principle it is possible to generalize the procedures presented in this chapter to concentrated solutions. An attempt at such a generalization has been reported by Jhon and Freed[13] in conjunction with a study on polymer migration.

There are many experimental works[79] on migration of a tagged polymer (or particle) in finite-concentration polymer solutions. Theoretical attempts have always been based on a continuum approach and are well summarized in a review paper by Leal.[80] Nonetheless, no theoretical attempt to examine this phenomenon from a molecular approach has been reported.

APPENDIX: REVIEW ON TURBULENCE†

The following is a radical condensation of the classical theory on turbulence. It seems necessary to explain something about the nature of turbulence itself, specifically the turbulence produced by high Reynolds number flow through a tube. The turbulent energy budget equation, derived by taking the time average of the Navier–Stokes equation and subtracting the kinetic energy equation for the mean flow, is

$$U_j \frac{\partial}{\partial x_j}\left(\frac{1}{2}\langle u_i u_i\rangle\right) = -\frac{\partial}{\partial x_j}\left[\frac{1}{\rho}\langle u_j p\rangle + \frac{1}{2}\langle u_i u_i u_j\rangle - 2\nu\langle u_i s_{ij}\rangle\right] - \overline{u_i u_j} S_{ij} - 2\nu\langle s_{ij}s_{ij}\rangle \qquad \text{(repeated indices implies a summation),} \tag{A.1}$$

where U_j and S_{ij} are the mean flow and mean strain rate, respectively; and u_i, p, and s_{ij} are the velocity, pressure, and strain-rate fluctuation, respectively. (*Note:* We use the same notations as Ref. 62 and 77.) Also, $\nu = \eta/\rho$ is the kinematic viscosity. The first term on the right-hand side is the divergence of three energy transport terms due to pressure work flux, Reynolds stress flux, and flux due to viscous stresses. These terms serve only to redistribute the energy from place to place and from one form to another. It can be shown that these terms add nothing to the total turbulent energy by applying the Gauss divergence theorem after integrating over the whole tube. The last two terms in Eq. (A.1) are the net contributors to the total energy. Clearly the first term produces energy through a synergism of the Reynolds stress and

†See Refs. 60–62.

the mean flow (mean strain), and the second term is the viscous dissipation due to the fluctuating strain field. If we presume (as is usually done) that the mean flow is stationary along the axis of the tube, then the term on the left-hand side is zero.

This points up the fact that the total energy generated from the mean flow is eventually dissipated by viscous stresses,

$$-\langle u_i u_j \rangle S_{ij} = 2\nu \langle s_{ij} s_{ij} \rangle. \tag{A.2}$$

Here we have assumed homogeneity, that is, the averaged quantities are not spatially dependent. By the following order-of-magnitude analysis we will show that the order of length at which turbulence production occurs is much larger than the order of length at which dissipation occurs. If u_+ is the rms fluctuation velocity, and L is the length scale of the largest eddies (roughly the pipe diameter), then S_{ij} is $O(u_+/L)$ because in order for the mean strain to contribute much of its energy to the turbulence, it must be able to interact with the fluctuations. Using this fact and the fact that $\langle u_i u_j \rangle \approx u_+^2$, we can rearrange Eq. (A.2) as

$$C_1 u_+ L S_{ij} S_{ij} = 2\nu \langle s_{ij} s_{ij} \rangle, \tag{A.3}$$

where C_1 is on the order of 1. Since $\mathrm{Re} \approx u_+ L/\nu$, and presuming that the Reynolds number is large, it is easy to see that $S_{ij}S_{ij}$, the mean strain, is much smaller than $\langle s_{ij} s_{ij} \rangle$, the fluctuating strain, and since the strain is related to the time scale of the eddies producing the strain, the time scale of the dissipative eddies is much smaller than that of the energy-producing eddies. This suggests that if polymer molecules are greatly affected by the presence of a large strain (most believe this is so), then high Reynolds number flow of a polymer solution may exhibit characteristics that are radically different from those of a Newtonian fluid.

It becomes important to identify the length scale over which the dissipation of the turbulent energy occurs. We will define η as this length scale which should be only a function of the fluid properties (that is, the kinematic viscosity ν) and the energy ε that needs to be dissipated. In addition, we expect that the Reynolds number characteristic for these eddies is low, say on the order of 1. By defining the characteristic velocity of these eddies as v, we obtain

$$\frac{\eta v}{\nu} = 1, \qquad \nu \left(\frac{v}{\eta} \right)^2 = \varepsilon. \tag{A.4}$$

This may be rearranged to give

$$\eta = \left(\frac{\nu^3}{\varepsilon}\right)^{1/4}. \tag{A.5}$$

We will calculate ε later, but it suffices to say that for large Reynolds numbers, η is several orders of magnitude below L. Thus it seems that energy is taken from the mean strain S_{ij} by large eddies having length and time scales similar to S_{ij}. However, dissipation primarily occurs on length and time scales much smaller than those of S_{ij} and which S_{ij} cannot act on. In short, *turbulence is basically the mechanism by which energy is absorbed from the mean shear, by length scales at which there is virtually no viscous dissipation, and is passed down to length scales where viscous dissipation predominates.* Clearly there can be no eddies much larger than L (the diameter of the tube), and we may also suspect that there is not enough energy around to support eddies much smaller than η, defined by Eq. (A.5). The eddy sizes between η and L effectively serve as energy transmitters. It is important to realize that eddies of size η are ambivalent to what is going on with eddies of size L and also that although large eddies have ample opportunity to interact with walls of the tube, small-scale eddies have a tendency to be isotropic even if the system itself is definitely not isotropic.

This fact is extremely convenient for the researcher who wishes to apply the molecular theory to drag reduction. We expect that the smallest eddies (the ones producing the largest strain rates) will have the largest effect on a molecule subjected to turbulence. This is convenient because a great deal of theory exists for isotropic turbulence, while few results (even experimental ones) are available for anisotropic turbulence. In fact, we have used isotropic turbulence assumptions to study the polymer conformation in Section IV.

This has set the stage for a unified view of turbulence from a qualitative perspective. Since it seems that eddies and their associated properties seem to be distinguished primarily by size and lifetime, a Fourier space representation of eddies appears advantageous. Looking more closely at $\langle u_i u_j \rangle$, we have

$$\langle u_i(0, 0)u_j(\mathbf{r}, t)\rangle = R_{ij}(\mathbf{r}, t), \tag{A.6}$$

where $\mathbf{r}$ is the distance between the measurements of u_i and u_j, and t is the time lag between the measurements. This differs from similar terms in the turbulent energy equations in which $\mathbf{r}$ and t are zero. Also we have assumed that R_{ij} is isotropic in space and stationary in time (the flow is not evolving in time). We hope to capitalize on the observation made previously, that the eddies of concern, the smallest ones, are roughly isotropic. Using the trace of R_{ij} and then the Fourier transform of the result, it is not hard to arrive at the Fourier space representation of the kinetic energy $E(k)$. Then we are able to draw the schematic illustration in Fig. A.1.

It remains to find how to characterize ε from the mean flow. Using the averaged equation of motion, we find

$$-\langle uv \rangle + \nu \frac{\partial u}{\partial y} = u_*^2 \left(1 - \frac{2y}{D}\right). \tag{A.7}$$

Here D is the tube diameter, v is the lateral fluctuating velocity, with u and v measured simultaneously, and u_* is the frictional velocity. It has been well established that the Reynolds stresses dominate for most of the interior of the inertial layer, so we can see that $-\langle uv \rangle \sim u_*^2$ in the inertial layer. If at each point inside the tube we can say that the energy production is equal to dissipation (and experiments bear this out[60]), then we have

$$\varepsilon \sim -\langle uv \rangle \frac{\partial U}{\partial y}, \tag{A.8}$$

which is the turbulence production term from Eq. (A.1). According to the so-called law of the wall (logarithmic velocity profile), $\partial U/\partial y = u_*^2/\nu K y_+$, where U is the local average velocity and $y_+ = yu_*/\nu$. Therefore it seems likely that, from Eq. (A.8),

$$\varepsilon \approx \frac{u_*^2}{Ky}$$

and therefore

$$\eta_+ \sim K^{1/4} y_+^{1/4}. \tag{A.9}$$

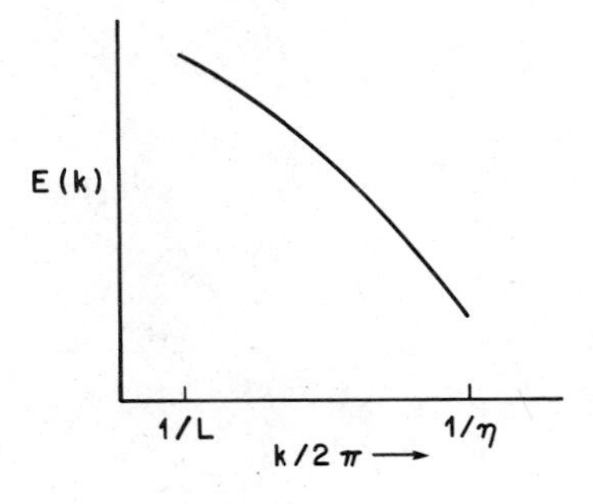

Figure A.1. Schematic representation of the turbulent energy distribution. k—Fourier counterpart of r, η is the smallest eddy size allowed by viscous effects.

In addition, it is fairly obvious that, as we get closer to the wall, the size of the largest eddy L is roughly proportional to the distance from the wall, and since $1/K = 1/2.5 \sim \mathcal{O}(1)$, we state

$$L_+ = Ky_+. \tag{A.10}$$

Here η_+ and L_+ are dimensionless parameters defined as $\eta_+ = \eta u_*/v$ and $L_+ = Lu_*/v$.

Acknowledgment

One of the authors (*MSJ*) would like to thank Professors Karl F. Freed (Section II), John L. Anderson (Section III), and Jacques L. Zakin (Section IV) for many discussions and for their generous help. The authors also thank Ms. Jeen-Ok Kim for her assistance in preparing this manuscript.

References

1. W. R. Schowalter, *Mechanics of Non-Newtonian Fluids*, Pergamon, Oxford, 1978.
2. R. B. Bird, R. C. Armstrong, and O. Hassager, *Dynamics of Polymeric Liquids*, vol. 1, *Fluid Mechanics*, Wiley, New York, 1977.
3. H. Yamakawa, *Modern Theory of Polymer Solutions*, Harper and Row, New York, 1971.
4. B. D. Coleman, *Viscometric Flows of Non-Newtonian Fluids*, Springer, Berlin, 1966.
5. R. B. Bird, O. Hassager, R. C. Armstrong, and C. F. Curtiss, *Dynamics of Polymeric Liquids*, vol. 2, *Kinetic Theory*, Wiley, New York, 1977.
6. R. H. J. Sellins and R. T. Moses, Eds., *Proc. 3rd Int. Conf. on Drag Reduction*, University of Bristol, July 2–5, 1984.
7. R. B. Bird, W. E. Stewart, and E. N. Lightfoot, *Transport Phenomena*, Wiley, New York, 1960.
8. R. B. Bird and C. F. Curtiss, *Phys. Today* **37**, 36 (1984).
9. M. M. Denn, *Process Fluid Mechanics*, Prentice-Hall, Englewood Cliffs, N.J., 1980.
10. S. F. Edwards and K. F. Freed, *J. Chem. Phys.* **61**, 1189 (1974).
11. K. F. Freed and A. Perico, *Macromolecules* **14**, 1290 (1981); A. Perico and K. F. Freed, *J. Chem. Phys.* **78**, 15 (1983).
12. K. F. Freed, in *Progress in Liquid Physics*, C. A. Croxton, Ed., Wiley, London, 1978, p. 343.
13. M. S. Jhon and K. Freed, *J. Polym. Sci. Polym. Phys. Ed.* **23**, 955 (1985).
14. N. G. van Kampen, *Phys. Lett. C* **24**, 171 (1976).
15. N. G. van Kampen, *Stochastic Processes in Physics and Chemistry*, Elsevier, New York, 1981.
16. Z. A. Akcasu, *J. Stat. Phys.* **16**, 33 (1977).
17. A. R. Altenberger, *J. Poly. Sci. Polym. Phys. Ed.* **17**, 1317 (1979).
18. J. G. Kirkwood and J. Riseman, *J. Chem. Phys.* **16**, 565 (1948).
19. R. Zwanzig, *Adv. Chem. Phys.* **15**, 325 (1969).
20. R. Zwanzig, J. Kiefer, and G. H. Weiss, *Proc. Nat. Acad. Sci. U.S.A.* **60**, 381 (1968).
21. J. T. Fong and A. Peterlin, *J. Res. Nat. Bur. Stand. B* **80**, 273 (1976).

22. J. Rotne and S. Prager, *J. Chem. Phys.* **50** 4851 (1969).
23. H. Fujita, *J. Polym. Sci. Polym. Phys. Ed.* **11**, 899 (1973).
24. R. C. Armstrong, and M. S. Jhon, *J. Chem. Phys.* **75**, 4160 (1981).
25. P. E. Rouse, Jr., *J. Chem. Phys.* **21** 1272 (1953).
26. B. H. Zimm, *J. Chem. Phys.* **24**, 269 (1956).
27. S. Middleman, *Fundamentals of Polymer Processing*, McGraw-Hill, New York 1977.
28. J. M. Dealy, *Polym. Eng. Sci.* **11**, 433 (1971).
29. J. Happel and H. Brenner, *Low Reynolds Number Hydrodynamics*, Prentice-Hall, Englewood Cliffs, NJ, 1965.
30. R. G. Cox and H. Brenner, *Chem. Eng. Sci.* **22**, 1753 (1967).
31. P. G. deGennes, *Scaling Concepts in Polymer Physics*, Cornell University Press, Ithaca, N.Y., 1979.
32. S. Daoudi and F. Brochard, *Macromolecules* **11**, 751 (1978).
33. T. D. Long and J. L. Anderson, *J. Polym. Sci. Polym. Phys. Ed.* **22**, 1261 (1984).
34. T. D. Long and J. L. Anderson, *J. Polym. Sci. Polym. Phys. Ed.* **23**, 191 (1985).
35. E. J. Hinch, *Phys. Fluids* **20**, 522 (1977).
36. T.Karis and M. S. Jhon, *Proc. Nat. Acad. Sci. U.S.A.* **83**, 4973 (1986).
37. H. Haken, *Rev. Mod. Phys.* **47**, 67 (1975).
38. H. E. Stanley, *Introduction to Phase Transitions and Critical Phenomena*, Oxford University Press, London, 1971.
39. M. Suzuki, *Adv. Chem. Phys.* **46**, 195 (1981).
40. R. H. Shafer, N. Laiken and B. H. Zimm, *Biophys. Chem.* **2**, 180 (1974).
41. R. H. Shafer, *Biophys. Chem.* **2**, 185 (1974).
42. J. H. Aubert and M. Tirrell, *J. Chem. Phys.* **72**, 2694 (1980).
43. J. H. Aubert, S. Prager, and M. Tirrell, *J. Chem. Phys.* **73**, 4103 (1980).
44. A. Dutta and R. A. Mashelkar, *Rheol. Acta* **22**, 455 (1983).
45. G. Sekhon, R. Armstrong, and M. S. Jhon, *J. Polym. Sci. Polym. Phys. Ed.* **20**, 947 (1982).
46. D. Lhuiller, *J. Phys. (Paris)* **44**, 303 (1983).
47. S. Yip and S. Ranganathan, *Phys. Fluids* **8**, 1956 (1965).
48. H. Mori, *Prog. Theor. Phys.* **33**, 423 (1965).
49. R. Zwanzig, in *Lecture in Theoretical Physics*, W. E. Britten, B. W. Downs, and J. Downs, Eds., Interscience, New York, 1961.
50. R. Armstrong and M. S. Jhon, *J. Chem. Phys.* **83**, 2475 (1985).
51. T. Cosgrove, B. Vincent, and T. L. Crowley, ACS Symp. ser. 240, *Polymer Adsorption and Dispersion Stability*, 1983; J. Klein and P. F. Luckham, *Macromolecules* **17**, 1041 (1984).
52. J. M. H. M. Schentjens and G. J. Fleer, *J. Phys. Chem.* **83**, 1619 (1979); **84**, 173 (1980).
53. P. Dejardin and R. Varoqui, *J. Chem. Phys.* **75**, 4115 (1981).
54. Ph. Myard and Ph. Gramain, *Macromolecules* **14**, 180 (1981).
55. Y. Cohen and A. B. Metzner, *Macromolecules* **15**, 1425 (1982).
56. C-H. Tsang, M. S. Thesis, University of California, Los Angeles, 1985.
57. J. J. Lee and G. G. Fuller, *Macromolecules* **17**, 375 (1984).
58. W. K. Idol, Jr., Ph. D. Dissertation, Carnegie Mellon University, Pittsburgh, 1985.

59. E. A. Dimarzio and R. J. Rubin, *J. Polym. Sci. Polym. Phys. Ed.* **16**, 457 (1978).
60. J. O. Hinze, *Turbulence*, 2nd ed., McGraw-Hill, New York, 1975.
61. G. K. Batchelor, *The Theory of Homogeneous Turbulence*, Cambridge University Press, Cambridge, 1953.
62. J. L. Lumley, *Phys. Fluids* **20**, S64 (1977).
63. R. Armstrong and M. S. Jhon, *J. Chem. Phys.* **77**, 4256 (1982).
64. R. Armstrong and M. S. Jhon, *J. Chem. Phys.* **79**, 3143 (1983).
65. R. Kubo, *J. Phys. Soc. J* **17**, 1100 (1962).
66. V. I. Klyatskin and V. I. Tatarskii, *Sov. Phys.-Usp.* **16**, 494 (1974).
67. J. L. Lumley, *Symp. Math.* **9**, 315 (1972).
68. N. Phan-Thein and R. I. Tanner, *Phys. Fluids* **21**, 311 (1978).
69. A. F. Horn and E. W. Merrill, *Nature* **312**, 140 (1984).
70. E. W. Merrill and A. F. Horn, *Polym. Commun.* **25**, 144 (1984); J. A. Odell, A. Keller, and M. J. Miles, *Polym. Commun.* **24**, 7 (1983).
71. I. Oppenheim, K. E. Shuler, and G. H. Weiss, *Stochastic Processes in Chemical Physics*, M.I.T. Press, Cambridge, Mass., 1977; G. H. Weiss, *Adv. Chem. Phys.* **13**, 1 (1966).
72. B. A. Toms, *Proc. Int. Cong. on Rheology* (Holland, 1948), North Holland, Amsterdam, 1949, pp. II-135–141.
73. J. B. Holt, *Oil and Gas J.* **19**, 272 (1981).
74. P. S. Virk, *AlChE J.* **21**, 625 (1974).
75. R. Armstrong and M. S. Jhon, *Chem. Eng. Commun.* **30**, 99 (1984).
76. J. L. Lumley, *J. Polym. Sci. Macromol. Rev.* **7**, 263 (1973).
77. R. Armstrong, Ph.D. Dissertation, Carnegie Mellon University, Pittsburgh, 1984.
78. P. S. Virk and H. Baker, *Chem. Eng. Sci.* **25**, 1183 (1970).
79. D. C. Prieve, M. S. Jhon, and T. L. Koenig, *J. Rheol.* **29**, 639 (1985).
80. L. G. Leal, *Ann. Rev. Fluid Mech.* **12**, 435 (1980).

INFLUENCE OF TRANSLATIONAL ENERGY UPON REACTIVE SCATTERING CROSS SECTION: NEUTRAL–NEUTRAL COLLISIONS

A. GONZÁLEZ UREÑA

Departamento de Química Física
Facultad de Ciencias Químicas
Universidad Complutense
Madrid, Spain

CONTENTS

I. INTRODUCTION

This chapter presents a general survey of the influence of translational energy upon reactive cross sections in neutral–neutral collisions. It will be mainly devoted to recent experimental and theoretical work essentially related to molecular beam investigations.

In a conventional kinetic experiment all the available excitation modes are populated as the temperature is raised, and it is difficult to isolate the specific roles of translational, vibrational, or rotational energy in promoting chemical reactions. For the past several decades different experimental techniques,[1] and in particular the molecular beam and the beam–photon interaction methods,[2,3] have permitted the study of chemical reactions with selective excitation of the reagents and for specific analysis of the products in vibrational, translational, etc., modes. This has provided an opportunity to test all the theories now available in the atom–molecule collision field.[4–6]

With regard to the influence of collision energy on reaction dynamics, many things can be expected a priori. Let us mention a few examples. It is well known that a translational excitation brings into a collision much more angular momentum than does a vibrationally excited molecule (for the same impact parameter). Obviously, different product states will be populated in the two cases. For complex mode reaction mechanisms (those involving long-lived intermediates) no energy selectivity can be expected since significant internal energy transfer should take place during the relatively long lifetime (~ 1 ps) of the complex. As the collision energy is increased to values where the collision time is much less than the lifetime of the complex, a more direct mode of behavior can be expected. From a potential energy surface (PES) point of view, two different potential energy surfaces should, in principle, give different collision energy dependences of the reactive cross section. Modern theories and simulation procedures can be a great help in correlating the observations with relevant potential features. Thus experimental results such as energy dependence should be a very sensitive probe of the unknown potential. In addition, since energy opens new channels as, for example, the onset of various chemical reactions, translational energy threshold measurements (including other effects such as tunneling and resonances) and their classical and quantum interpretation will clarify the nature of the chemical interactions.

Therefore, many questions justify the main purpose of this chapter.

1. To what extent, if any, can translational energy promote chemical reactions?
2. What can be learned about the potential energy surface from such influence?
3. What can be learned from translational excitation about reaction dynamics that is not available from other excitation modes?
4. Can translational excitation probe the unknown part of the potential energy surface, that is, elucidate the transition state structure?

We shall attempt to answer these kinds of questions by presenting recent experimental and theoretical results. Accordingly the chapter has been organized as follows.

Section II is dedicated to experimental procedures and some definitions. It is not intended to present a very detailed review of experimental techniques, but to give a general description of the main experimental procedures to measure the influence of translational energy upon reactive scattering and more specifically upon the excitation function.

Section III reviews the basic model and formulae. It is beyond the scope of this chapter to present the full reaction dynamic theories. The main purpose of this section is to summarize the main models and working expressions and to provide a basis for the review and discussion of the main results, presented in Sections IV to VI.

This review is not intended as a general treatise on molecular reaction dynamics, but is dedicated to collision energy dependence. Fortunately, there are excellent general reviews[1-24] that will assist the reader who wants to complement Sections II and III and to clearly understand many aspects of Sections IV to VI.

II. EXPERIMENTAL: ON THE DETERMINATION OF THE EXCITATION FUNCTION

A. Excitation Function

For a bimolecular reaction A + BC → AB + C the reaction cross section is given by[25]

$$\sigma_R(E_T) = \frac{F_{AB}}{n_A n_{BC} v_r \Delta V}, \tag{1}$$

where F_{AB} is the (integrated) total flux of the AB product, v_r the relative reagent velocity, ΔV the (beam intersection) reaction volume, n are the number densities, and the subscripts indicate the species concerned. Because of the spatial dispersion of products, σ_R is not measured directly in molecular beam experiments, but must be inferred from the product angular distribution. In addition, due to difficulties in establishing the absolute magnitudes of n_A, n_{BC}, ΔV, and the detector efficiency, there is usually a considerable margin of error in determining the reaction cross section accurately. On the other hand only relative values of the reaction cross section need to be measured to determine the influence of the translational energy upon the reactive cross section, that is, the shape of the excitation function $\sigma_R(E_T)$.

The total flux is independent of the coordinate system and we can introduce the flux of AB scattered per unit time per unit solid angle in the center-of-mass direction (θ, ϕ) as

$$F_{AB}(\theta, \phi) = n_A n_{BC} \sigma_R \Delta V P(\theta, \phi).$$

The product $\sigma_R P(\theta, \phi)$ is called the differential (solid angle) cross section with dimensions of area per unit solid angle, $\sigma_R P(\theta, \phi) = d^2\sigma/d^2\omega$, where $d^2\omega = \sin\theta \, d\theta \, d\phi$. For a spherically symmetrical potential there is no ϕ dependence of the scattered intensity per unit solid angle, and one can obtain the differential (polar) cross section integrating over the azimutal angle, that is,

$$\frac{d\sigma}{d\theta} = \int \sin\theta \frac{d^2\sigma}{d^2\omega} \, d\phi$$

$$= 2\pi \sin\theta \frac{d^2\sigma}{d^2\omega},$$

which represents the fractional contributions to σ_R from any final polar angle θ. Hence the total reaction cross section can now be obtained,

$$\sigma_R = \int_0^\pi \frac{d\sigma}{d\theta}\, d\theta = \int_0^\pi 2\pi \sin\theta \frac{d^2\sigma}{d^2\omega}\, d\theta$$

$$= \int_0^\pi \sigma_R(\theta) 2\pi \sin\theta\, d\theta.$$

Of course, the differential reactive cross section $\sigma_R(\theta)$ is a function of the collision energy and should be written as $\sigma_R(E_T, \theta)$. The collision energy dependence of this quantity should be defined as the differential (solid angle) excitation function.

Other types of differential reaction cross sections arise when the product translational energy distribution is considered, that is, $P(E'_T)$. Now $\sigma_R P(E'_T)$ is the differential reaction cross section for scattering into a given final translational energy.

The integral reaction cross reaction can be written[2,3]

$$\sigma_R(E_T) = \int_0^E dE'_T \frac{d\sigma}{dE'_T}, \qquad 0 \leqslant E'_T \leqslant E,$$

where E is the total available energy for the products. Similarly this differential cross section is a collision energy dependent function, and we can also define the differential (product energy) excitation function $\sigma_R(E_T, E'_T)$, which will measure the collision energy dependence for scattering into a given final translational energy.

More detailed differential cross sections are usually deduced from measurements of the product flux–velocity angle distributions in the laboratory system.[35] These data are transformed to the center-of-mass system to obtain the detailed differential cross section in center-of-mass velocity space, that is, a quantity proportional to $d^3\sigma(\theta, w')/d^2\omega\, dw'$, where w' is the magnitude of the center-of-mass recoil velocity of the products. (See Section IV for examples of these contour maps.) The aforementioned differential cross sections are in fact convolutions of this triple differential cross section.

In general we can define the n-differential excitation function $\sigma_R(E_T, \alpha_1, \alpha_2, \ldots, \alpha_n)$ as a measure of the collision energy dependence for scattering into a given final set $(\alpha_1, \alpha_2, \alpha_3, \ldots, \alpha_n)$ of product states. It is given by

$$\sigma_R(E_T, \alpha_1, \alpha_2, \ldots, \alpha_n) = \sigma_R P(E_T/\alpha_1, \alpha_2, \ldots, \alpha_n),$$

where $P(E_T/\alpha_1, \alpha_2, \ldots, \alpha_n)$ is the normalized product distribution into n different states at fixed collision energy E_T.

From an experimental point of view the initial expression can be decomposed into three factors,

$$\sigma_R(E_T) = \frac{1}{n_A n_{BC} \bar{v}_r} \frac{1}{\Delta V} F_{AB}.$$

The first is basically related to the known reagent conditions, including reactants, densities, and average relative velocity. The second factor, the beam intersection volume, is completely determined by beam geometry and collimation. Thus it is an energy-independent factor, and rather than being measured, it is usually neglected since only relative values are required to determine the excitation function. (For a discussion on the determination of the absolute reaction cross section see Ref. 10.) Finally the total flux F_{AB}, is given in terms of laboratory measurables,

$$F_{AB} = \int_{-\pi}^{+\pi} \int_{-\pi/2}^{+\pi/2} I(\Theta, \Phi) \cos \Phi \, d\Phi \, d\Theta,$$

where $I(\Theta, \Phi)$ is the flux collected by the detector located at (Θ, Φ) within the detector solid angle $d^2\Omega = \langle \cos \Phi_d \rangle \, d\Phi_d \, d\Theta_d$.

The total flux is the crucial quantity to be measured in order to obtain the excitation function. A typical procedure for determining the excitation function consists in maintaining one beam fixed (that is, beam velocity and density fixed), while the velocity of the second beam is changed, such as by seeding the heavier component as described in the next subsection, to achieve different collision energies. Then relative values of the total flux are measured and normalized via Eq. (1) to obtain the kinetic energy dependence of the reactive cross section, provided that the average relative velocity of the reagents is known. If the spread in relative velocity is large and the cross section a strong function of E_T, an iterative procedure is generally used to extract $\sigma_R(E_T)$ from $\sigma_R(\bar{v}_r)$.[47]

B. Reagent Preparation and Changes in Translational Energy

One widely used technique has been the seeded nozzle beam technique for the BC reactant. In this technique the reactant BC is added to a larger amount of a lighter inert carrier gas, such as He or H_2, and then the mixture is expanded through a supersonic nozzle to produce a narrow translationally "cold" velocity distribution of the BC molecules. By varying the seed–carrier mixture, among other factors, collision energies up to 3.0 eV can be achieved under favorable conditions for neutral–neutral interactions.

Supersonic nozzle beams of atoms and radicals seeded in an inert buffer gas have been increasingly used in molecular beam scattering experiments.[17] Alkali atoms seeded in hydrogen, halogen atoms produced by thermal dissociation, oxygen and halogen atoms from a radiofrequency discharge source and, more recently, by microwave discharge have provided available intense supersonic beams suitable for studying the translational energy dependence of the reaction cross section. In addition to the seeded beam procedure a (mechanically) rotor-accelerated beam technique has been used to accelerate atomic beams. With subsequent electron bombardment excitation this method produces beams of electronically excited rare gases to velocities $\leqslant 1.7$ km/s.[58]

A novel method in molecular beam arrangements uses independently rotatable sources which permit the collision energy in the center-of-mass frame to be varied smoothly by changing the beam intersection angle without having to change the molecular beam source conditions.[26] This experimental setup has been used to measure the state-to-state vibrational excitation of I_2 in collisions with He and can therefore be extended to determine excitation functions under favorable kinematic conditions (see Section II.C.3).

Finally an important method, which has been used to produce fast neutral alkali beams, is the charge exchange beam source. Its capability depends upon the fact that the cross section for the resonant charge transfer process

$$A^+_{\text{fast}} + A \rightarrow A_{\text{fast}} + A^+$$

is much larger than the momentum transfer cross section. This technique has been used to study the collision energy dependence of ion-par function and collisional ionization.[1,27]

Laser Techniques. A novel method of producing translational fast metal atoms consists of the vaporization of a metal film by a short-pulse laser.[28] The dense cloud of vaporized material so produced expands and becomes collisionless. This method has been used in a beam–gas arrangement to determine the kinetic energy dependence of the chemiluminescent oxidation of zinc atoms by nitrous oxide producing electronically excited ZnO^*.[28]

If a diatomic molecule is photofragmented by the absorbtion of monochromatic radiation, the excess of energy upon dissociation appears as translational energy of the separating atoms. This translational excitation is shared between the two atoms in the inverse ratio of their masses. This method has been used to produce fast H, D, and T atoms from XH dissociation (X being halogen) and applied to threshold energy measurements of hydrogen atom abstraction. Recently this nonbeam method has been coupled to laser-induced fluorescence detection of the product reaction under single-collision conditions in order to determine the internal state distribution of the product as a function of translational energy.[29]

C. Product Detection: Experimental Procedures for Determining the Excitation Function

Many difficulties arise in the detection of the products when the (integral) total or partial reactive cross section is measured, especially because of integration over the laboratory angles of the reactively scattered products in neutral–neutral reactions. A survey of current techniques for determining excitation functions is given in Table I.

Typical procedures to measure the product angular and velocity distributions use rotatable flux or density detectors working with the single-pulse method of time of flight (TOF) and more recently with pseudorandom correlation time-of-flight methods. The improved efficiency of data retrieval by using these methods together with the use of intense supersonic beams leads to a higher reactive scattering signal and well-defined kinematics. This permits a direct inversion of laboratory data to obtain full contour maps of the differential reaction cross section as a function of collision energy. For technical details we refer the reader to Refs. 13 and 18. Our main goal now is to summarize the procedures that have been used so far to obtain the collision energy dependence of the reaction cross section.

1. Optical Potential Method: Analysis of Nonreactive Scattering†

In the early studies carried out for M + RX → MX + R (M being alkali atoms, X halogen, and R radicals), a surface ionization detector was used and the relative translational energy of the reagents was established by velocity selection of the atom beam. Typically most of the scattered atoms result from elastic collisions, but at larger angles the intensity of atoms was less than one can expect from elastic scattering measurements in systems of similar size because of the onset of the reactive scattering at small impact parameters. From this missing intensity at any angle the differential reaction cross section $\sigma_R(E, \theta)$ was calculated,

$$\sigma_R(E, \theta) = \sigma(E, \theta)_{\text{elastic}} - \sigma(E, \theta)_{\text{observed}},$$

and therefore summation over all angles gave the reaction cross section

$$\sigma_R(E_T) = 2\pi \int_0^{\pi} \sigma_R(E_T, \theta) \sin\theta \, d\theta.$$

†See Ref. 31.

Survey of Current Techniques for Determining Excitation Functions[a]

Method (product detection)[b]	Quantity measured	Advantage(s)	Disadvantage(s)
Surface ionization (FD)	Laboratory angular and velocity distribution of products	Flux detection; high-efficiency detection; in-plane and out-of-plane detection	Applicable only to low ionization potential products.
UV/VIS Chemiluminescence (FD)	Electronic spectra with resolvable vibrational bands (v_{ij}); total photon yield	Spectroscopic precision; resolution of different electronic states of, products reaction	Low photon yields. Interpretation of spectrum to obtain state-to-state cross sections is difficult. It also requires knowledge of molecular spectroscopic constants. Product of dark channel reaction cannot be measured.
Chemi-ionization (FD)	Total ion yield	Easy product collection; versatility	Low cross section. High collision energy often required to exceed thresholds. Low intensities, difficult to measure differential cross sections.
Laser-induced fluorescence (DD)	Electronic spectra with resolvable vibrational bands	Spectroscopic precision; dark channel reaction cross section can be measured; versatility due to available tunable lasers	Since it is a density detection method the reaction cross section can only be, determined if laboratory velocity distribution of products is known. This difficulty disappears for favorable kinematic conditions.
Mass spectrometry (electron bombardement) (DD)	In-plane laboratory angular and velocity distributions for product reactions	Versatility and high resolution	Same as laser-induced fluorescence method, namely density detection constraints. In addition most of the measurements are constrained to in-plane scattering. Total cross sections can be obtained if good velocity distribution of products is available or if confined at centroid.

[a]Only beam methods are included.
[b]FD—flux detection; DD—density detection.

This method provides an upper limit to the integral reactive cross section because actually this subtraction procedure also includes the inelastically scattered products in the reactive yield. As the collision energy increases, more inelastic channels are opened and the procedure loses validity (see, for example, Fig. 2).

2. *Total Flux Integration Method.*†

From Eq. (1) it becomes apparent that by in-plane and out-of-plane integration of the laboratory angular distribution one can determine the total flux and then the excitation function $\sigma_R(E_T)$. This procedure has been widely used in the molecular beam technique of the type shown in Fig. 1, essentially for alkali metal reactions where products with low ionization energy are formed.[33–54]

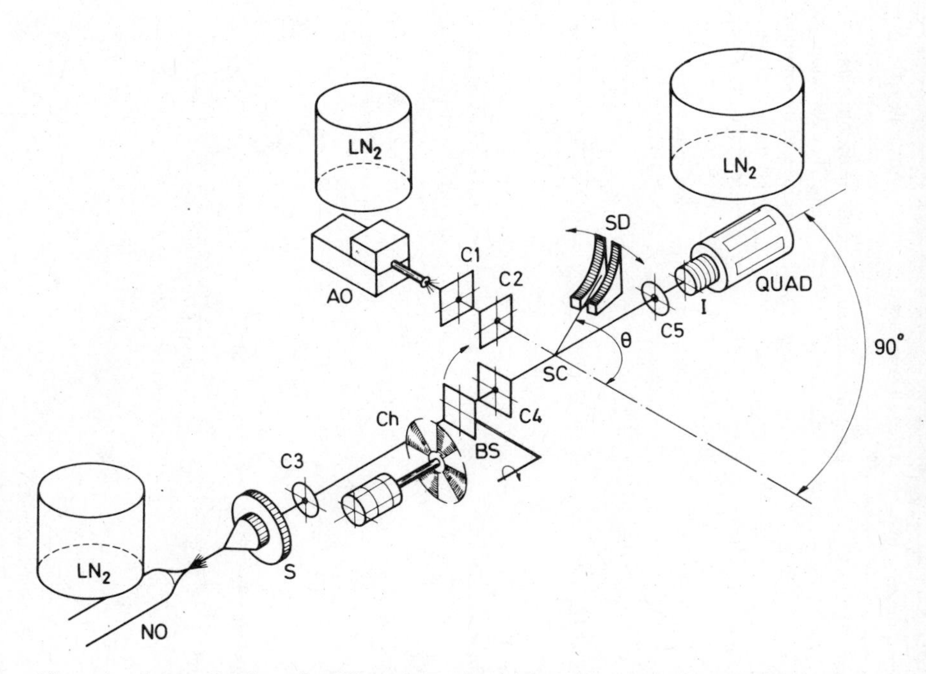

Figure 1. Schematic view of a differential molecular beam apparatus for the measurement of the excitation functions. AO and NO—alkali and nozzle ovens; Cl–C5—beam collimators; Ch—chopper; BS—beam stop; I and QUAD—ionizer and quadrupole mass filter; LN_2—liquid nitrogen trap; θ—in-plane scattering angle (out-of-plane angle ϕ not shown). The alkali beam direction defines the $\theta = 0$ values, as usual. See Ref. 38 for more details.

†See Refs. 37–75.

Figure 2 shows the first study of this method carried out by Bernstein and co-workers[25] for the reaction $K + CH_3I \rightarrow KI + CH_3$. Low energy data obtained by the optical model methods are also shown for comparison.

The scattered product intensity is relatively easy to detect in chemi-ionization reactions where the ion yield can be extracted independently of the scattering angle by proper ion optics, or in chemiluminescence reactions where the (total) product photon flux can also be collected directly. In these situations the total flux $F_{AB} \propto S_c, S_{cI}$, where S_i stands for the total (integrated) ion or photon yield.[55-75]

Furthermore a novel method[55] for determining the absolute chemi-luminescence cross section without knowledge of the absolute sensitivity has been described for reactions involving electronically excited reactants such as metastable alkaline earth. Consider the reaction $M^* + N_2O \rightarrow MO^* + N_2$. It has been shown that the absolute chemiluminoscence cross

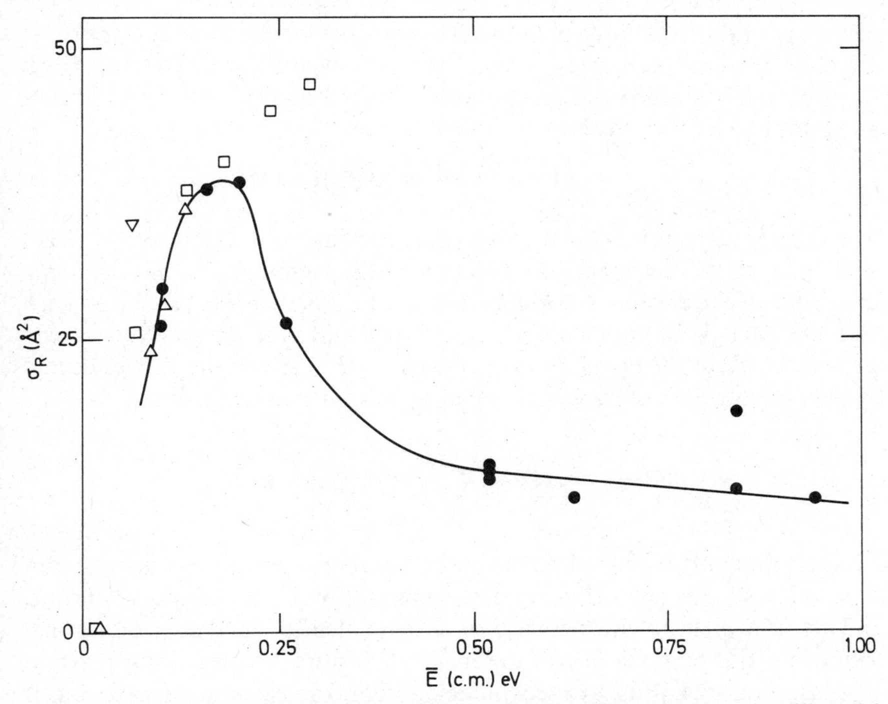

Figure 2. First excitation function measured from a direct integration of the reactive differential cross section for the reaction $K + CH_3I \rightarrow KI + CH_3$. Experimental results shown by filled circles from Ref. 25. Open symbols are from optical potential analysis of nonreactively scattered K measured by several authors.[25,31] Solid line is drawn through filled circles for clarity.

section is given by

$$\sigma_{\text{chem}} = \frac{S_{\text{MO}*} A f_J}{S_{\text{M}*} n_{\text{N}_2\text{O}} \bar{v}}$$

where $S_{\text{MO}*}$ is the wavelength integrated signal from MO* in a certain spectral region or band system, $S_{\text{M}*}$ is the metastable emission intensity, f_J its fraction at a particular J level, A the atomic transition probability, $n_{\text{N}_2\text{O}}$ the number density of the reactant, and $\bar{v}$ the average velocity. This procedure shows much promise in the study of absolute excitation functions for reactions with excited species, eliminating important difficulties associated with the precise knowledge of reactant flux, the size of the emitting volume, the solid angle subtended by the detector, and so on.

For reactions involving hydrogen atoms an interesting method[50] uses MoO_3 film as a detector. It was found that hydrogen atoms are chemisorbed on MoO_3 with nearly unit efficiency to give "molybdenum blue."[51] Martin and Kinsey[51b] were the first to extend the sensitivity and increase the specificity of this technique with the use of tritium labeling and radioactivity detection in their pioneering molecular beam experiments. Recently Kwei and Lo[50] have studied the reaction of H and D atoms with T_2 molecules using this kind of tritium atom detection scheme.

3. *Total Density Integration Method*†

If a density detector is used, as in the laser-induced fluorescence (LIF) method,[76] or in the mass-spectroscopic (MS) technique,[87] *only with the knowledge of the product laboratory velocity distribution* $P(v'(\Theta, \Phi))$ *at each angle* (Θ, Φ) (where the detector is located) *would it be possible to estimate the product flux and therefore to determine* $\sigma_R(E_T)$. Since the flux collected by the detector, $I(\Theta, \Phi)$, is related to the product density $n(\Theta, \Phi)$ by

$$I(\Theta, \Phi) = n(\Theta, \Phi) \int_0^\infty v' P(v'/(\Theta, \Phi))\, dv',$$

it is clear that only with the knowledge of the laboratory velocity distribution at each (Θ, Φ) can one get σ_R by direct integration in the laboratory frame.

This crucial problem, namely, to convert density to flux product distributions, still remains one of the major difficulties in measuring reaction cross sections by LIF or MS techniques, and only in exceptional cases has it been solved successfully (see below). The only attempt to measure the

†See Refs. 76–95.

laboratory recoil distribution by LIF has been made by measuring the Doppler profile at different scattering angles, which allows the full three-dimensional velocity distribution to be recovered. This technique was introduced by Kinsey in 1977 and is known as Fourier transform Doppler spectroscopy.[96] The first application carried out in Kinsey's laboratory was a study[97] of the exothermic reaction $H + NO_2 \rightarrow OH + NO$, where the velocity angle contours of the OH product flux were obtained by measuring Doppler profiles at 20 intervals over a wide angular range for the $R_1(17)$ line of the (0, 0) band of the OH $A^2\Sigma^+ - X^2\Pi$ transition. Unfortunately this method has not been used extensively, perhaps due to its difficulty to obtain (laboratory) product velocity distributions.

In certain cases favorable mass combinations, as in light atom exchange reactions, facilitate this procedure. For example, if $m_c \ll m_{AB}$, then $v_{AB} \simeq v_{c.m.}$, where $v_{c.m.}$ is the center-of-mass velocity, which can be calculated from the known reagent conditions. Under these circumstances the n_{AB} density is proportional to its (integrated) fluorescence intensity or mass spectrometric signal. In addition one can take advantage of these favorable kinematics and calculate the reactive cross section involving angular momentum conservation. This follows from the fact that in these cases (that is, $H + HL \rightarrow HH + L$, where H and L represent heavy and light atoms) the reagent orbital angular momentum L is effectively carried over into product rotation $J' \simeq L$ since the light atom does not carry away much angular momentum as compared to the heavier one. This condition holds as long as the initial reactant molecular (rotational) angular momentum is small, a condition that is quite well satisfied in seeded beam conditions.

Therefore one can write for the product AB rotational energy

$$E'_{\text{rot}}(b) = J'^2/2\mu_{AB}r_{AB}^{0\,2} \simeq L^2/2\mu_{AB}r_{AB}^{0\,2}$$

$$= (\mu_{A,BC}vb)^2/2\mu_{AB}r_{AB}^{0\,2} \simeq E_T(b/r_{AB}^0)^2,$$

where μ is the reduced mass, v the initial relative velocity at an impact parameter b, and r_{AB}^0 the AB equilibrium distance. The mean rotational energy is defined as

$$\bar{E}_{\text{rot}}(E_T) = \int_0^{b_{\max}} q(b)E'_{\text{rot}}(b)\,db,$$

with the partition function of the impact parameters given by

$$q(b) = 2b/b_{\max}^2.$$

If $P(b)$ is a step function [that is, $P(b) = 0\,(1)$ for $b > (<)\, b_{\max}$] and with the approximation $\mu_{A,BC} \simeq \mu_{AB}$, one obtains by integration of the above equation,

$$\bar{E}'_{rot}(E_T) = E_T b_{max}^2 / 2 r_{AB}^{0\,2}$$

$$= E_T \sigma_R(E_T) / 2\pi r_{AB}^{0\,2}$$

or

$$\sigma_R(E_T) = 2\pi r_{AB}^{0\,2} \bar{E}'_{rot}(E_T)/E_T,$$

which allows us to estimate approximate relative cross sections from experimental $\bar{E}'_{rot}(E_T)$ distributions.

The center-of-mass cross section has the advantage (for nonoriented beams) of depending only on the single center-of-mass scattering angle θ. This permits a reliable evaluation of the integral as long as significant ranges of θ are spanned, even though laboratory data may be taken only on a single plane ($\Phi = 0$), and the cross section so obtained will probably have additional uncertainties introduced in the laboratory to center-of-mass transformation.

Therefore for any system it is possible to estimate the reaction cross section at a given energy if the product translational energy and center-of-mass angular distribution have been deduced from a full analysis of the laboratory angular distributions in addition to the measurement of the in-plane density yield and beam densities. This procedure has been applied to the $Hg + I_2 \rightarrow IHg + I$ excitation function by Bernstein and co-workers,[87a] and details of its reaction dynamics will be presented later.

4. *Inversion Procedures*†

It is well known that the rate constant for a reaction can be obtained by convoluting the excitation function with the Maxwell–Boltzmann distribution, where k_B is Boltzmann's constant and μ the reduced mass of the reagent,[14]

$$k(T) = (\pi\mu)^{-1/2}(2/k_B T)^{3/2} \int \sigma_R(E_T) E_T \exp(-E_T/k_B T)\, dE_T,$$

which is based on the assumption that $\sigma_R(E_T)$ depends only on the collision energy. This equation is the basic link between microscopic (σ_R) and macroscopic $k(T)$ kinetics,[98] and essentially it shows the rate constant as the Laplace transform of $E_T\sigma_R(E_T)$. From an experimental or calculated cross

†See Ref. 14, 98.

section one can build rate constants with a high degree of accuracy since the overall temperature dependence of rate constants is determined by the Boltzmann factor in the temperature ranges typically studied. The opposite process, that is, to derive the excitation function from rate data as a function of temperature, is more difficult and not so accurate as the former because minor changes in $k(T)$ may cause a large variation in $\sigma_R(E_T)$.[14]

An interesting method has been developed by Roberts and co-workers[99] to invert molecular rate data with effective optical potentials. By making reasonable assumptions concerning the general features of the effective potential, one can invert the total reaction cross section as a function of collision energy to obtain reaction probabilities as a function of both impact parameter and energy once the excitation function is known. Reaction probabilities $P(E', b)$ can be generated via the working equation

$$P(E', b) = P(E) = (\pi b_0^2)^{-1}\left[\sigma_R(E) + E\frac{d\sigma_R(E)}{dE}\right],$$

where $E = E'(1 - b^2/b_0^2)$ and b_0 is some critical distance (the model parameter) where the strong interaction for the reaction takes place. This simple optical inversion has been applied to the $K + CH_3I$ system, and the result showed good agreement with the trajectory calculations of La Budde et al.[192] The model can also be extended to generate differential cross sections and other dynamic properties.

5. *Molecular Beam Excitation Function Data*

Table II lists most of the excitation function data obtained from molecular beam experiments. It is not intended to cover all the existing data, so there are omissions. The goal is to present enough examples of each group of different experimental procedures to show the state of the art of the collision energy dependence of the reaction cross section. Translational energy parameters, such as, those energy values where $\sigma_R(E_T)$ shows some relevant feature as threshold, maximum, and so on, are also presented. For clarity the shape of $\sigma_R(E_T)$ has been classified into different groups as follows:

1. These $\sigma_R(E_T)$ values show only one translational feature, such as post-threshold rising (with or without threshold) or monotonically decreasing functionality. They would be classified as $+1$ and -1, respectively.
2. These excitation functions show a mechanistical bimodality associated with two translational features, such as post-threshold plus post-maximum evolution in $\sigma_R(E_T)$ or low energy decline plus subsequent rise after a minimum in $\sigma_R(E_T)$. They would be classified as $+2$ and -2, respectively.

TABLE II.
Excitation Function Data from Molecular Beam Experiments

Reaction and detection method	Translational energy parameters (kJ/mol)			Shape class	Reference
	E_0	E_{max}	E_{min}		
Laser-induced florescence					
$Sr + FH \rightarrow SrF + H$	25.4	—	—	+1, +2?	79
$Ba + FH \rightarrow BaF + H$	9.6	29.2	—	+2	80
$Ba + HBr \rightarrow BrBa + H$				−1	84
$Ba + ClH \rightarrow BaCl + H$				−1	84
$H (D) + Br_2 \rightarrow HBr (DBr) + Br$				−1	82
$F(^3P_{3/2}, {}^2P_{1/2}) + HBr (DF) \rightarrow HF (DF) + Br(^2P_{3/2}, {}^2P_{1/2})$				−1	81
Surface ionization					
$K + CH_3 \rightarrow KI + CH_3$	3.18	17.4	—	+2	25
$K + CH_3Br \rightarrow KBr + CH_3$[a]	23.2	—	—	+1	44
$K + C_2H_5I \rightarrow KI + C_2H_5$				−1	38
$K + C_2H_5Br \rightarrow KBr + C_2H_5$	11.58	33.6	?	+2, 3?	48, 49
$K + ClH \rightarrow KCl + H$	6.37	41.5	—	+2	39, 46, 54
$K + FH \rightarrow KF + H$	<63	—	—	+1	54
$K + Br_2 \rightarrow KBr + Br$				−1	40
$Na + CH_3I \rightarrow NaI + CH_3$	~5	13.5	—	+2	53
$Rb + CH_3I \rightarrow RbI + CH_3$	—	—	87	−2, 3	37, 42
$Rb + CH_3Br \rightarrow RbBr + CH_3$[a]	19.3	—	—	+1	44
Mass spectrometry					
$O + H_2S \rightarrow HSO + H$	14.2	—	—	+1	88a
$O + C_2H_4 \rightarrow C_2H_3O + H$	5.3	—	—	+1	88b

$O + C_2H_2 \rightarrow OC_2H + H$	8.3	—	—	+1	88b
$Hg + I_2 \rightarrow HgI + I$	108	300	—	+2	87a
$F_2 + \begin{pmatrix} I_2 \\ ICl \end{pmatrix} \rightarrow \begin{pmatrix} FI + IF \\ FI + FCl \end{pmatrix}$	25.5	—		+1	95a
	84	—		+1	95a
$F_2 + \begin{pmatrix} I_2 \\ ICl \\ HI \end{pmatrix} \rightarrow \begin{pmatrix} IIF \\ ClIF \\ HIF \end{pmatrix} + F$	16.7	—	—	+1	95a
	25.8	—	—	+1	95a
	46	—	—	+1	95a
$Eu + O_2 \rightarrow EuO + O$	14.5	—	—	+1	91
$F_2 + \begin{pmatrix} n - H_2 \\ D_2 \end{pmatrix} \rightarrow \begin{pmatrix} HF(v = 1, 2, 3) \\ DF(v = 1, 2, 3, 4) \end{pmatrix} + H$				+1 (total)	151
				+1 (total)	152

Chemiluminescence

A N_2O, NO, NO_2, O_2, O_3 reactions

$Sm + N_2O$ (cold) $\rightarrow SmO^*$ (blue) $+ N_2$	5	25	50	3	66c
$Sm + N_2O$ (hot) $\rightarrow SmO^*$ (blue) $+ N_2$	—	—	50	−2	66a
$Sm + N_2O$ (hot) $\rightarrow SmO^*$ (red) $+ N_2$	—	—	>80	−2	66a
$Sm + N_2O$ (cold) $\rightarrow SmO^*$ (red) $+ N_2$	5	25	>80	3	66a
$Ba + N_2O$ (cold) $\rightarrow BaO^* + N_2$	<8	10	>40	3	66b, 66c
$Ba + N_2O$ (hot) $\rightarrow BaO + N_2$	—	—	>40	−2	67
$Pb + N_2O \rightarrow PbO^* + N_2$	56.8	—	—	+1	28b
$Zn + N_2O \rightarrow ZnO^* + N_2$				+1	28a
$B + N_2O \rightarrow BO^* + N_2$	<96	—		+1	28c
$Ho + N_2O \rightarrow HoO + N_2$	<48	—		+1	28c
$Ba + NO_2 \rightarrow BaO^* + NO$	—	—	27	−2	67, 199
$NO + O_3 \rightarrow NO_2^* + O_2$	13.4			+1	57, 65a, 64a
$Ba + (NO_2) \rightarrow BaO^* + NO + NO_2$	0	—	—	+1	199
$Y + O_2 \rightarrow YO(A, B) + O^b$				+1	69
$La + O_2 \rightarrow LaO(A, B, C) + O^b$				+1	69
$Sc + O_2 \rightarrow ScO(A) + O^b$				+1	69
$O_2(^1\Delta_g)$ + methyl vinyl ether	41.8	—	—	+1	66d

TABLE II (*Cont.*)

Reaction and detection method	Translational energy parameters (kJ/mol)			Shape class	Reference
	E_0	E_{max}	E_{min}		
	Chemiluminescence				
	A N_2O, NO, NO_2, O_2, O_3 reactions				
$O_2(^1\Delta_g)$ + 1,1-diethoxyethylene	~30	—	—	+1	66d
$O_2(^1\Delta_g)$ + *N*,*N*-dimethylisobutenylamine				0	66d
	B. Rare gas reactions				
$Xe(^3P_{2,0}) + CH_3Br \rightarrow XeBr(B, C) + CH_3$	7	60	—	+1	60
$Xe(^3P_{2,0}) + CH_3I \rightarrow XeI(B) + CH_3$	<6	—		+2	59
$Xe(^3P_{2,0}) + Br_2 \rightarrow XeBr^* + Br$				−1	58,59
$Xe(^3P_{2,0}) + Cl_4C \rightarrow XeCl^* + Cl_3C$				−1	58,59
$Xe(^3P_{2,0}) + ICl \rightarrow XeCl^* + I$				−1	58,59
$Xe(^3P_{2,0}) + CF_3I \rightarrow XeI^* + CF_3$				−1	58,59
$Xe(^3P_{2,0}) + HCl \rightarrow H + XeCl$	15.4			−1	58,59
	C. Halogen reactions				
$F_2 + \begin{pmatrix} I_2 \\ ICl \\ Br_2 \end{pmatrix} \rightarrow \begin{pmatrix} IF^* \\ ClF^* \\ BrF^* \end{pmatrix} + \begin{pmatrix} FI \\ FI \\ FBr \end{pmatrix}$	17.5	—	—	+1	
	25.8	—	—	+1	95
	47.2	—	—	+1	
$Ba + F_2 \rightarrow BaF^* + F$				−1	22
$Ba + Cl_2 \rightarrow BaCl^* + F$				−1	22
$Ca + F_2 \rightarrow CaF^* + F$				−1	22
$Ca + F_2 \rightarrow GaF(^3\Pi) + F^b$	25.8	33.4		+2	73
$In + F_2 \rightarrow InF(^3\Pi) + F^b$	3	—	—	+1	73

Chemi-ionization					
$Ca + F_2 \rightarrow CaF^+ + F^-$				−1	70
$Ca + Cl_2 \rightarrow ClCa^+ + Cl^-$	36.8±9.6	—	—	+1	70
$Ca + Br_2 \rightarrow CaBr^+ + Br^-$	82	—	—	+1	70
$Ca + NF_3 \rightarrow CaF^+ + NF_2^-$	48.3	—	—	+1	56
$Ba + F_2 \rightarrow BaF^+ + F^-$				−1	136, 22
$Ba + Cl_2 \rightarrow BaCl^+ + Cl^-$	—	21	—	−2	136
$Ba + Br_2 \rightarrow BaBr^+ + Br^-$	<14	23	—	+2	22
$Ba + SF_6 \rightarrow BaF^+ + SF_5^-$	96.5	—		+1	56
$Sr + F_2 \rightarrow SrF^+ + F^-$				−1	70
$Sr + Cl_2 \rightarrow SrCl^+ + Cl^-$	0.00±9.6	—		+1	70
$Sr + Br_2 \rightarrow SrBr^+ + Br^-$	33.8±14.5	—		+1	70
$Sr + SF_6 \rightarrow SrF^+ + SF_5^-$	145	—		+1	56
$Sr + NF_3 \rightarrow SrF^+ + NF_2^-$	57.9			+1	56

[a]Only in-plane yield measurements.
[b]Relative rate constant measurements.

3. These $\sigma_R(E_T)$ values include extra topological features, in addition to each of the previous classes, such as a shallow minimum in addition to category + 2.

Obviously as the collision range being studied is extended, any (less elaborate) category such as + 1 may change to category + 2 or − 1 to − 2, and so on.

III. THEORY: SIMPLE MODELS AND THEORIES TO CALCULATE EXCITATION FUNCTIONS

A. Model Calculations

1. Product Energy Disposal as a Function of Collision Energy

Excellent references have described a great number of detailed models for product angular and velocity distribution, and therefore it is beyond the scope of this chapter to repeat them here. This section provides a simplified description of the more current models in neutral–neutral collisions just to give the necessary background for interpreting the collision energy dependence of the main reaction attributes.

Direct Models for Energy Disposal.† One of the most frequently used models is that of the spectator stripping model for the reaction A + BC → AB + C. The assumption is made that no momentum is transferred to atom C. It can be shown that the final product translation is given by $E' = \cos^2 \beta \, E_T$, where $\cos^2 \beta = m_A m_C / m_{AB} m_{BC}$ and β is the so-called skewing angle. In a momentum representation the spectator model can be expressed as $p_c = \gamma_i p_i$, where p_c is the partition of the incident momentum $p_i (\gamma_i = m_C / m_{BC})$.

Other models assume that reaction proceeds via a rebound mechanism where the exoergicity is released as A recedes from the BC molecule. (This repulsive release is the basic idea of the photodissociation model described below.) One can then write $p_c = \gamma_i p_i + I$, where I is the impulse imparted during the repulsive release of energy $\mathbf{R}$. Then

$$\mathbf{R} = I^2 / 2 \mu_{BC} \qquad \text{and} \qquad E'_T = (\cos \beta \sqrt{E_T} + \sin \beta \sqrt{\mathbf{R}})^2.$$

In addition one can establish that the momentum of C can also change due to A–C(A–B followed by B–C) collisions. Then one obtains[110,111]

$$E'_T = [(\mu'/\mu)^{1/2} \sqrt{E_T} + \sin \beta \sqrt{\mathbf{R}}]^2$$

where $\mu'(\mu)$ stands for the product (reactant) reduced mass.

†See Refs. 9, 11, 14, 100–111.

A different approach takes advantage of incorporating some variable product as the mass which is introduced via angular momentum conservation of simple hard-sphere collisions. In this model the mean product translational E'_T is given by[103]

$$\bar{E}'_T = \frac{3(E_T + Q_{max})^2 - y^2 E_T^2}{6(E_T + Q_{max}) - 3yE_T}, \qquad E_T < Q_{max}/(y - 1),$$

$$= \tfrac{2}{3}(E_T + Q_{max}), \qquad E_T > Q_{max}/(y - 1),$$

where $y = \mu/\mu'$ and Q_{max} is the reaction exothermicity.

Many of the more detailed models for product energy disposal in chemical reactions belong to the impulsive model category where an instantaneous release of force in B–C is assumed. Polanyi has reviewed (see Ref. 9, Fig. 8) those that are based on the notion that the products are retreating from the barrier crest. One of the most widely used impulsive models is that of Herschbach,[11] which has found extensive application in electron jump type reactions. The central assumption is to consider a repulsive energy of magnitude **R** abruptly released in B–C. This estimate of repulsive energy is then given by

$$\mathbf{R} = \mathrm{EA(B)} - D(\mathrm{BC}) - \mathrm{EA(BC)},$$

where EA(B) is the electron affinity of B, D(BC) the dissociation energy of BC, and EA(BC) the electron affinity of BC. In Herschbach's model the distribution of repulsive energy is obtained from the quasi-diatomic reflection approximation often used for photodissociation and dissociative attachment processes. The Franck–Condon reflection of the BC vibrational phase distribution leads to a $P(\mathbf{R})$ distribution,

$$P(\mathbf{R}) \sim \psi^2(\mathbf{R}),$$

where ψ^2 is the vibrational wave function of the target molecule. Therefore for the ground state $P(\mathbf{R})$ takes a Gaussian form centered at $\mathbf{R}_0$,

$$P(\mathbf{R}) \sim \exp(\mathbf{R} - \mathbf{R}_0)/\Delta\mathbf{R}.$$

This expression can be used (1) in the DIPR-DIP model (see below) to obtain, the product angular translational vibrational distribution; (2) to build an idealized potential energy surface on which to run trajectories; (3) to obtain product translational distributions invoking linear momentum conservation,

$$P(E'_T) = P(\mathbf{R})\frac{d\mathbf{R}}{dE'_T},$$

where $d\mathbf{R}/dE'_T$ is given by

$$\frac{d\mathbf{R}}{dE'_T} = \frac{\mu_{AB,C}}{\mu_{B,C}}$$

(applied to the A + BC → AB + C reaction scheme); and (4) to obtain the excitation function by using Magee's[104] well-known formula for the crossing radius R_c (the ionic–covalent crossing distance) via

$$R_c\,(\text{Å}) = \frac{14.4}{\text{IP(A)} - \text{EA(BC)}},$$

where IP(A) is the ionization potential of A and EA(BC) the vertical electron affinity of BC, both in electron volts. Since EA(BC) is a function of the energy release (B–C internuclear distance), R_c can be convoluted over the collision energy to obtain the excitation function

$$\sigma_R(E_T) = \pi\langle R_c\rangle^2 = \pi\int_{R_{min}}^{R_{max}} R_c^2(\mathbf{R})P(\mathbf{R})\,d\mathbf{R},$$

where both limits of the integral are energy dependent.

More elaborate models to calculate other product distributions for direct reactions have been used extensively (see the literature), and there are excellent reviews[14] describing their characteristics. Some of the most used are the DIPR and DIPR-DIP models.[11] The former is an acronym for "direct interaction with product repulsion" and has been applied for many reactions occurring via electron jump, that is, $A + BC \rightarrow |A^+ + BC^-| \rightarrow AB + C$. The basic physics of the model is that the product distribution is essentially determined by the repulsive energy release between B and C while A and B are still separated. Under these circumstances the model predictions for product distribution reproduce trajectory results on the full surface quite well.[9] A simple version of the model[105] gives the translational energy of the product as

$$E'_T = f_m E(1 - 2q_v q_T + q_v^2),$$

where $f_m = AC/(A + B)(B + C)$ and q_v and q_T are model parameters.

Here the capital letters mean the mass of the particle. Once the product translational energy is obtained from the recoil of C, the rotational excitation R', from angular momentum conservation, and the vibrational excitation V', from energy conservation, can also be calculated. When allowance is made for a distribution of repulsive energy release, that is, $P(\mathbf{R})$, the DIPR model can be extended further, depending on the way the distribution is obtained. Although the $P(\mathbf{R})$ distribution can be obtained from experiments by velocity analysis, an important version uses the photodissociation model of Herschbach.[11] This analogy to photodissociation leads to the DIP (distributed as in photodissociation) extension to the original DIPR model, named DIPR-DIP model. It has accounted quite successfully for product distributions in reactions of alkali atoms plus halogen molecules.[100] Recently it has been extended to chemiluminescent reactions where the product aligment as a function of internal state was also calculated.[107d] These models provide exact descriptions of the actual dynamics in the repulsive (bond-breaking) and attractive (bond-forming) interactions if the repulsive energy release is *sequential* and *separable* on the regions of the surface sampled by the trajectory. These two conditions mean first that the BC repulsion must precede the A–B attraction, and second that there is no interaction between C and AB during the B–C repulsion and, on the other hand, no interaction between C and AB during the AB attraction.

Levine and co-workers have used the maximum entropy procedure of information theory[108,112], to obtain the product state distribution and its dependence on the initial translational energy. By using as a constraint a model requiring the least transfer of momenta to the nuclei during the collision (that is, a principle similar to that of Franck and Condon, but for molecular collisions), they obtained for the translational energy disposal[109]

$$P(E'_{\mathrm{T}}) = P^{0}(E'_{\mathrm{T}}) \exp[-\lambda_{\mathrm{T}}(E'^{1/2}_{\mathrm{T}} - \varepsilon^{1/2})^{2} - \lambda_{0}]$$

only for the case RRHO (diatomic + atom product), where

$$P^{0}(E'_{\mathrm{T}}) = \tfrac{15}{4} E'^{1/2}_{\mathrm{T}}(E - E'_{\mathrm{T}})/E^{5/2}$$

is the prior expectation of statistical distribution,[112] E the total energy available to the product, λ_{T} the translational surprisal parameter, and ε the value of E'_{T} at the peak of $\ln[(P(E'_{\mathrm{T}})/P^{0}(E'_{\mathrm{T}}))]$, that is, the surprisal. The value of ε can be obtained either from experiments or from model calculations. The above functionality has been shown to reproduce quite well the product energy recoil distribution in many direct (rebound) reaction mechanisms.[110,111]

2. *Excitation Function Models and Post-Threshold Laws*†

The simplest starting point is the classical model where a collision may be characterized by its impact parameter b and the reaction cross section is given by[2]

$$\sigma_R(E_T) = \int_0^{b_{max}} 2\pi P(b) b \, db,$$

where the reaction probability (opacity function) decreases to zero for $b > b_{max}$. This approach leads to the venerable textbook cases of the line-of-center and centrifugal barrier models[2] if one calculates b_{max} from the maximum value that can surmount a step barrier or a centrifugal barrier. These examples are models 1 and 3 displayed in Table III, where we have summarized the most used models for excitation function data. Many of the expressions are not repeated here for brevity. Table III also lists a few comments and recommendations on model applications and validity.

These two examples belong to the two-body collision category within the dynamic model entry. Further extensions are also considered, such as steric models and three-particle models.

Post-Threshold Laws‡ A special section of Table III is dedicated to post-threshold laws. The initial consideration comes from quantum mechanical threshold laws. In the absence of Coulomb forces at large separations, Wigner has developed the limiting threshold laws for a two-channel system when the momentum in either the initial or the final state of the system becomes very small.[113]

Let k_i and l_i be the momentum and the orbital angular momentum quantum number in channel i. The threshold limit of the scattering amplitude for the exoergic transition $a \to b$, $E_a < E_b$, is

$$\lim F_{ab} \sim k_a^{l_a - 1/2}, \qquad a \to b.$$

The corresponding limit for the endoergic process is

$$\lim F_{ab} \sim k_a^{l_a + 1/2}, \qquad b \to a.$$

These threshold laws are quite general for any kind of bimolecular inelastic scattering event. In the $E_a \to 0$ limit we would have the s-wave limit $l_a = 0$,

†See Refs. 102–138.
‡See Refs. 113–134.

TABLE III
Survey of Current Models for Excitation Functions

Model and reference	σ_R	Comments	Recommendation[a]
	Simple dynamic models and post-threshold laws		
1. Line of centers[2]	$\pi R_c^2\left(1-\frac{E_0}{E}\right)$	$V(R)=\begin{cases}E_0, & R\leqslant R_c\\ 0 & R>R_c\end{cases}$	+1
2. Eu and Liu[117]	$\pi R_c^2\left(1-\frac{E_0}{E}\right)^{1/2}$	Strictly based on scattering considerations	
3. Centrifugal barrier[2]	$\sim\left(\frac{C_n}{E_T}\right)^{2/n}$	Long-range potential $V(R)\cong -C_nR^{-n}$. Only trajectories surmounting the centrifugal barrier lead to reaction.	−1
4. Modified hard spheres[103]	Low energy $\sim\sigma_{LC}\left(1-\frac{yE_T}{2(E_T+Q_{max})}\right)$ High energy $\sim\frac{\sigma_{LC}}{2y}\left(1+\frac{Q_{max}}{E_T}\right)$	$y=\mu/\mu'$; Q_{max} = reaction exothermicity; the product mass (μ') is introduced via angular momentum conservation; σ_{LC} is line-of-centers value.	+2
5. Shin's model[118]	$\pi R_c^2 P(a)$; $P(a)=f(E_T, E_v, \text{masses})$	Three-particle (hard-sphere) interaction. See Ref. 118 for expression of reaction probability.	−1
6. Harris and Herschbach[135]	$\pi R_c^2\begin{cases}\frac{n}{n-2}\left(1-\frac{d}{E_T}\right)\frac{(D-d)}{(E_T-d)^{2/n}}, & n>2\\ \frac{d-\varepsilon}{E_T}, & n=2\end{cases}$	Particular example of centrifugal model using potential $V(r)=d-(d-\varepsilon)(a/R)^n$ for $R\leqslant a$, and $V(a)=\varepsilon$, $V(\infty)=d$.	+2

TABLE III

Model and reference	σ_R	Comments	Recommendation[a]
Simple dynamic models and post-threshold laws			
7. Dynamic model[122]	$\pi R_c^2 \begin{cases} \left(1 - \dfrac{E_0}{E_T}\right) \\ A\left(1 + \dfrac{Q_{max}}{E_T}\right)\dfrac{1}{1 - \gamma E_T} \end{cases}$	A and γ are constants for a given system. Dynamic approximation based on total angular momentum conservation.	+2 +3
8. Levine and Bernstein[116]	$\sigma_0 \left(\dfrac{E_0}{E_T}\right)^{2/s} \left(1 - \dfrac{E_0}{E_T}\right)^{1-2/s}$	Based on microscopic reversibility. Hard-sphere limit is obtained for $s = \infty$. Long-range interaction $V(r) = -C_s r^{-s}$ was assumed for exoergic channel.	+1
9. Steric models[132–134]	$\sim \dfrac{(E_T - E_0)^2}{E_T}$	Based on orientation dependence of barrier to reaction. See Ref. 134 for more elaborated steric models.	+1
10. Wigner laws[113]	$\sim E_T^{-1/2}$, exoergic case $\sim E_T^{1/2}$, endoergic case	Quantum mechanical threshold laws valid for any kind of bimolecular inelastic scattering event in the limit $E_T \to 0$ and s wave.	−1 +1
11. Classical threshold law[114]	$(E_T - E_0)^{5/4}$, $q < 1$ $E_T - E_0$, $q > 1$	Classical expression valid for attractive r^{-4} potential. See text for details and q value.	+1
Statistical models			
13. Eu[119] (atom–diatom)	$AE_T\left(1 + \dfrac{Q_{max}}{E_T}\right)^{2.5}$	Valid in high-energy limit of reaction cross section. Practical applications use n as a free parameter instead of 2.5.	−1 −2
14. Kaplan and Levine[108]	$AE_T^{-1/2}$	Statistical approximation for highly exothermic reactions.	−1

15. Microcanonical transition state theory[126–130,140]	$\frac{(E_T - V^{\ddagger})^n}{E_T}$	Valid for internally cold reagents $n = s + r/2 - 1$. See Ref. 128. s and $\dot{r}$ are vibrational and rotational degrees of freedom of transition state when harmonic approximation is used. See Ref. 130 and Table IV for explicit expressions in atom–diatom case.	+1 −2
16. Transition state cross section for truncated rotator-oscillator model[138]	AE_T^{-1}	Valid for high-energy limit. A is a function of product parameters and spectroscopic constant.	−1
		Electron jump models	
17. Harpooning model[2,137a]	$\pi R_c^2 P$	P is the adiabatic transition probability leading to sudden electron jump at R_c.	−2
18. Modified harpooning model[136]	$\sigma_{harp} P(E)$	σ_{harp} is previous model. $P(E)$ is probability of a second diabatic jump. This model has been used for chemi-ionization reactions.	−2
19. Gislason and Sachs[137b]	$\pi R_c^2 \left\{ \begin{array}{ll} 1 + \frac{\varepsilon}{E_T}, & 0.5 \leqslant E_T \leqslant \infty \\ \frac{2}{2^{1/3}} \left(\frac{\varepsilon}{E_T} \right)^{1/3} & 0 \leqslant E_T \leqslant 0.5\varepsilon \end{array} \right\} + \sigma_{NA}$	Electron jump model where ε and R_c are Lennard–Jones parameters, σ_{NA} is a nonadiabatic contribution calculated by Landau–Zener model.	−1 −2?
20. Modified electron jump model[102]	$\sigma_R = \sigma_{LC} + \sigma_{RC} + \sigma_D$	The three contributions are barrier surmounting, recrossing, and diabatic jump. See text for more details and Ref. 202	+3

[a]The recommendation follows the classification given in Table II, for example, +1 stands for post-threshold rise in $\sigma_R(E_T)$, etc.

and thus the cross section $\sigma \sim (F)^2$ will be given by

$$\lim_{k_a \to 0} \sigma_{a \to b} \sim k_a^{-1}, \qquad \text{exoergic case}$$

and

$$\lim_{k_a \to 0} \sigma_{b \to a} \sim k_a, \qquad \text{endoergic case.}$$

Even though this quantum mechanical threshold law is sometimes useful for collisions yielding free electrons in the final state, it holds in a region smaller than 0.01 eV for typical atomic and molecular proccesses.[115a]

In a classical context the masses of the interacting particles are large enough that quantization of the final orbital angular momentum is not crucial at high energy. In addition many of the classical threshold laws assume that dynamic factors do not change rapidly with energy, and therefore only kinematic factors in the initial state and the average state-to-state transition probabilities of the reaction are considered. The classical threshold law is obtained by summing over a set of final partial waves, or integrating over final impact parameters. Obviously the upper limit of the sum depends on the large-range potential. For an attractive r^{-4} potential (ion–molecule reaction) the threshold law is given by[114]

$$\sigma_R \sim \begin{cases} (E_T - E_0)^{5/4}, & q < 1 \\ E_T - E_0, & q > 1, \end{cases}$$

where

$$q = \mu_{AB}(r^0_{AB})^2/\mu_{A,BC}(r^\ddagger)^2.$$

Here the symbols correspond to the reaction A + BC → AB + C, and $r^\ddagger$ is the distance between the products at the position corresponding to the barrier in the effective potential energy in the final state.

Well-defined threshold laws have been found in electron impact spectroscopy and in electronic excitation of alkali atoms by collision with molecules.[115c] For example, threshold laws of the form $(E - E_0)^n$ fit quite satisfactorily the energy dependence of the cross section for electronic excitation of K atoms by collision with nitrogen ($n = \frac{5}{2}$) and CO ($n = \frac{3}{2}$). It was stated that the n values show an intriguing qualitative correlation with the N_2^- and CO^- "shape resonances" found in electron impact spectroscopy. There the theory of threshold laws suggests $n = l + \frac{1}{2}$, where l is the orbital angular momentum (in units of $\hbar$) of the added electron. No such nice correlations have been found in neutral–neutral interactions for threshold laws, even in those mechanisms occurring via electron jump. Levine and Bernstein[116] deduced a post-threshold law for an endoergic reaction of the form

$$\sigma_R(E_T) = A\left(1 - \frac{E_0}{E_T}\right)(E_T - E_0)^{-2/s}$$

by using a long-range potential $V(r) = C_s r^{-s}$ coupled with the centrifugal barrier model for the exoergic channel and then invoking the microreversibility principle. On the otherhand when a complex mechanism is observed, the most common interpretation of the parameter n is the number of active degrees for the internal motion of the transition state. Statistical theories predict a reaction cross section,[126–131]

$$\sigma_R \sim \frac{(E_T - V^{\ddagger})^{s+r/2-1}}{E_T},$$

where s and r are the number of active vibrations and rotations of the transition state, located at the potential barrier $V^{\ddagger}$.

A concave-up behavior has also been predicted by a hard-sphere steric model developed by Smith and extended by Levine and Bernstein. This functionality is of the quadratic form $\sigma_R \sim (E_T - E_0)^2/E_T$ and is in qualitative agreement with the results found from early trajectory calculations.[133]

Many other post-threshold laws have been reported in the literature. The reader should refer to Table III for those excitation function models recommended with +1 (post-threshold laws), which for simplicity are not repeated here.

The next group in Table III contains electron jump models. Simple nonadiabatic models[137b] have appeared to account for the $M + X_2$ and $M + RX$ (and N_2O) reactions where M stands for alkali and alkaline earth metal, R is a radical and X a halogen.

Many of these treatments consist of a simple two-body collision model plus an extra correction due to the nonadiabatic jump.

Recently a modified electron transfer model considering two reaction channels has been used for several direct reactions.[202] The reaction is initiated in the adiabatic channel through an electron transfer at the avoided crossing of the covalent and ionic configurations. The model accounts for reflection from the inner repulsive potentials and the possibility of a diabatic contribution to the reaction initiated by a transition during the incoming or outgoing phases of the trajectory. The total reactive cross section is given by

$$\sigma_R(E_T) = \sigma_{LC}(E_T)[1 - P_c(E_T)] + \sigma_D(E_T)$$

where σ_{LC} accounts for barrier surmounting (it is identified with the line-of-center model), P_c is the probability of nonreactive reflection from the repulsive wall, and σ_D is the partial cross section for transition to the upper surface. (The Landau–Zener model is used for this contribution.) The model has been applied to M + RX → MX + R reactions among other examples, and some results will be presented and discussed in the next section.

Statistical models are required for complex reaction dynamics, and the next section is dedicated to statistical theories. For direct reactions a simple microcanonical transition state model calculation has been reported, which leads to an excitation function of the form[130]

$$\sigma_R \sim A\left(\frac{E_T + S}{E_T}\right)^n.$$

Explicit values for A and n have been calculated for the "loose" atom–diatom case. A novel feature was introduced by considering a movable location of the transition state along the reaction coordinate. In this treatment S was the potential energy at the transition state location taking the reactant value as the energy origin. Table IV summarizes several values of n and S for atom–diatom reactions.

Obviously at this point more detailed treatments necessarily require a better knowledge of the actual potential energy surface. In this case one can still use two approaches. Either we can use some information from the potential energy surface and carry out a statistical calculation (transition state, RRKM, etc.), or we can go directly to a dynamic calculation (one dimensional or three dimensional) by quasi-classical or quantum methods. We really need a theory, and some of the more important are presented in the next section.

B. Statistical Theories

1. *Microcanonical Transition State Theory*†

Traditionally transition state theory (TST) and RRKM theory have been the two more important theories applied to calculate rate constants. Transition state theory is a statistical mechanical theory of chemical reaction rates based on two fundamental assumptions: the local-equilibrium and the

†See Refs. 24 and 140.

TABLE IV.
Reaction Cross Section Expressions for Different Transition State Models
$\sigma_R = A(E - E_0^\ddagger)^n/E_T$

Model[a]	Location[b]	A^c	n	$E - E_0^\ddagger$
Two-body model	R = 1	$A_1 = \sigma_0$	1	$E_T - \varepsilon_0$
Two-body model	P = 2	$A_2 = \sigma_0 y$	1	$E_T + Q_{max}$
A-D(RRHO)	R	$A_3 = \sigma_0 D/3$	3	$E_T + \bar{E}_I - \Delta E_0^\ddagger$
A-D(RRHO)	P	$A_4 = y\sigma_0 A_I' D/3A_I$	3	$E_T + Q_{max}$
A-D(RR)	R	$A_5 = \sigma_0 D\hbar\omega$	2	$E_T + \bar{E}_I - \Delta E_0^\ddagger$
A-D(RR)	P	$A_6 = 4A_4/A_I' B_e'$	2	$E_T + Q_{max}$
A-D(HO)	R	$A_7 = A_5' B_e/\hbar\omega$	2	$E_T + \bar{E}_I - \Delta E_0^\ddagger$
A-D(HO)	P	$A_8 = A_6 B_e'/\hbar\omega'$	2	$E_T + Q_{max}$
Three-body model		$A_9 = C[2\mu(3)R^{\ddagger 2}/\hbar^2]^{5/2}$	2.5	$E - E_0^\ddagger$

[a]A-D = atom–diatom; RRHO = rigid-rotor-harmonic oscillator; RR or HO = rigid rotor or harmonic oscillator model for diatom involved in transition state.

[b]Location R, or 1 refers to a "loose" transition state close to a reactant configuration; Location P, or 2, the same, but close to product configuration. See Ref. 130 for details.

[c]$\sigma_0 = \pi R^{\ddagger 2}$; $y = \mu'/\mu$; and $D = (\bar{E}_I \Delta E_{int})^{-1}$; C is defined in Ref. 130. Primed quantities stand for products.

nonrecrossing assumptions. The first establishes that transition state species that originate as reactants are in local equilibrium with them. The second assumption is that any system passing through the transition state does so only once. Since transition state theory focuses on the properties of the complex in thermal equilibrium, it cannot be easily used to predict state-to-state cross sections. Marcus[127] has developed a microcanonical version of the transition state theory which, together with his RRKM generalization to yield state-to-state cross sections, provide a central body of theories quite comparable to those designed especially for describing complex reactions, such as the (phase space) theories proposed by Nikitin[125] and Light.[123] This discontinuity between direct and complex mechanism theories has been elegantly connected by Miller with the unified statistical theory (UST), although this has not yet been applied to realistic systems.[129]

Clearly the validity of the first and second assumptions of the transition state theory depends on the location of the transition state. This was at the origin of one of the most interesting developments of the transition state theory as it is the variational criterion for choosing the transition state as indicated by Miller and others.[129] Whereas conventional transition state theory refers to placing the transition state at a saddle point on the potential energy surface, generalized transition state theory (GTST) deals with arbitrary locations of it. Obviously variational transition state theory

(VTST) refers to GTST when the transition state is determined variationally, looking for the optimum transition state that corresponds to a minimum sum of states (microcanonical ensemble).

Our main goal is to provide a clear view of how the excitation function can be calculated following the microcanonical variational transition state theory.

The conventional transition state theory rate constant $K(E)$ for the microcanonical ensemble of A, BC pairs at total energy E is given by

$$K_{\mu\mathrm{CTST}}(E) = \frac{N(E - V^{\ddagger})}{h\rho(E)},$$

where h is Planck's constant and $\rho(E)$ the reactant density of states per unit energy and volume. $N(E - V^{\ddagger})$ is the total number of internal states within the active modes, excluding the reaction coordinate, with energy less than or equal to $E - V^{\ddagger}$, where $V^{\ddagger}$ is the potential energy of the transition state. In variational transition state theory the dividing surface is chosen so as to minimize the product flux. Garret and Truhlar[140] take this dividing surface to be a straight line in configuration space, interesecting the reaction coordinate s at right angles at the point s^*, which is called microcanonical variational transition state. Then the rate constant $k_{\mu\mathrm{v}}$ is given by

$$k_{\mu\mathrm{v}}(E, s^*) = \frac{N[E - V(s^*)]}{h\rho(E)},$$

where $N[E - V(s^*)]$ is the minimum total number of internal states along the reaction coordinate s, $N[E - V(s^*)] = \min N[E - V(s)]$, and s^* is the location of this minimum in $N[E - V(s)]$ and therefore in $k(E, s)$. Now given the rate constant of the microcanonical variational transition state theory, the corresponding reaction cross sections are

$$\sigma_{\mathrm{R}}(E_{\mathrm{T}}) = \left(\frac{\mu}{2E_{\mathrm{T}}}\right)^{1/2} \frac{N(E_{\mathrm{T}} s^*)}{h\rho(E)},$$

where μ and E_{T} are the reduced mass and the translational energy of the reactant, respectively. At this stage, to calculate the reaction cross section by this variational method, the potential energy surface is required to evaluate the number $N(E, s)$ along the reaction coordinate and then to find its minimum value. This procedure has been widely used by Truhlar and coworkers in the collision reaction, A + BC and an explicit expression for the reaction cross section is available for the Morse oscillator signal rotor approximation for BC.[131]

Even in the simple case of having found a good transition state, one may expect that the microcanonical variational transition state theory would not provide exact expressions for rate constants because it still involves the local-equilibrium assumption and additional approximations are required to calculate the flux through the transition state. Furthermore the variational search for the best transition state is usually performed with constraints for practical reasons.

2. *RRKM and Phase Space Theory*†

The RRKM theory is parametrized in terms of long-lived collision complex and critical configuration in that the quasi-equilibrium between activated species and the transition state is the main assumption, that is, the following scheme is adopted for the overall reaction:

$$A + BC \rightarrow (ABC^{\ddagger}) \rightarrow ABC^*$$

$$ABC^* \rightarrow (ABC^{+}) \rightarrow AB + C,$$

where ABC^* is the long-lived complex and ABC^+ ($ABC^{\ddagger}$) the transition state for the exit (entrance) channel. The theory was designed for calculating rates and leads to the following expression for the unimolecular rate constant:

$$k_E = \sum_{E_{vR}^{+}=0}^{E^{+}} P(E'_{vR})/hN(E^{*}_{vR}),$$

where the numerator is the sum of all levels of the critical configuration (excluding the reaction coordinate) with energy E_{vR}^{+}; $N(E_{vR}^{*})$ is the density of states of the complex at energy E_{vR}^{*}, and h is Planck's constant. The random lifetime assumption implicit in RRKM theory relates the average lifetime of the complex to k_E by $\tau = (k_E)^{-1}$. The lifetime of the complex could be deduced from product angular distributions obtained from molecular beam experiments, which provide a test of the energy randomization assumption. In addition the molecular beam method provides data on the translational energy of the reaction products formed via a collision complex (already detected by the forward–backward peaked angular distribution), and again these results may be used to test the RRKM theory.

†See Refs. 123–126.

In the loose transition state no coupling of the radial and internal coordinates exists in the exit channel, and therefore the final state distribution of the product mirrors that of the transition state. Obviously in this case RRKM theory can be used without further approximation for discussing the energy distribution of the products.

In the loose transition state model no added assumptions are needed because AB and C rotate freely, and so their vibrational–rotational motion is uncoupled from the radial–orbital motion in ABC and in the rotation to the products AB + C. Marcus and co-workers[127,128] have developed a microcanonical transition state theory for this case, where the product translational energy is given by

$$P(E'_{\mathrm{T}}) = A(E'_{\mathrm{T}})N^{+}_{\mathrm{vr}}(E_{\mathrm{tot}} - E'_{\mathrm{T}})$$

with

$$A(E'_{\mathrm{T}}) = \begin{cases} 1, & l_{\mathrm{m}} \geqslant l_{0\mathrm{m}} \\ (l_{\mathrm{m}}/l_{0\mathrm{m}})^2, & l_{\mathrm{m}} < l_{0\mathrm{m}}. \end{cases}$$

These expressions are valid for loose transition states when $l_0 \gg j_0$ and $l \gg j$. Here $l_{0\mathrm{m}}$ and l_{m} are the maximum values of the initial and final orbital angular momentum quantum numbers, respectively, and j is the final rotational quantum number.

In general the energy level density may be replaced by its classical limit for most complexes,

$$N^{+}_{\mathrm{vr}}(E_{\mathrm{tot}} - E'_{\mathrm{T}}) \approx (E_{\mathrm{tot}} - E'_{\mathrm{T}})^{m-1}$$

where $m = r_{\mathrm{p}}/2 + s_{\mathrm{p}}$. The subscript p refers to the product, r_{p} being the number of active rotations of the products and s_{p} that of active vibrations. Furthermore the A factor can be evaluated for several long-range potential cases. Table V displays the A expressions for a three-atom loose transition state.[131]

For a tight complex the situation is more complicated. Extensive energy transfer between different degrees of freedom may occur as the system moves from the transition state to the products. As a result the product distribution need not be statistical, even if it were so at the transition state. At this point one needs added assumptions as a dynamic model or a full dynamic trajectory calculation. In the view of Marcus exit channel effects may distort the product translational energy distribution, such as the so-called statistical adiabaticity connecting the bending motion of the transition state with the

TABLE V
$A(E'_T)$ Factors for Several Entrance and Exit Channel Model Potentials.

	Channel[a]						
Case	Entrance	Exit	q	K_1	K_2	K_3	v
1	LRP	LRP	$\frac{n-2}{n}$	$y^{n/(n-2)}\left(\frac{C_n}{C'_n}\right)^{2/(n-2)}$	0	0	1
2	SB	LRP	$\frac{n-2}{n}$	$y^{n/(n-2)}\frac{n-2}{n}\left(\frac{2}{nC_n}\right)^{[2/(n-2)][(n-2)/n]}$	ε_0	0	$\frac{n}{n-2}$
3	SB	SB	1	$y\left(\frac{R_c}{R'_c}\right)^2$	ε_0	ε'_0	1
4	LRP	SB	1	$\frac{y}{R_c'^2}\left(\frac{n}{n-2}\right)^{(n-2)/2}\left(\frac{nC_n}{2}\right)^{2/n}$	0	ε'_0	$\frac{n-2}{n}$

SB = step barrier potential: $V(R) = \varepsilon_0$ for $R \leqslant R_c$ and $V(R) =$ for $R \geqslant R_c$. LRP = Long-range potential: $V(R) = -C_nR^{-n}$. The case when no barrier is present can be obtained using the step barrier and ε_0 or ε'_0 zero. General expressions: $A(E'_T) = (E'_T/B'_m)^q$ for $E'_T < B'_m$ or 1 for $E'_T \geqslant B'_m$; $y \equiv \mu/\mu'$. $B'_m = K_1(E_T - K_2)^v + K_3$. Primed magnitudes stand for products.

free rotation of the products. This effect has accounted for a shift to higher values of the translational energy of the products with respect to the statistical distribution.[128b] The relative translational energy distribution for a tight complex is given by

$$P(E'_T) \sim (E_{tot} - E'_T)^{s+r/2-1} A'_t(E'_T),$$

where s and r are the number of active vibrations and rotations in the complex. The expressions $A'_t(E'_T)$ are more complicated and can be found in Ref. 128.

In the next section the main assumptions and the expression of the phase space theory are presented, and we shall indicate its equivalence to the loose transition state described.

Phase Space Theory. In the statistical theory of Light and co-workers[124] a complex intermediate is considered to be formed under the "strong complex" assumption which states that the mode of decay of the collision complex is independent of its formation, except via conservation of total energy and angular momentum. This theory considers the transition state at the top of the barrier to radial motion formed from the effective orbital angular momentum potential and the long-range attractive potential. The states are then characterized by their asymptotic properties and the transition state is called a loose. In the usual formalism of Pechukas et al.[124] the state-to-state cross section is given by

$$\sigma(n, j, l, n_0, j_0, l_0, E) = \frac{\pi \hbar^2}{2\mu_0 E_0 (2j_0 + 1)} \sum_J (2J + 1) P^J(njl; n_0 j_0 l_0 E),$$

where for fixed energy E and total angular momentum J the reactant (product) state has quantum numbers $j_0 n_0 l_0 (jnlJ)$, and where the sum on J is restricted by $|j_0 - l_0| \leqslant J \leqslant |j_0 + l_0|$; E_0 is the initial translational energy, and $P = 1/N(E, J)$ if both the initial and the final states are accessible from the complex, and zero otherwise. $N(E, J)$ is the total number of states accessible from the complex of total energy E and total momentum J. This expression may be summed in various ways to give a less complete vibrational or rotational distribution. For the particular case where the reactant orbital momentum is large with respect to the rotational angular momentum ($l_0 \gg j_0$), one gets, taking $j_0 \simeq 0$, the following simplification:

$$\sigma(n, l, n_0, l_0, E) = \frac{\pi \hbar^2}{2\mu_0 E_0} \sum_{l_0 = 0}^{L_{\max}} (2l_0 + 1) \sum_{l=0}^{L_{\max}} P(n, l, n_0, l_0),$$

where $L_{\max}$ is the maximum orbital angular momentum quantum number of the reactant leading to reaction. In the loose transition state case the probability P is independent of the internal states of A and BC, and the average reaction cross section is given by

$$\sigma_R(E) = \frac{\pi \hbar^2}{2\mu_0 E_0} \sum_{l_0 = 0}^{L_{\max}} (2l_0 + 1) P_{l_0},$$

where P_{l_0} is the reaction probability for a colliding pair A and BC having an initial orbital angular momentum quantum number l_0. P_{l_0} is a step function of the energy excess $E_0 - B_{l_0}$ (B_{l_0} being the centrifugal barrier) and is given by

$$P_{l_0} = \begin{cases} 1, & E_0 > B_{l_0} \\ 0, & E_0 < B_{l_0}. \end{cases}$$

These conditions lead to the well-known expression

$$\sigma_R = \frac{\pi \hbar^2}{2\mu_0 E_0} (L_{\max} + 1)^2,$$

which typically reduces to

$$\sigma_R = \frac{\pi\hbar^2}{2\mu_0 E_0} L_{max}^2,$$

ignoring the difference between L_{max} and $L_{max} + 1$.

In this strong coupling complex two models are used to obtain the cross section for complex formation. (1) For reaction without activation energy one uses the centrifugal model to obtain L_{max} or the maximum impact parameter b_{max} via the approximation $\hbar l_0 = \mu v b$. (2) For reactions with an energy threshold E_0 one uses a line-of-centers model taking $b_{max} = D(1 - E_0/E_T)^{1/2}$. Finally much of the phase space applications depend on a precise quantization of the energy levels of the reaction products which requires a degree of parmetrization similar to conventional RRKM. Phase space theory is the same as transition state theory if the transition states for both channels are loose and when angular momentum restrictions are fully taken into account.

3. *Excitation Function Based on the Born Approximation*

Eu[119] has used the Born approximation to calculate the energy dependence of the reaction cross section. He derived an analytical form for the state-to-state cross section valid for the high energy limit, which was found to decrease as E_T^{-1} as E_T increases. Those cross section values were summed over all final product states to obtain the following expression for the reaction cross section:

$$\sigma_R = AE_T \left(1 + \frac{Q_{max}}{E_T}\right)^{2.5},$$

where A is a function of the Morse potential parameters of the constituent diatomics for A + BC (that is, AB, AC, BC) and Q_{max} is the reaction exothermicity. This formula is not expected to be valid near threshold, but applies well to the post-maximum energy range of the cross section for the exoergic $K + CH_3I \rightarrow KI + CH_3$ reaction if CH_3 is considered as a single particle. In many applications the exponent 2.5 is considered as a free n parameter.

The Born approximation has also been used by Kafri et al.[111] to predict the product state vibration–rotation distribution in atom–diatom reactions for many reactions leading to functional forms similar in type to those shown in the information-theoretical model described.

C. Dynamic Treatments†

When the potential energy surface is available, it is possible to calculate classical trajectories by solving numerically the equations of motion of the nuclei. There are excellent reviews[5] describing the methodology of this kind of calculation. As far as the excitation function is concerned, the procedure can be summarized as follows. At a given relative velocity and impact parameter, trajectories are calculated for various sets of initial parameters chosen by the Monte Carlo averaging procedure. If the total number of trajectories is $N(b)$ and $N_r(b)$ is the number leading to reaction, then the reaction probability $P(b)$ is

$$P(b) = \lim_{N \to \infty} \frac{N_r(b)}{N(b)}.$$

In practice the upper limit for N is established when $P(b)$ tends to a well-defined constant value. This procedure is repeated for other impact parameters to obtain the reaction cross section $\sigma_R(E_T)$ from

$$\sigma_R(E_T) = \int_0^{b_{max}} P(b, E_T) 2\pi b \, db.$$

Then the entire procedure is repeated for a series of different energies to give the excitation function.

The validity of classical trajectories has been clearly stated, and it is well known that quantum effects (such as tunneling and interferences) play a role, sometimes a very important one, in molecular collisions.

Recently a great effort has been made in the development and applications of the so-called quasi-classical trajectory (QCT) method. This method is important for the study of reactive scattering problems and therefore for the calculation of the influence of the translational energy upon reactive cross sections. Essentially the quasi-classical trajectory method represents the molecular collisions by using the laws of classical mechanics, even though they are prepared in discrete internal energy states corresponding to the quantum states of the molecule. Therefore due to this classical evolution the quantization relaxes as the trajectory begins. The final state quantization has to be made by approximate methods such as a histogram, which causes the internal motion of the product molecule to be smeared out over a continuous range. This is the reason why cross sections calculated by quasi-classical trajectories do not satisfy the detailed-balance and time-reversal requirements.[141]

†See Refs. 5, 141–143.

In a different version of the method one quantizes the actions in the final state rather than the initial one. Thus by running trajectories on the reverse reaction and applying a detailed balance, one can calculate the reactive cross section of the forward reaction. This latter procedure is called quasi-classical trajectory reverse histogram method (QCTRM) whereas the former is called quasi-classical trajectory forward histogram method (QCTFM).[143]

In the past several years there has been an enormous increase in our sophistication concerning an accurate dynamic calculation of reactive collisions from quantum mechanics. In addition approximate quantum treatments for predicting reaction probabilities and cross sections have been developed, such as the J_z conserving, Born approximation, and overlap methods, as elegantly reviewed by Wyatt.[142]

All these techniques have been applied to calculate collinear $H + H_2$, $F + H_2$, $Cl + H_2$, $I + H_2$, $O(^3P) + H_2$, $H + F_2$, $H + Cl_2$, and $Li + FH$ reactions, and in higher dimensionality for the $H + H_2$ reaction. These studies on state-to-state cross sections have provided a clear picture of detailed quantum effects, quantized whirlpools, and so on, and some of these results will be presented later. It should be pointed out that although quasi-classical trajectory results are unable to reproduce such quantum effects, they give a good account of the average trend in the collision energy dependence of collinear state-to-state reaction cross sections at energies well above the reaction threshold, since it is near threshold that these quantum effects tend to be more pronounced.

IV. DIFFERENTIAL EXCITATION FUNCTION: RESULTS AND DISCUSSION

A. Established Representations of Direct Reactions

In many cases the character of the angular distribution of a direct-mode reaction may well be correlated by simple dynamic models, such as spectator stripping or DIPR-DIP approximation.

Figure 3 shows a typical representation of the diatomic reflection approximation used in the electron jump mechanism $A + BC - A^+ + BC^- \rightarrow AB + C$. Morse and Wentworth potentials are often used for the covalent and ionic potentials, respectively.[52] Within this approximation and following the photodissociation model the product average translational energy is given by

$$\bar{E}'_T = \frac{\int_{r_{\min}}^{r_{\max}} E'_T \psi_0^2(r)\, dr}{\int_{r_{\min}}^{r_{\max}} \psi_0^2(r)\, dr}$$

where $\psi(r)$ is the ground state harmonic oscillator function and $r_{\min}$ and

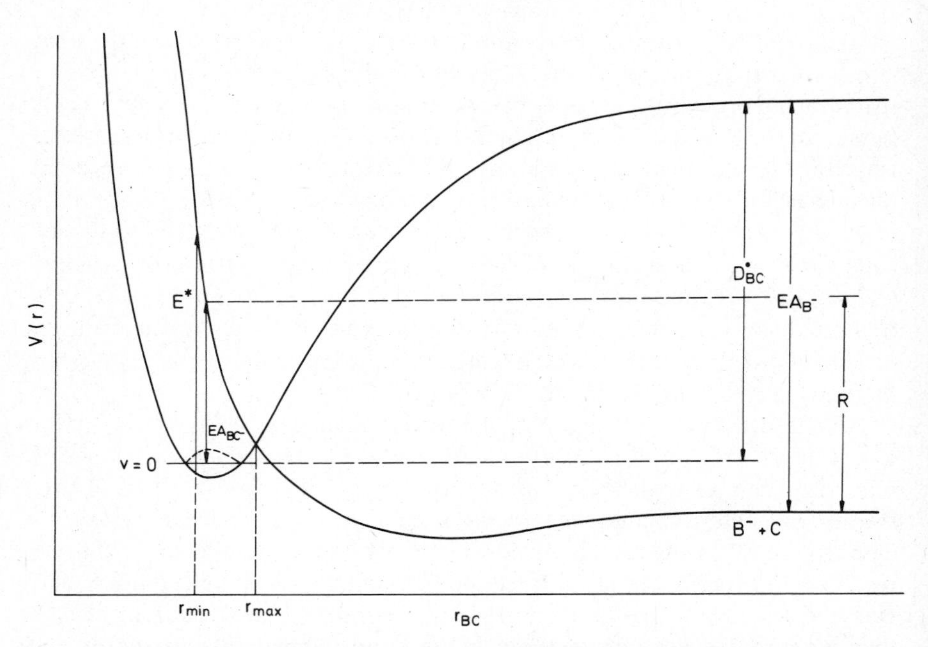

Figure 3. Typical representation of the diatomic reflection approximation between BC and BC^-. For a better illustration the r_{min}, r_{max} limits are shown. The arrow indicates a particular electron jump for the BC equilibrium internuclear distance.

r_{max} are defined as follows: r_{max} is given by the condition $V_{covalent}$ (Morse) $= V_{ionic}$(Wentworth); r_{min} is given by the minimum distance allowed for the electron jump to occur, that is, $V_{ionic}(r_{min}) - V_{cov}(r_{min}) = E^*$, where E^* is the energy consumed for the electron jump bounded by the condition $E_T \leqslant E^* \leqslant E_T + Q_{max}$.

In this model calculation the E'_T values are generated from the momentum conservation condition described in Section III.A.1.

Once the $P(E'_T)$ or $P(\mathbf{R})$ distribution were available, extensive calculations of product energy disposal have been carried out for the M + RX family. (See Ref. 48 for details.) Good agreement between experimental and theoretical values of $\bar{E}'_T$ versus E_T were obtained when $E^* = E_T$, that is, only the collision energy is used for the electron jump. This indicates that the electron jump occurs before the R group goes away from X^-, supporting the well-established opinion that the energy available for the products appears as translational energy via the RX^- repulsion. (See below for this translational adiabaticity.)

One of the great advantages of this DIPR-DIP model is its ability to predict angular and internal distributions of the products. Once $P(E'_T)$ or

$P(\mathbf{R})$ has been generated, the angular distribution in the center of mass can be calculated in a straightforward manner.

Figure 4 shows the center-of-mass angular distribution of KI obtained for $K + C_2H_5I \rightarrow KI + C_2H_5$. Both the theoretical and the experimental results agree quite well for this rebound type mechanism.

A key feature concerning the product angular distribution is the trend of increasing forward scattering of the product which contains the new bond as the collision energy is increased. This effect has even been shown for strongly backward reactions.[9] This is in agreement with the prediction of the DIPR model and many trajectory studies and in fact has led to the definition of a stripping threshold energy T_s as the collision energy above which 90% of the molecular products is scattered into the forward hemisphere. Polanyi[9] has suggested that the magnitude of T_s could be used as an index of the repulsive energy release **R** for these reactions where B and C repel. Only for a few cases, mainly from theoretical trajectories, have the T_s values been determined, as for example in the reaction $Rb + CH_3I \rightarrow RbI + CH_3$, where $T_s > 5$ eV, from a six-atom trajectory study.[203]

From a general point of view increasing either the translational or the vibrational energy leads to an extension of the range of impact parameters

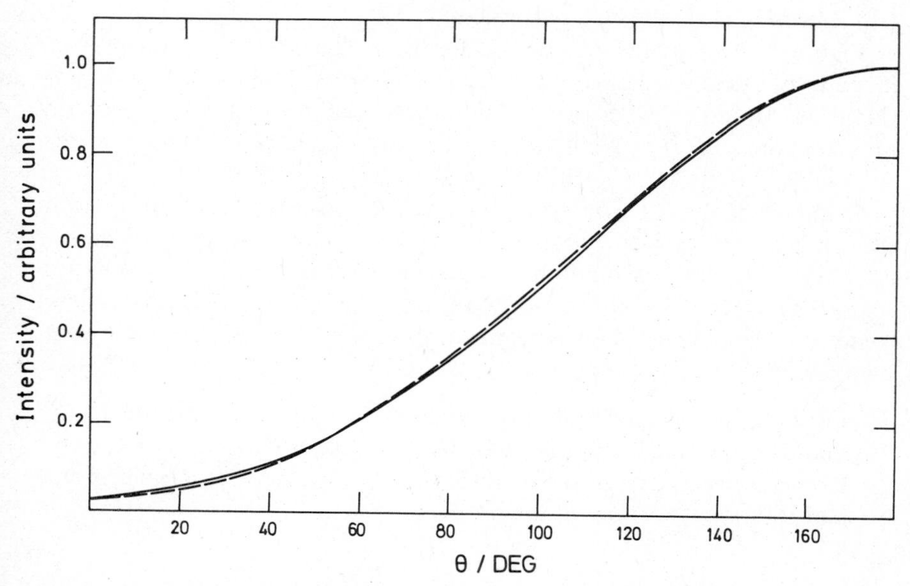

Figure 4. Experimental versus computed angular distribution for the reaction $K + C_2H_5I \rightarrow KI + C_2H_5$. Solid line—center-of-mass angular distribution from Ref. 48; dashed line—center-of-mass angular distribution obtained by the DIPR-DIP model. See Ref. 48 for details.

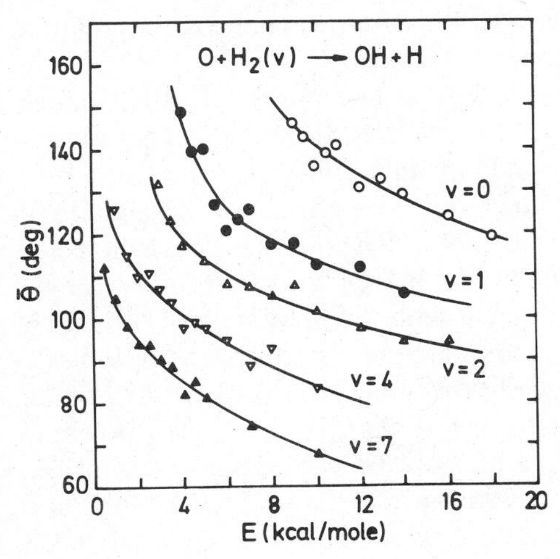

Figure 5. Average scattering angles $\bar{\theta}$ as a function of the collision energy E_T for the reaction $O + H_2(v) \rightarrow OH + H$. Symbols indicate different vibrational states for H_2. From quasi-classical trajectory calculations of Ref. 150.

resulting in reaction, and therefore one may expect an accompanying decrease in scattering angles. This effect has been shown by a recent three-dimesnional quasi-classical trajectory calculation carried out for the nearly thermoneutral reaction of oxygen atoms[150] with hydrogen molecules by using a LEPS form of potential energy surface.

Figure 5 displays the average scattering angles of the product $\bar{\theta}$ as a function of of collision energy for the reaction $O + H_2(v = 0, 1, 2, 4, 7) \rightarrow OH + H$. As shown, $\bar{\theta}$ decreases with the increase of either collision energy or vibrational energy. Similar behavior was obtained in other systems, such as $O + H_2$, $Cl + H_2$, $H + H_2$, and $F + H_2$.[141,150]

In analyzing the reaction dynamics a well-known correlation comes from the work by Polanyi and co-workers on the dynamic problem of barrier crossing. Trajectory calculations on the so-called surface I (and II), representing an early (late) barrier location, have shown the following general trends in the differential reaction cross section for all the mass combinations.[9,14]

1. For reagent energy close to the threshold, reaction transposition occurs, that is, $\Delta E_T \rightarrow \Delta E'_v$ on surface I and $\Delta E_v \rightarrow \Delta E'_T$ on surface II.
2. Excess collision energy above the barrier ΔE_T leads to extra translational and rotational energies in the products

$$\Delta E_T \sim \Delta E'_T + \Delta E'_{rot}.$$

This effect has been termed "induced repulsive energy release,"[14] indicating a shifting toward reaction through more compressed configuration. Furthermore the reagent vibrational energy in excess of that required for the reaction is converted into product vibration energy, such as $E_v \rightarrow E'_v$, which can be understood in terms of a contribution from "induced attractive energy release,"[14] explained by a shift toward reaction through more extended configuration when $E_v \rightarrow E'_v$. A similar tendency has been shown even for the four-center exchange reaction AB + CD → AC + BD.

Examples of these rules have been widely evidenced by many experimental and trajectory studies.[14,141] Recently transposition effects have been reported by the laser-induced fluorescence study of the product vibrational distribution of BaBr formed in the reaction Ba + CH_3Br as a function of collision energy. Figure 6 shows both $P(E_v)$ distributions at several collision energies.[85]

Figure 7 shows the energy disposal as a function of average collision energy for several reactions of the M + HX family obtained by laser-induced fluorescence. Note that the product BaCl (from Ba + HCl) and BaBr (from Ba + BrH) vibrational energy depends very little on the reagent collision energy. A major portion of the excess collision energy is shared by product translation and rotation, in good agreement with the above general rule. The case of the Ba + FH reaction is poorly satisfied, particularly at low collision energies. It was speculated that complex trajectories may couple the reagent modes, vibration and translation, causing the disposal of excess

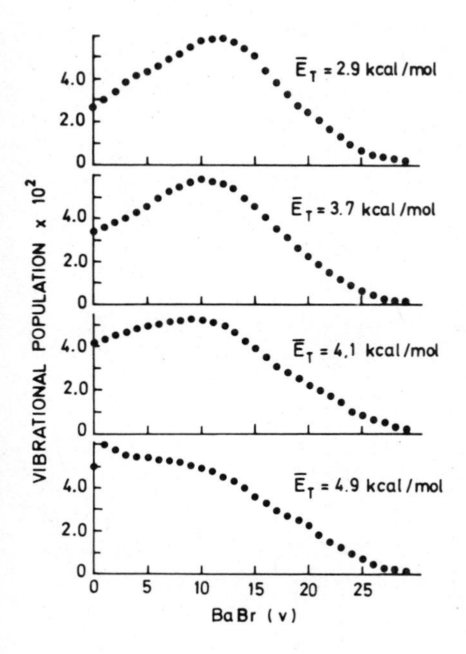

Figure 6. Vibrational population N_v plotted against vibrational quantum number for different collision energies E_T of BaBr formed in the reaction Ba + CH_3Br. Results from Ref. 85.

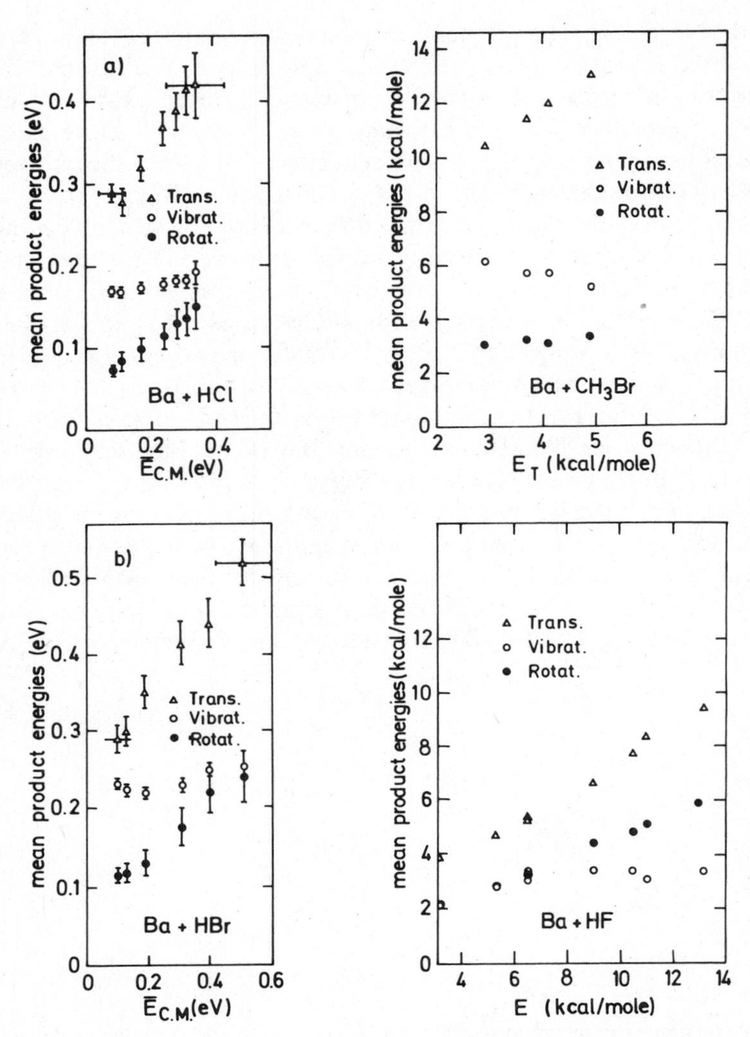

Figure 7. Rotational, vibrational, and translational energy disposal as a function of collision energy for several reactions as indicated. The results were obtained from the laser-induced fluorescence studies of Refs. 79, 80, 84, and 85.

energy in the product modes to be more partitioned. The more exothermic reaction Ba + CH_3Br shows a drastic decrease of vibrational excitation as E_T increases. For this reaction a dual-path reaction mechanism was proposed to explain the observed shift in product vibrational excitation as the collision energy increases. At low collision energy an indirect mechanism takes place in which BaBr is produced through an initial attack of Ba on CH_3 and followed by migration to the Br end (due to the poor electron acceptor capability of the CH_3 radical). This migratory encounter may produce BaBr in highly excited vibrational states. On the other hand the direct mechanism follows an impulsive breakage of the C–Br bond, resulting in little excitation of the products in clear analogy to the electron jump mechanism of the M + RX → MX + R reaction family. Here more than 70 to 80% of the excess collision energy is transferred to product translational energy, as shown in Fig. 8, which is typical of the repulsive character of the potential energy surface already mentioned. In Fig. 9 a detailed example of energy disposal is shown together with various model calculations. The common feature of an abrupt energy release, present in all the models, captures the essence of the (rebound) electron jump dynamics.

Detailed cross sections have also been reported for several reactions by either the LIF[80] or the MS[151] method. Figures 10 and 11 illustrate various examples. In Fig. 10 the resolved reaction cross section of the product (vibrational) state is displayed for the Ba + FH reaction as a function of collision energy. Note that the product vibration first rises and then falls off. Furthermore as the collision energy increases, the breadth of the vibrational distribution also increases.

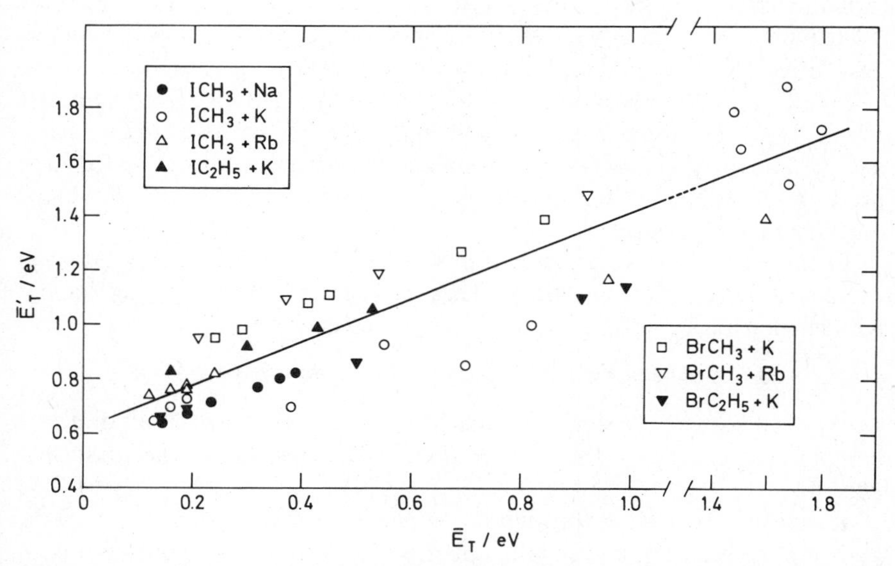

Figure 8. Translational energy disposal for several members of the M + RX family, as indicated. Solid line is drawn for a better illustration. Note the translational adiabaticity following the Polanyi rules. See text for further comments.

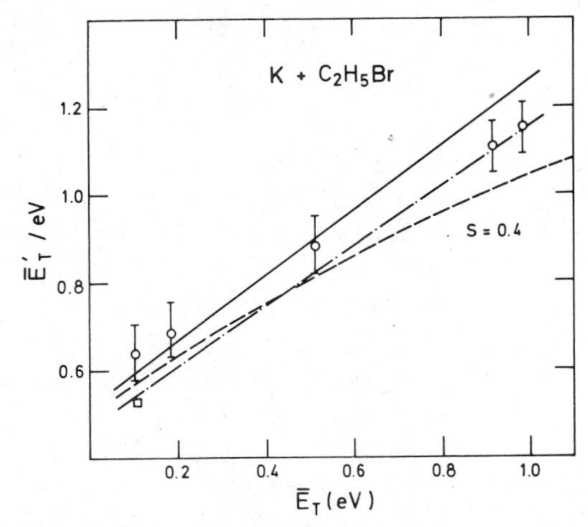

Figure 9. Specific display of the energy disposal for the K + C_2H_5Br reaction. Symbols—experimental results from Ref. 48; lines—different models as follows: solid line—calculated according to collision model given by Ref. 103; dash-dot line—from DIPR-DIP analysis of Refs. 48 and 49; broken line—from impulsive model of Levine and co-workers.[108–111]

Figure 10 (right) shows the detailed cross section for InF($^3\Pi_0^+$) formed from the reaction In + F_2 at two collision energies $E_T = 3.1$ and 7.9 kcal mol.[73] Essentially the average energy dispossed in vibration changes very little, following the same rules as described earlier in this section.

Finally detailed cross sections for HF($v = 1, 2, 3$) formed from the F + $n - H_2$ reaction are displayed in Fig. 11 as a function of the collision energy. The total reaction cross section and the partial cross section for each product vibrational state increase as the collision energy is raised with E (not shown in the figure).

In the normalized cross section plot the branching ratio $(v = 3)/(v = 2)$ decreases with the collision energy. These results are in qualitative agreement with chemical laser and infrared chemiluminescence results.

B. Collision Energy Dependence of Complex Reactions

As it is well known, classical reactions have been classified as either direct or complex, depending on the ratio of the collision lifetime to the rotational period.

Reactions proceeding through a complex, live many rotational periods, and the center-of-mass angular distribution exhibits symmetry about $\theta = \pi/2$; but the actual observation of this forward–backward symmetry does not provide us with any value for the actual lifetime, except that it is longer than a rotational period, typically 1 ps. Suppose that a reaction takes

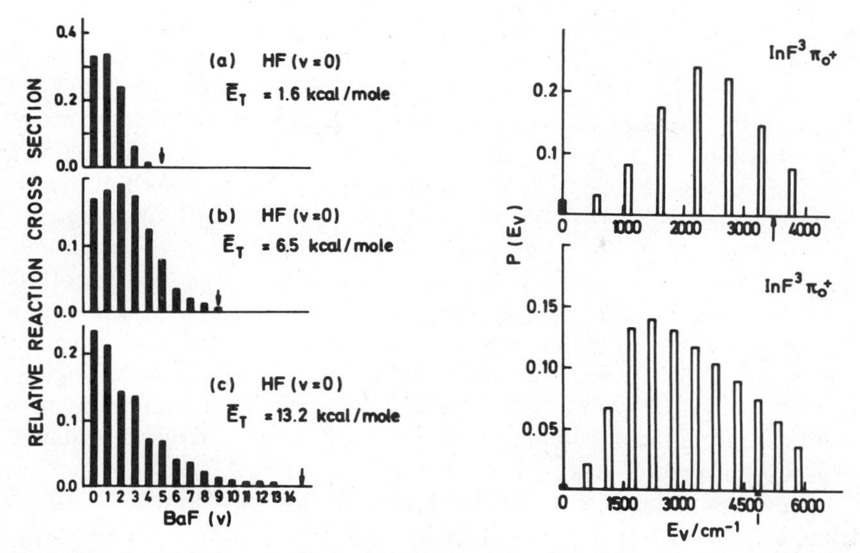

Figure 10. Left: Relative cross section $\sigma_R(v')$ for producing BaF(v') from the Ba + FH reaction as a function of collision energy. Results are from Ref. 79. An arrow marks the highest BaF vibrational level energetically accessible for each reagent condition. Right: Detailed cross section for InF($^3\Pi_0^+$) formed in the reaction In + F_2 at two collision energies E_T = 3.1 kcal mol (top) and E_T = 7.9 kcal mol (bottom). Adapted from Ref. 73.

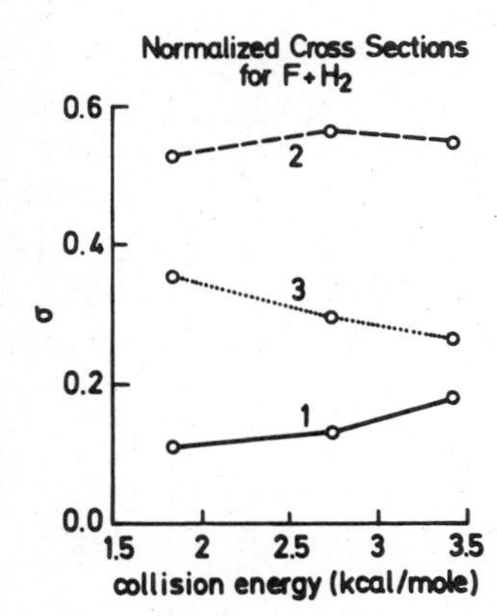

Figure 11. Relative detailed cross section for FH(v') formed in F + H_2 as a function of collision energy. The final v' of the FH product is indicated. Results from Ref. 151.

place via a long-lived complex at collision energy that is much lower than the binding energy of the complex. As the collision energy is increased, one expects some asymmetry to appear in the center-of-mass angular distribution as the lifetime of the complex becomes comparable to or shorter than the rotational period, because even though the complex forms, it decomposes before one full rotational period elapses. This process has been reported as the osculating complex model proposed by Herschbach and co-workers.[11,153] They have shown that deviations from forward–backward symmetry begin to appear when the complex lifetime becomes comparable to the rotational period.

Figure 12 shows two center-of-mass angular distributions for the IF product formed in the $F + CH_3I$ reaction at two different collision energies.[92] A comparison of the falloff functions of the osculating model for various ratios of complex lifetime to rotational period could give the lifetime of the complex if a reasonable estimate of the rotational period is available from a knowledge of the complex moment of inertia and the total angular momentum. When this can be approximated to the initial orbital angular momentum, we can write $\tau_{rot} = 2\pi(I/L)$, and therefore we estimate a lower limit of τ_{rot} from the maximum value of L, L_{max}, obtained from an assumed long-range attraction plus centrifugal term.

In addition to an estimate of the complex lifetime, further insight into the dynamics of unimolecular decomposition can be obtained from molecular beam experiments when the collision energy of the reagents is increased to values on the order of the well depth of the intermediate complex, so that its lifetime becomes comparable to a rotational period. An important ingredient of statistical theories is that energy randomization takes place since intramolecular energy transfer is assumed to be rapid compared to the chemical reaction time. In fact this should hold true only when the lifetime of the complex is expected to be long compared to a rotational period, which should occur if the binding energy of the complex is large compared with the excess energy available for complex dissociation. On the other hand if the total energy available for reaction is much larger than the stability of the complex, its lifetime may be reduced to the extent that energy randomization cannot occur in this time period. The reaction will then be more "direct," especially for systems containing a small number of atoms. Consequently

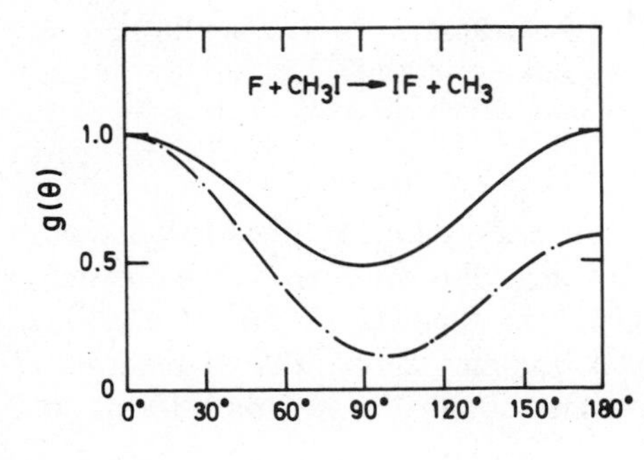

Figure 12. Center-of-mass angular distribution of FI from $F + CH_3I$. —— $E = 2.60$ kcal/mol; –·– $E = 14.1$ kcal/mol. From Ref. 92.

one can expect to find a change in the complex decomposition from statistical to nonstatistical as the collision energy is increased.

The dynamics of the reaction O + XY (X and Y being halogens) exhibit complex reaction dynamics and have been studied as a function of translational energy by Grice and co-workers.[88c,88d,88e] Most follow a long-lived collision complex mechanism at low collision energy, which has been assumed as arising from a hollow on the triplet potential energy surface in the O–X–Y configuration. The trend in the well depth for these reaction is in line with the electronegativity ordering rule proposed by Herschbach, whereby a triplet O–X–Y complex gains stability as the electronegativity of the central halogen atom decreases.[154] This would explain the long lived complex for the O + IBr reaction, whereas the O + Br_2 reaction dynamics show only a short-lived collision complex. Furthermore, emergence of rebound dynamics at higher initial translational energy has been reported for O + I_2 and O + Br_2. This transition from complex to direct trajectories has been probed in many cases, both experimentally and theoretically, and is to be expected because of the reduction in the lifetime of the collision complex.

As an example Fig. 13 shows contour maps of the differential reaction cross section for OBr from O + Br_2 at different collision energies. Both maps show a peak in the forward direction. However, there is much greater scattering into the backward hemisphere as the collision energy is increased. In addition the high collision energy backward scattering exhibits a higher OBr product velocity than the scattering into the forward. These results illustrate the collision energy dependence of the O + Br_2 reaction dynamics. Whereas at low E_T the reaction proceeds via a short-lived collision complex with a lifetime approximately equal to its rotational period, at higher energy there is increased scattering into the backward hemisphere with higher product translational energy. It was shown[88c] that the scattering at low E_T and the forward component at high E_T retain the same dynamics, consistent with randomization of energy over the internal modes of the collision complex. It is interesting to note that the O + IBr reaction (which has the same exoergicity) follows a long-lived complex even at 80 kJ/mol of total energy, as one would expect from a greater well depth with respect to the reaction product. In general for the O + XY systems three factors have been proposed to account for this collision energy dependence of the observed reaction dynamics.

First, the potential energy hollow becomes progressively more displaced into the exit valley along the series of reactions O + ICl, IBr, Br_2, I_2. Also it may be influenced by the well depth relative to the reaction exoergicity, as is the case for the O + I_2 reaction with respect to O + Br_2 because of the greater exoergicity of the former. Finally one can also expect that the greater the moment of inertia of the halogen molecule, the lower is the reaction

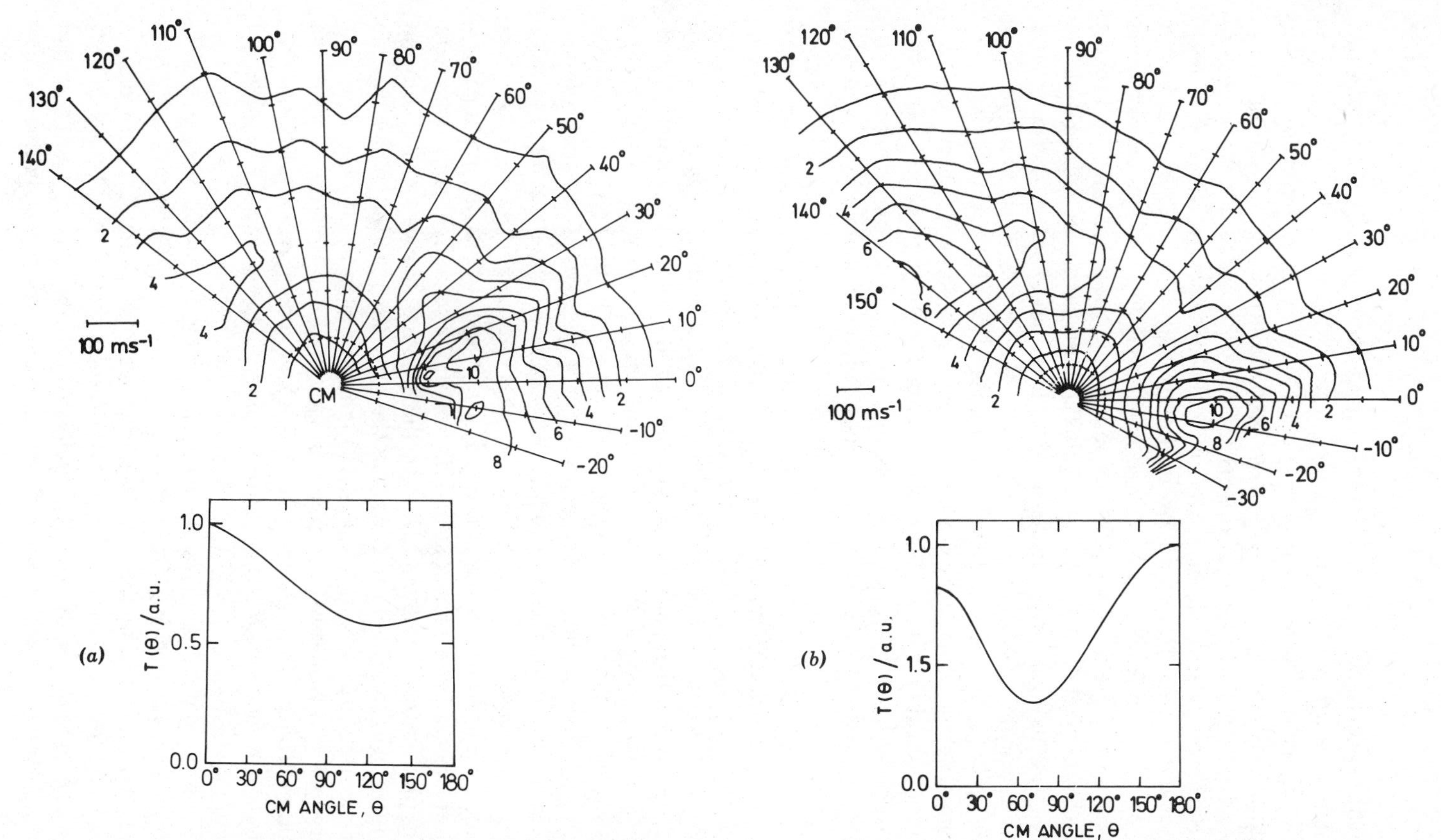

Figure 13. Polar contour maps of OBr flux from O + Br_2. (*a*) Collision energy E_T = 15 kJ/mol. Bottom shows angular distributions (that is, the integral of the left panel over the velocity distribution). (*b*) Same as (*a*), but at collision energy E_T = 40kJ/mol. Note the increased scattering into the backward hemisphere as the collision energy increases. Adapted from Ref. 88c.

probability in larger impact parameter collisions. This is due to the increased difficulty of the halogen molecule to follow the orbital angular motion of the attacking atom in high-energy collisions at large impact parameters, thereby reducing the forward scattered component.

A trajectory calculation by Fitz and Brumer has shown[155] a clear transition from complex to direct mechanism. The study used a symmetrical potential energy surface with a deep potential energy well. For a collision trajectory study in a thermoneutral reaction it was shown that at high translational energy only those complexes persisted which were formed by trajectories directed toward dissociation, that is, exciting the symmetric stretch. On the other hand in those complexes formed by trajectories directed out, the reactive or nonreactive channels do not persist to high relative translational energies.

One clear example of this dynamic transition from complex to direct mechanism as the translational energy increases is the $Hg + I_2 \rightarrow HgI + I$ endothermic reaction, studied by Bernstein and co-workers[87a] over the range of translational energies from 96 to 357 kJ/mol. Figure 14 displays the observed solid-angle differential cross section as a function of relative energy. At low energies above threshold, 111 kJ/mol, there is a backward–forward symmetry of the angular differential cross section, in agreement with a previous work on the same reaction, indicating the formation of a long-lived complex via insertion of Hg into the I–I bond. Also, from 140 to 210 kJ/mol the behavior typical of an osculating complex is observed with a progressively shorter lifetime (3 to 2 ps). At energies above 220 kJ/mol the center-of-mass angular distributions become strongly backward peaked, showing the emergence of a rebound mechanism via the collinear approach of the Hg atom to the I–I bond, which at energies above 310 kJ/mol accounts for most of the observed scattering. This change in mechanism is explained in terms of the potential energy surface for the perpendicular and collinear approach and the existence of a late barrier in the last of the potential energy surfaces.

Also in the case of the spin-forbidden reaction $SO_2 + Ba \rightarrow SO + BaO$ there seems to be a change from complex to direct rebound mechanism as energy increases, though less substantiated as in the previous example. Up to 29 kJ/mol of translational energy this reaction occurs via complex formation, but at 45 kJ/mol BaO is scattered preferentially backward via a direct reaction mechanism.[87b]

Before finishing this section we should keep in mind the difficulty in interpreting the product translational energy distribution for complex reactions whose exit channel transition state is tight, such as the reactions involving methyl elimination, measured by Rice, Lee, and co-workers,[128]

$$F + CH_3RC = CR'R'' \rightarrow (CH_3FRC - CR'R'')^* \rightarrow CH_3 + FRC = CR'R''$$

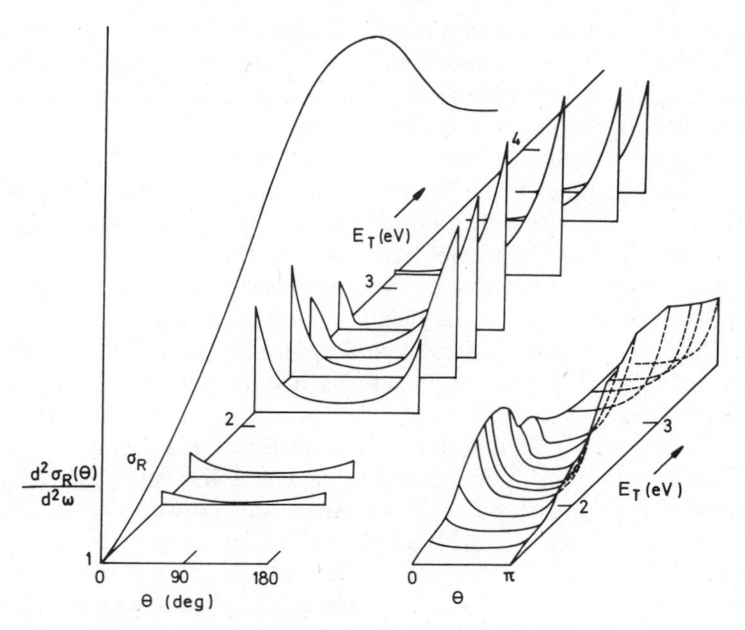

Figure 14. Solid angle differential cross section for the reaction Hg + I_2 → HgI + I. Zero degrees in the center of mass corresponds to the Hg beam. Note the forward–backward symmetry at lower energy, which gives way to preferential forward peaking as the collision energy increases and then becomes backward at higher energy. Solid line on the left shows the integrated reaction cross section as a function of collision energy. The insert on the lower right is a three-dimensional surface representing the center-of-mass angular distribution from threshold to 3.75 eV. From Ref. 87a.

where R, R′, and R″ are alkyl groups. In these cases, as pointed out by Rice et al.,[128] coupling of radial and internal coordinates distorts the final energy distribution, which is shifted to higher values of translational energy than that calculated from phase space or loose microcanonical transition state theory.[128]

C. Reaction Probability Versus Collision Energy: Oscillatory Behavior

One quantum effect which contributes an oscillatory structure of the cross section extending over a wide energy range arises from the interference of reactive and nonreactive scattered waves. Similar to the Ramsauer–Townsend effect in elastic scattering,[144] one-dimensional reactive scattering calculation as well as single partial waves in two- and three-dimensional calculations show a pronounced oscillatory structure of the reaction probability as a function of incident energy.[145]

In the ideal case of collinear heavy + light → heavy atom reactions, the absence of rotational excitation, multiple encounters, etc. supports the experimentally found rule of translational energy conservation

$$\bar{E}_T \simeq \bar{E}'_T$$

and makes it rigorous. Therefore vibrational adiabaticity holds for symmetric collinear reactions such as

$$I + HI(v) \begin{cases} I + HI(v') \\ HI(v') + I \end{cases}$$

This vibrational adiabaticity causes collinear reactions of the above type to become resonant two-state scattering processes. In a similar way as in excitation transfer or symmetric charge transfer in atom–atom collisions, the transition probability for these processes can be calculated by[156,157]

$$P_{vv} = \sin^2[\eta_v^+(E) - \eta_v^-(E)],$$

where $\eta_v^{\pm}(E)$ are the phase shifts, computed for the two one-channel problems that are the symmetric (+) and antisymmetric (−) combinations of the two states. Then at a given kinetic energy one can calculate reaction probabilities by computing phase shifts from the scattering of a particle (three-body reduced mass) by the vibrationally adiabatic potential.

Semiclassical WKB approximation for the two phase shifts can be used to evaluate the reaction probability. The high energy result is[156a]

$$P_{v-v} \simeq \sin^2(a + bE_T^{1/2}),$$

where a and b are energy independent constants. The concept that the resulting probability oscillates as a function of collision energy is in agreement with exact quantum results. Figure 15 shows the reaction probability versus $E_T^{1/2}$ for the reaction I + MuI($v = 0$) → IMu($v = 0$) + I. It has been pointed out that the oscillations are a manifestation of interferences between waves scattered by the symmetric and antisymmetric curves correlating the same vibrational level of the diatom as one goes through the short range of the three-body potential. It is also interesting that an oscillatory reaction probability has been observed in a strictly classical trajectory calculation.[149]

Marcus and co-workers have developed an approximation for the transfer of an H atom in an asymmetric reaction. The ground vibrational state transition probability for the reaction IH + I → I + HI is illustrated in Fig. 15.[157a] Figure 15*c* shows state-to-state quantum-mechanical collinear

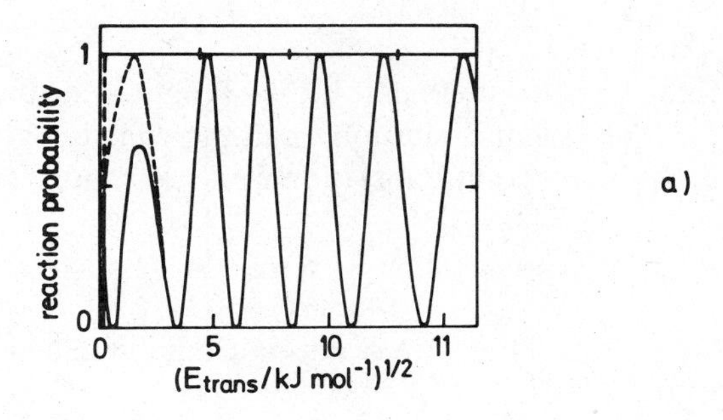

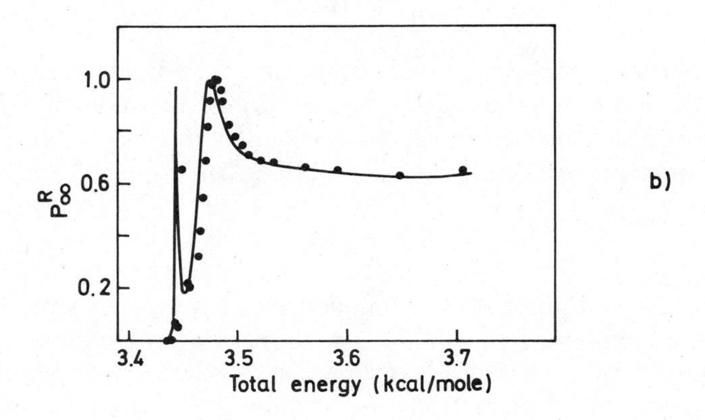

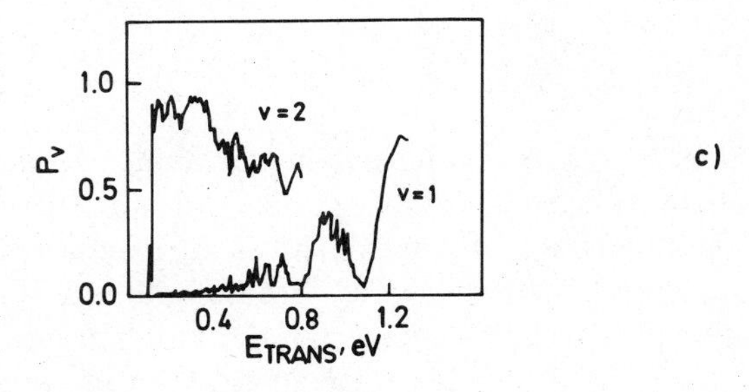

Figure 15. (*a*) Reaction probability $P_{v=0,v=0}$ versus translation energy $E_T^{1/2}$ for the reaction I + MuI(v = 0) → IMu(v = 0) + I. ——, semiclassical WKB approximation[156a] and exact quantum results[156b]; ---, semiclassical calculation[157c] shown only when appreciable difference occurs. (*b*) Ground vibrational state transition probability for the reaction IH + I → I + HI.——, approximate method of Ref. 157a; filled circles are the accurate numerical results of Ref. 156b. (*c*) Collinear reactive probabilities versus translational energy for the reaction Li + FH(v = 1, 2) → LiF + H. Adapted from Ref. 157b.

reactive probabilities for the reaction Li + HF → LiF + H as a function of translational energy. Note that the resonance structure is more pronounced for the case $v = 1$, where the initial vibrational state is below the saddle point.[157b]

D. Reactive Resonances and Adiabatic Barriers

While there is only little direct experimental evidence for the "fuzzy edges" of the $\sigma_R(E_T)$ problem where current theories are insufficient, theory as well as nuclear and atomic physics give us indications of quantum effects that one may expect in total cross sections.

The existence of shape and Feshbach scattering resonances through which access is gained to long-lived compound states is well established through three-dimensional quantum calculation[146,160] as well as through classical experiments in nuclear physics[147] and electron–atom collision physics.[148] The formation and unimolecular decay of metastable molecules has been described in terms of isolated and overlapping scattering resonances. While the experimental observation of resonances in heavy particle collisions is made difficult by their extreme narrowness ($\Delta E \sim \hbar/\Delta\tau \times 10^{-3}$ eV for $\Delta\tau \sim 4 \times 10^{-12}$ s), one must always expect them to be present.

Building potential energy surfaces, vibrationally adiabatic curves can be derived, which in fact are effective potentials for the translational motion of reactants or products described by a single quantum number for the vibrational action. Even in the case of no wells on the potential energy surface, the (vibrationally) adiabatic curves can show wells and barriers as the potential energy surface perpendicular to the translational coordinate widens and narrows, respectively. (These effects would be even more pronounced for curves of higher vibrational quantum numbers.) These wells can support quasi-bound states similar to shape resonances. Reactive scattering through this temporarily bound state can therefore create reactive resonances.

For surfaces in which wells exist the trapping can be explained intuitively. However, for many systems such a well does not exist.[146] The mechanism of the trapping is of a more sophisticated dynamic nature, involving Feshbach processes.[146] In this cases the system and the energy can be trapped in "internal" coordinates. The system can only separate into products when enough energy flows back into a "reactive" degree of freedom. In shape resonances energy can only be trapped in the same degree of freedom that leads to reaction. The question then arises: can dynamic resonances affect the collision energy dependence of the reaction cross section? The answer is positive, but it seems very difficult to distinguish such influence from the main direct scattering contribution. To understand these difficulties we only need to consider the following two effects to compare collinear and three-dimensional resonances.[151]

The first is the contribution of collision with nonzero orbital angular momentum L to the reaction. If an $L = 0$ resonance occurs at energy E, then at approximately $E + BL(L + 1)$ a quasi-bound state can be formed by a collision of orbital angular momentum L, where B is the rotational constant of the complex.

The second effect, less important, is that the energy of the quasi-bound state will be higher in the three-dimensional case by the additional zero-point energy complex of the bending modes. Consequently the larger number of partial waves involved in reactive scattering allows the resonance to be accessed over a wide energy range, appearing as broad, smooth features in the excitation function, and many difficulties can therefore be expected in distinguishing such resonances from direct scattering.

Note that although the reaction probability $P(b, E_T)$ as a function of impact parameter b and energy would be an oscillatory function of energy for fixed b (and also an oscillatory function of b for fixed E), the averaging over b, such as $\sigma_R(E_T) = \int 2\pi b\, db\, P(b, E)$ would tend to quench most of the oscillatory structure. So the expectation is that the energy dependence of the reactive cross section at a fixed scattering angle will show the type of oscillatory structure found in collinear results.

Accurate quantum mechanical reactive cross section calculations have been done[159–160] for the $H + H_2$ reaction, but only for the total angular momentum quantum number $J = 0$. These calculations have been extended through that first resonance. The results are shown in Fig. 16. It indicates that going from one-dimensional to three-dimensional the position of the first resonance is shifted upward in energy in steps of about 0.050 eV, which is close to the zero-point energy of the bending mode of the transition state. Note that for the reaction $H + H_2(v = 0) \rightarrow H_2(v' = 1) + H$ shown in Fig. 16*b*, where the results are summed over all the final product rotational states, three-dimensional effects do not wash out the first resonance. This is therefore a case where resonances present in collinear models do not disappear in the three-dimensional world. *It is still premature to exclude broad and easily observable structures at higher energies due to interference effects.*

Several calculations of the adiabatic curves for the M5 surface in the $F + H_2$ system[161–162] have shown that the lowest energy quasi-bound state of FH_2 is confined within the strong coupling region by the $H_2(v = 0)$ adiabatic barrier on the reactant side, and the $HF(v = 3)$ barrier on the product side. Moreover, quantum mechanical scattering calculations have predicted quantum effects in the form of dynamic resonances.[161] Thus collinear reactive scattering calculations have obtained sharp resonances in the reaction probability as a function of collision energy.[158–160]

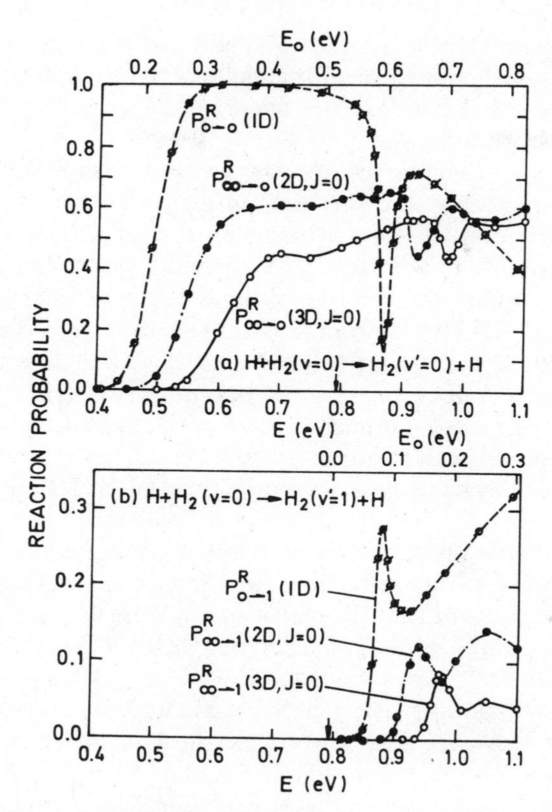

Figure 16. Collinear (one-dimensional), coplanar (two-dimensional), and three-dimensional reaction probabilities for the H + H_2 exchange reaction as functions of the total energy E and initial relative translational energy E_0. $P^R_{0\to0}$ and $P^R_{0\to1}$ are the collinear reaction probabilities from the $v = 0$ state of the reagent to the $v' = 0$ and $v' = 1$ states, respectively, of the product. $P^R_{00\to0}$ and $P^R_{00\to1}$ are the two- or three-dimensional (as specified) reaction probabilities for the total angular momentum $J = 0$ state of the reagent to the $v' = 0$ and $v' = 1$ states, respectively, of the product, summed over all product rotation states within a given vibrational manifold. From Ref. 160.

Quantum calculations carried out on the M5 surface of the F + H_2 system predict that as the collision energy is raised from 2 to 3 kcal/mol, the $v = 2$ distribution of FH formed by H + H_2,[175–177]

$$F + H_2(v = 0) \rightarrow FH(v) + H$$

shifts from backward-peaked to sideways-peaked, whereas the $v = 3$ distribution remains backward peaked.

Lee and co-workers have carried out a high-resolution crossed molecular beam study for this reaction[151] in an attempt to observe the effects of dynamic resonances. Differential cross sections and kinetic energy distributions were obtained for each vibrational state.

The HF product angular distribution for F + para-hydrogen at 1.84 kcal/mol is shown in Fig. 17. The Newton diagram is also included. The laboratory distribution shows considerable structure. Note that peaks in the angular distribution (indicated by the dotted lines) occur at laboratory angles where the laboratory velocity vector is nearly tangent to a Newton circle, that is, $\Theta_{lab} = 28$ and 45° from backward $v = 3$ and $v = 2$, respectively; for $\Theta_{lab} = 8°$, forward scattering belonging to $v = 3$ was observed over the range of collision energies from 0.7 to 3.4 kcal/mol. This striking feature has not been seen in previous experimental work or in any scattering calculation by classical or quasi-classical trajectory studies and has been explained, in terms of dynamic resonances, as a major feature of the HF($v = 3$) angular distribution.

Calculations based on the M5 surface indicated that collisions with the highest collision energy of 3.4 kcal/mol would correspond to impact parameter $b \leqslant 1$ Å. It does not seem likely that such intense forward peaks would result only from collisions at impact parameters as low as 1 Å. Indeed classical calculations[162] carried out for F + D_2 angular distribution at 6 kcal/mol showed no product at $\Theta \leqslant 45°$, even at impact parameters $b = 1.45$ Å. As pointed out by Lee and co-workers, the forward scattering can be

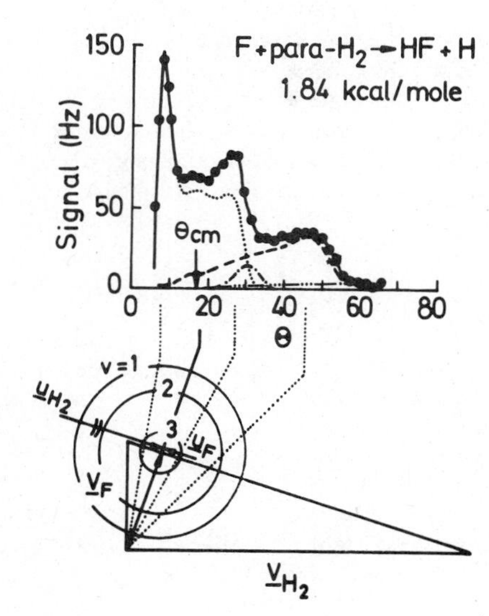

Figure 17. Laboratory angular distribution for F + p-H_2 at 1.84 kcal/mol and Newton diagram. Both data and calculated Laboratory distributions are shown. ●, data, ——, total calculated, —···—, $v = 1$; ---$v = 2$; ···$v = 3$; -·- $v = 3'$. From Ref. 151.

enhanced if the resonance lifetime is comparable to the rotational period of the complex.

These authors have also pointed out that the discrepancy between observations and quantum predictions can be attributed to inadequacies in the potential energy surface, suggesting several modifications to the M5 surface in these calculations.

It has been shown that the $v = 3$ product forward scattering is enhanced as the collision energy is raised, which is to be expected in light of the basic properties of dynamic resonances. Since at higher collision energies the resonance is accessed by higher L, it would result, among other factors, in a quasi-classical state with shorter rotational periods. If the lifetime of the quasi-bound state remains the same, the complex would rotate more before dissociation takes place and therefore increase the amount of HF product scattered into the forward hemisphere. Furthermore at sufficiently high energy, the trend toward increased forward scattering should reverse as the contribution from resonances becomes less important.

This picture is quite consistent with the treatment of resonant periodic orbits by Pollak and others.[163–166] Within the strong coupling region these bound classical trajectories (subjected to semiclassical quantization conditions) match very closely the resonance energies.

Pollak and Pechukas[163] have shown by using the dynamic criterion that trajectories leaving the transition state should not return. The best transition state in the microcanonical (fixed-energy) version of the theory corresponds to a periodic trajectory that vibrates back and forth between the two, equipotential at a given energy. Such trajectories, called pods (periodic orbit dividing surfaces), may be found numerically and can be viewed as unstable bound states of the triatomic system across the interaction region of the potential energy surface. These pods have been classified as attractive or repulsive, depending on whether a trajectory released from the same total energy contour at an infinitesimal displacement from one of the pod's turning points will turn either toward or away from the pod when it reaches the opposite turning point.

In general one may find pods having integer semiclassical quantum numbers v,

$$\int_{\text{pods}} \vec{p}_s \vec{ds} = \left(v + \frac{1}{2}\right) h\nu$$

where $\vec{p}_s$ and $\vec{ds}$ refer to the mass-weighted Jacobi coordinates (x, y). They are the momentum vector and the length element, respectively, in (x, y) space along the pods, and the integral is taken over a complete period of the periodic orbit. Pollak has shown that repulsive quantized pods define the vibrationally adiabatic barrier for reaction, that is, the location and length of the maximum of a vibrationally adiabatic potential energy surface, which closely corresponds to the dynamic barrier observed in quantum scattering calculations, as the resonance effects we are discussing.

Classical trajectory calculations predict small variations in the product angular distributions among the reaction $F + H_2$ and its isotopic variants. However, recent state-selected differential cross sections measured for the reactions $F + D_2$ and F + HD carried out by the crossed molecular beam method[152] have shown that whereas the DF product from both reactions was predominantly backward scattered, the HF angular distribution from F + HD showed considerable forward scattering ($v = 3$). These results support the same conclusion drawn in the $F + H_2$ study, since they agree with the predicted dependence of dynamic resonance effects on isotopic substitutions. Moreover, the experimental product angular distribution for DF($v_f = 1, 2, 3, 4$) measured at two different collision energies was followed by a quantum infinite-order sudden approximation (IOSA) carried out on the Muckerman V surface.[166b] This study showed that at the lower energy all four distributions were essentially backward peaked, but for the higher energy only the three lowest states yielded angular distributions, in qualitative agreement with the experimental ones. The highest state yielded a pure backward distribution where the forward experimental peak was missing.[166b]

In conclusion, the vibrationally state-resolved differential cross section for $F + H_2$ described represents one of the most detailed experimental studies in reaction dynamics to date, and the explanation of the results in terms of dynamic resonances provides a direct link between the details of the unknown part of the potential energy surface and the asymptotic scattering states observed in the experiment. Attempts to study the transition state via emission and absorption have not yielded a great deal of information since these experiments typically involve electronic transitions between two unknown surfaces. Therefore the observation of resonances should provide a more sensitive probe of the strong coupling region of the potential energy surface yielding information on these quasi-bound states. It has been stated that the "observation of reactive resonances is equivalent to performing vibrational spectroscopy of the transition state."[151]

V. EXCITATION FUNCTION AND ITS SHAPE: CAN IT BE A LEARNING ROUTE INTO DYNAMICS?

The answer is yes, but care should be taken in interpreting its shape evolution. Ambiguity may be present, and important dynamic information can be lost by the angular integration leading to the total reaction cross section. It is best to perform a simultaneous analysis of the differential excitation function, when available, since it provides much more detailed information. In general the collision energy dependence of $\sigma_R(E_T)$ has an important dynamic interpretation near threshold, and it is easy to interpret when the exchange reaction is the only channel accessible over the energy range studied, and only one potential energy surface is involved along the relevant reaction processes. For example, a maximum in $\sigma_R(E_T)$ can, in principle, be attributed either to a dynamic recrossing mechanism or to the onset of collision-induced dissociation. Nevertheless the significant body of data now available makes it easy to isolate some common translational features, family trends which could be interpreted in terms of relevant topological factors of the potential energy surface or idealized collision models. The following sections are dedicated to discuss some of the more interesting examples.

A. Family Trends and Bond Characterization in the Excitation Function

The traditional picture concerning the overall characterization of the excitation function describes $\sigma_R(E_T)$ as increasing (positive) for endoergic reactions and decreasing (negative) for exoergic reactions. This typical behavior is displayed in Fig. 18 and follows the simple functionalities described by Levine and Bernstein.[116] Further features arise when exoergic reactions show energy thresholds in which cases the cross section may or may not present a post-maximum decline. Moreover endothermic or thermoneutral reactions may exhibit a slight maximum[54] or high concave-up curvature[64] in the post-threshold range. We will concentrate our attention

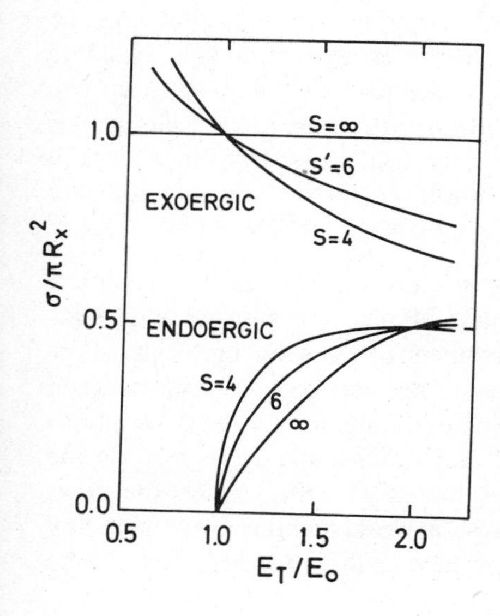

Figure 18. Exoergic and endoergic cross sections for various asymptotic long-range potentials in the exothermic channel. $s = 4$, ion–molecule interaction; $s = 6$, neutral–neutral interaction; $s = \infty$, Arrhenius or finite-range interaction. Plotted are the models represented by cases 1 and 3 of Table III. From Ref. 116.

in the following sections on discussing many of the available results. Therefore in the present section we will present an overall view of the main systematic trends already observed for well studied families of reactions. This will provide us insight for discussions in the following sections.

One of the better studied families so far is the alkali plus alkyl halide reactions. Figures 19 and 20 show a composition of several excitation functions which have been grouped into different trends for better illustration. The following comments can also be extended to the electronically excited rare gas reactions [as, for example, Xe($^3P_{2,0}$)], since a clear analogy between these and alkali metal reactions has been shown to occur over a wide energy range.[49]

The following family trends can be noted.[49]

a. Attacking Atom Effect.[43,58,59] Energy thresholds for reactions with the same molecule seem to decrease in the sequence Na > K > Rb > Cs(Xe*). This E_0 evolution has been explained by prestretching associated with interaction outside the vertical electron jump distance, which would reduce the negative electron affinity, favoring the attack of a heavier atom at a given translational energy.

For a reaction involving an outer harpooning, that is, in the long-range interaction potential, some expectation about the threshold comes from Magee's formula[104] $R_c = 14.4/(\mathrm{IP} - \mathrm{EA})$, where a lower threshold (higher cross section) is to be expected as the ionization potential diminishes. This trend is in good agreement with experimental findings.

b. Alkyl Group Effect.[31c,43] Some alkyl group effects have been reported as follows. (1) The heavier the alkyl group, the broader is the backward cone of the MX angular distribution. (2) More energy from the available exoergicity goes into internal energy of the products as the alkyl group size increases. (3) As the size of the alkyl group increases, lower reactivity is found. In particular a comparison of the reaction cross sections of methyliodide and ethyliodide with potassium atoms was carried out at a collision energy of 0.4 eV. The measured branching ratio was Γ(methyl/ethyl) = 1.4.[38]

c. Halogen Effect.[43,65b] It has been pointed out that whereas the excitation function for the alkyl iodide reactions with alkalis and metastable rare-gas atoms and for dissociative electron attachment by the same molecules show very low thresholds, in some cases there is a steep decline in the zone between 0.2 and 0.5 eV of collision energy. The excitation function for the corresponding alkyl bromide processes shows higher E_0 values and an increase in the post-threshold zone, typically from 0.1 to 0.3 eV, after which the cross section levels off or in some cases shows a weak maximum without any sharp post-maximum decrease. An important analogy to this effect can be

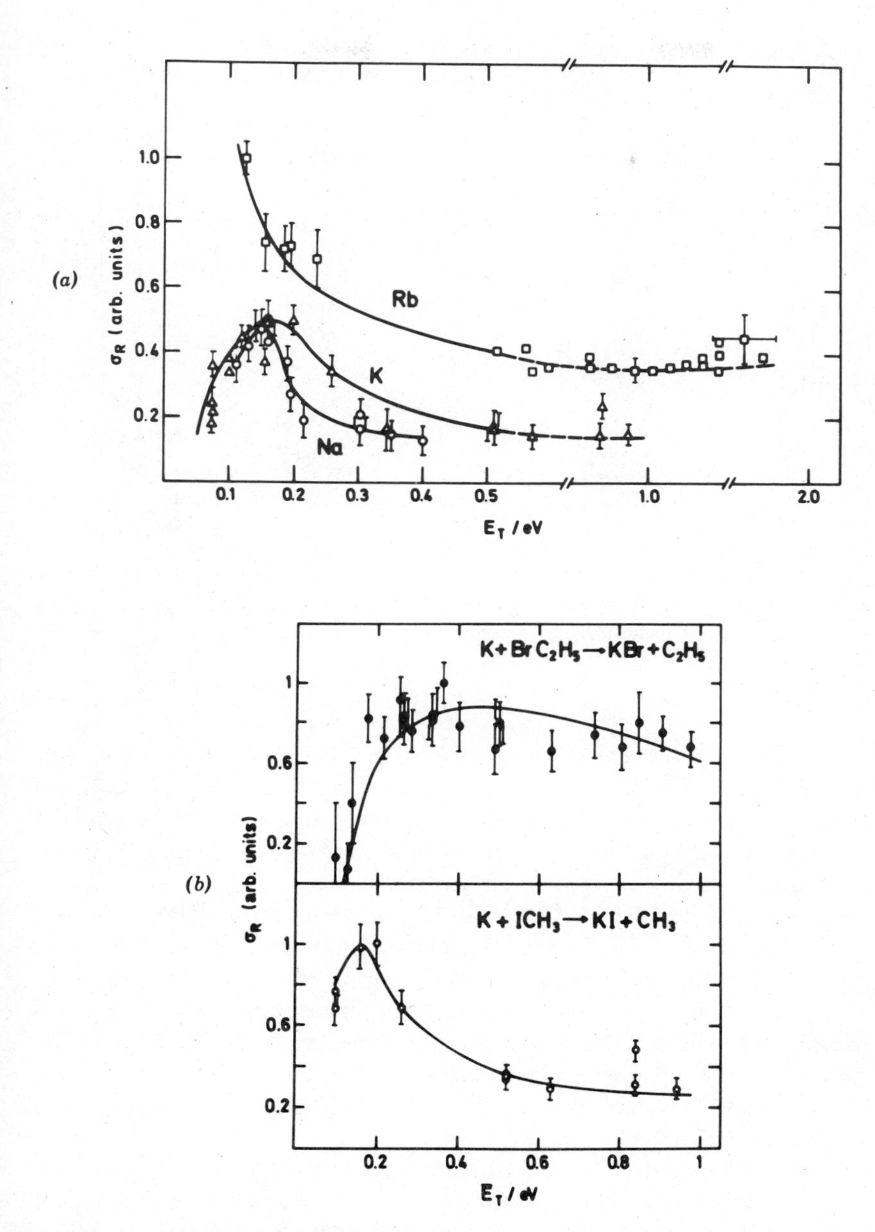

Figure 19. Composition of several M + RX → MX + R excitation functions as indicated. (*a*) Attacking atom effect. (*b*) Halogen effect. See next figure for references.

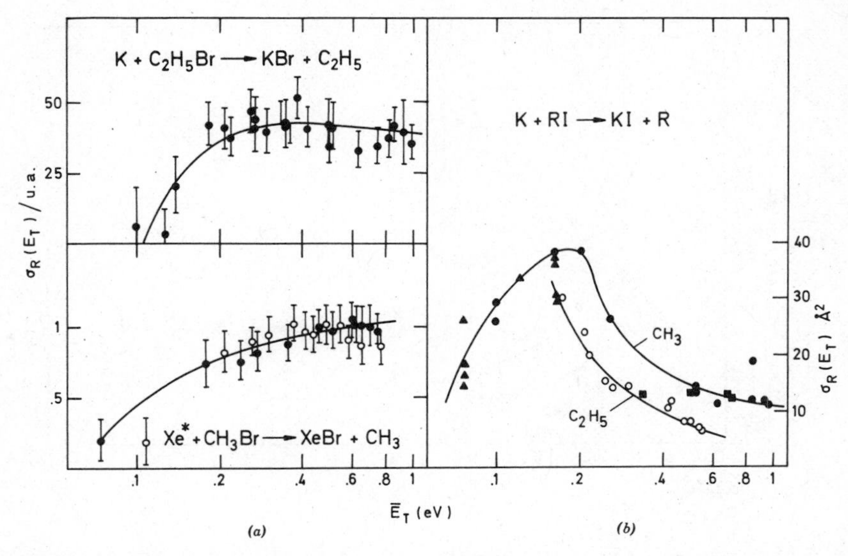

Figure 20. Same as Fig. 19. (*a*) Alkali and metastable rare gas comparison. (*b*) Alkyl group effect. Adapted from Refs. 48, 49, and 38.

found in the potential energy surface for the reactions K + XH → KX + H (X being F, Cl, H) calculated by Zeiri and Shapiro by using a VB semiempirical method.[64b] In these calculations they have shown that the barrier height goes down and the saddle point moves from the exit to the entrance valley with increasing halogen atomic numbers.

If the electron transfer mechanism is the first step of the M + RX → MX + R reaction dynamics, it is interesting to compare both reactive and dissociative electron attachment processes. Figure 21 shows such a comparison. The points correspond to reactive scattering measurements; the curves correspond to dissociative electron capture experiments. Note the main difference, as the collision energy increases, associated with the C–X bond, which is why we have grouped separately the C–I bond processes and those involving the C–Br bond.

The main conclusion to be drawn from an inspection of Fig. 21 is that there is a close analogy between reactive processes and electron attachment processes with regard to the low-energy part of their excitation functions. It also confirms the difference between the C–I and the C–Br shapes (different bond-breaking reaction dynamics) and once again the analogy between the reaction of alkali metals and that of metastable rare-gas atoms over a wider energy range.

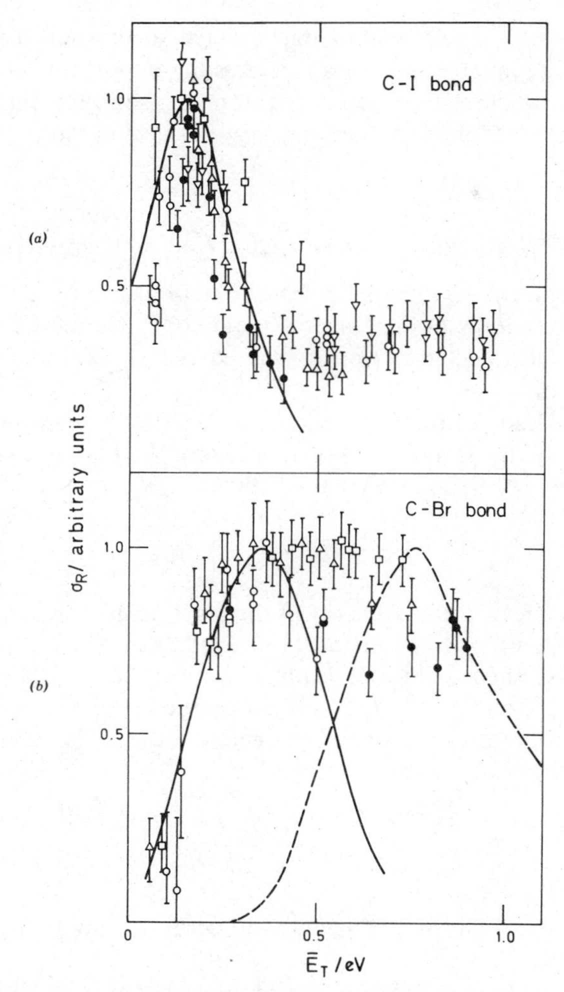

Figure 21. Experimental excitation function for reactive and dissociative electron attachment processes. (*a*) ○, $CH_3I + K \rightarrow KI + CH_3$; △, $C_2H_5I + K \rightarrow KI + C_2H_5$; ▽, $CH_3I + Rb \rightarrow RbI + CH_3$; □, $CH_3I + Xe(^3P_{2,0}) \rightarrow XeI + CH_3$; ●, $Na + CH_3I$. The solid curve corresponds to the I^- ion current versus electron collision energy reported in Stockdale et al.[168] for the process $CH_3I + e^- \rightarrow I^- + CH_3$. (They give a maximum at 0.15 eV and a FWHM of 0.30 eV.) In the case of the reaction $CH_3I + Rb \rightarrow RbI + CH_3$ we plotted σ_R values for E_T lower than 1.0 eV. It was shown that σ_R increases with E_T for $E_T > 0.9$ eV. (*b*) □, △, $CH_3Br + Xe(^3P_{2,0}) \rightarrow XeBr^* + CH_3$ with 330 and 490 K nozzle beam (CH_3Br) temperature, respectively; ○, results for the reaction $C_2H_5Br + K \rightarrow KBr + C_2H_5$. Both curves correspond to dissociative electron capture; the solid curve results for the process $CH_3Br + e^- - Br^- + CH_3$ (values reported by Stockdale et al.; maximum at 0.35 eV, FWHM 0.40 eV); dashed line is the excitation function of Christophorou et al.[169] for $C_2H_5Br + e^- \rightarrow Br^- + C_2H_5$. In both parts the curves have been scaled to 1 at the maximum; the points have been scaled for best to the solid curves. Adapted from Refs. 48 and 53. See references cited therein.

The failure of the electron attachment shape to reproduce the reactive one as the collision energy increases indicates the important role associated with the presence of the electron-donating atom (the bond-forming process on exit channel topology of the potential energy surface) in molecular reaction dynamics.

B. Translational Energy Thresholds: What Can Be Learned About Them?

The translational energy threshold for a reaction can be defined as the minimum kinetic energy of the reagents required for reaction, and its value can lead to important information on the energetics and dynamics of the chemical reaction.

While the meaning of the translational energy E_0 is clear, the Arrhenius activation energy E_a is not simply a measure of microscopic collision energetics. It has been shown that the activation energy is given by[170,171]

$$E_a = \langle E^* \rangle - \langle E \rangle,$$

where $\langle E \rangle$ is the mean thermal energy of the reacting mixture, $\langle E \rangle = \frac{3}{2}RT$ if the internal degrees of freedom are neglected, and $\langle E^* \rangle$ represents the average energy of those collisions leading to reaction. This latter quantity depends on the threshold energy and on temperature as well as on the functional form of the reaction cross section $\sigma_R(E_T)$. For an excitation function of the form

$$\sigma_R \sim \frac{(E_T - E_0)^n}{E_T}$$

an analytical expression for the activation energy E_a has been derived,[171]

$$E_a = E_0 + RT(n - \tfrac{1}{2}).$$

1. *Energetic and Molecular Properties of the Reagents.* Measurements of translational energy threshold provide an excellent means of determining the electron affinity of the target molecule BC collisional ionization processes,

$$A + BC \rightarrow A^+ + BC^-$$

since the threshold $E_0 = \mathrm{IP(A)} - \mathrm{EA(BC)}$. Typically, what is done is to measure the threshold for the above processes for several attacking atoms (alkali atoms, for instance) whose ionization potentials are well known, leading therefore to several independent values of the electron affinity of BC.

In addition molecular electron affinities can also be obtained from endoergic charge transfer reaction thresholds using negative atomic ion beams, such as in

$$X^- + BC \rightarrow X + BC^-,$$

the endoergicity of which is given by E_0 = EA(X) – EA(BC). The measured threshold energy will be equal to E_0 provided that there is a negligible activation barrier for the above reaction, and therefore EA(BC) can be calculated directly from threshold measurements. A consistency check can be made comparing electron affinity data from collisional ionization and charge transfer experiments.

On the other hand, for an exoergic reaction occurring via electron transfer mechanism, such as $M + RX \rightarrow M^+ + RX^- \rightarrow MX + R$, the observed energy threshold is associated with the negative vertical electron affinity of the molecule. Since the potential curve calculated for the molecular anion may lie above that of the reactant molecule for the equilibrium internuclear distance, an intersection zone situated above the zero point level, as shown in Fig. 3, can create a low translational threshold. As proposed by Wu,[44c] the threshold is given by the energy difference between the intersection of the covalent and the ionic curves and the ground vibrational state of the molecule. Table VI lists calculated and experimental energy threshold values for several reactions. The calculated values were obtained from the energy difference between the crossing point $V_{AB} = V_{AB^-}$ and the ground state energy level in the Morse well.

The agreement is satisfactory, but if the negative vertical electron affinity of the molecule were the only important requirement in the threshold energies, it would be independent of the attacking atom for a given molecule

TABLE VI
Calculated versus Experimental Energy Threshold Values and Related Parameters for Several Reactions[a]

				$E_0 \times 10^2$ (eV)	
System	ν(C–X) (cm^{-1})	r^0(C–X) (Å)	D^0(eV)	Calculated	Experimental
$CH_3Br + K$	611	1.94	2.94	20	24
$C_2H_5Br + K$	562	1.94	2.84	10	12
$CH_3Br + Rb$	611	1.94	2.94	20	20
$CH_3I + K$	533	2.14	2.45	1.6	3.3
$CH_3I + Rb$	533	2.14	2.45	1.6	3.2

[a]Adapted from Ref. 49.

reaction, which is not exactly the case. These differences may be due to prestretching associated with reagent interactions outside the vertical electron jump distance that would reduce the negative electron affinity and, among other effects, would favor the heavier attacking atom at any given collision energy, as is generally observed. Also this model neglects the influence of the bond "forming" process (that is, the influence of the exit channel of the potential energy surface) in molecular reaction dynamics, and therefore one may expect important deviations from the observed features of $\sigma_R(E_T)$ as the collision energy increases.

An interesting evidence for this electron transfer mechanism, including the correlation between the electron affinity of the molecule and the reaction threshold, has been reported in the chemiluminescent $Ba + N_2O \rightarrow BaO^* + N_2$ cross section.[64] A significant cross section enhancement was observed through N_2O bending vibration by heating the N_2O reactant from 283 to 613 K. It was pointed out that the N_2O electroaffinity of the linear N_2O is negative by 0.15 to 1 eV, and it increases (becomes positive) on bending, when the mode population (010) rises from 9 to 27% over the temperature range mentioned. Therefore as the electroaffinity increases with N_2O bending, so does the electron transfer distance and with it the chemiluminescent cross section due to the threshold reduction. Figure 22 shows this

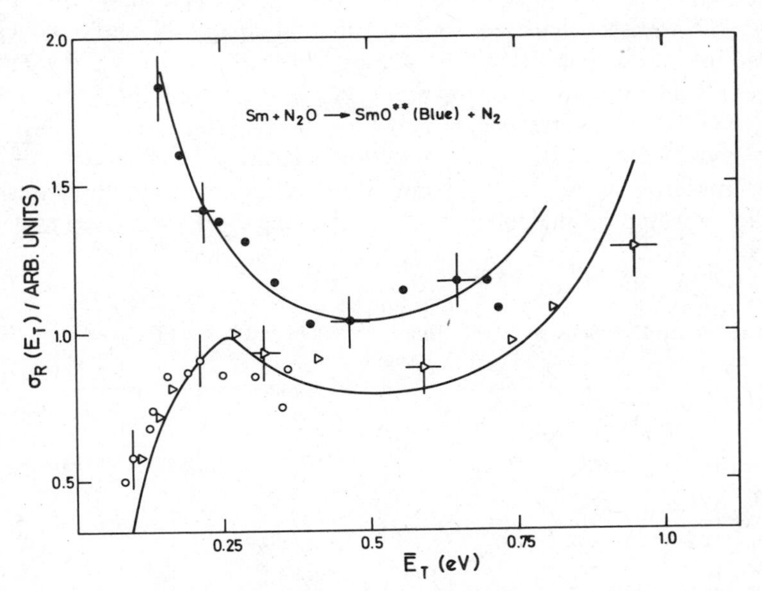

Figure 22. Excitation function for blue channel of the Sm + N_2O reaction. Solid symbols, vibrationally hot (T_v = 613 K) N_2O; open symbols, vibrationally cold (T_v = 218 K) N_2O. solid lines are calculated from the dynamic model in Ref. 122. See details therein.

cross section enhancement for the $Sm + N_2O \rightarrow SmO^* + N_2$ reaction cross section.[65]

For complex reactions the determination of the translational threshold may give important information on the estimation of the binding energy of radical or stable intermediates. An interesting example is the determination of the binding energy of CH_3IF from the threshold behavior of the endoergic reaction $F_2 + CH_3I \rightarrow CH_3IF + F$, from which an energy threshold of 11 kcal/mol was observed. Since the F_2 bond energy is 36 kcal/mol, a lower limit of 25 kcal/mol was deduced for the CH_3IF binding energy. This method shows great promise and the capability of the molecular beam technique to provide new information on the stability and binding energy of radical molecules.[92]

2. *Energy Threshold and Potential Energy Surfaces: Effects of Dynamics and Kinematics.* Since the threshold energy for a reaction can be defined as the minimum kinetic energy of the reagents required for reaction, it may depend on other energy modes of excitation such as vibration. In fact, the question of which energy mode is more effective in promoting chemical reaction is a crucial one in interpreting molecular reaction dynamics.[2,3]

One clear interpretation is based on a rule by Polanyi and co-workers on the effect of barrier location on the potential energy surface.[9,14] Two well surface types designated I and II were introduced. While surface I is typical of an early barrier (exoergic reactions), surface II corresponds to a late barrier type, associated with an endothermic reaction. They showed that in surface I translational energy is more effective than vibrational energy, and vice versa, for surface II. In a condensed expression this can be written as $\sigma_R(T) > (<) \sigma_R(V)$ for surface I (II), where T and V stand for energy allocated in translational and vibrational motion, respectively. The interpretation is straightforward; the energy is more effective when it is allocated in the necessary coordinate to surmount the barrier, that is, translation (A–BC) for early barrier or vibration (AB–C) for late saddle point. A great number of experimental and trajectory calculations have been reported documenting such a general rule, in good agreement with the predictions. An example is that recent trajectory calculations for the collinear LiFH carried out on an ab initio potential energy surface have shown that vibrational energy is so vital for the reaction that at lower energy the reaction probability is zero (0.002) unless almost all the energy required to cross the barrier is in the form of reagent vibration: "the vibrational threshold is equal to the barrier height for this endothermic reaction."[200]

3. *Orientation Dependence of the Reaction Threshold.* In general the threshold energy may depend on the angle of attack due to the A–C interaction (A–B–C) for fixed AB and BC distances. Figure 23 shows the barrier

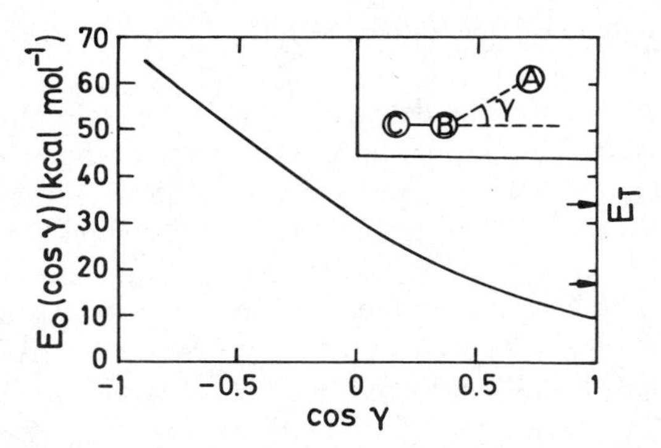

Figure 23. Barrier height E_0 versus cos γ for ab initio H_3 potential energy surface. Adapted from Ref. 133. Right-hand ordinate is the translational energy E_T taken from the floor of the potential energy surface in the asymptotic region. Arrows indicate E_T values at which comparisons are made with trajectory computations.

height E_0 versus cos γ for the ab initio H_3 potential energy surface,[133] where γ is the angle of attack defined in the inset. A well-established model dependence of the threshold energy upon the angle of attack is given by[132,133]

$$E_{th} = E_0 + E'(1 - \cos\gamma),$$

which is valid for an atom–diatom reaction. A recent inversion has been made for the Rb + CH_3I reaction and yielded the reaction probability as a function of the angle of attack.[173] See Section V.C for more details.

4. *Quantum Reaction Thresholds and Quantum Effects.* An interesting point can be stated about the quantum origin of the reaction threshold and whether or not quasi-classical trajectories can reproduce the quantum results. The question is the following. If one assumes that motion perpendicular to the reaction coordinate is vibrationally adiabatic, then the reaction threshold energy is simply equal to the sum of the classical barrier E_c plus the difference between the saddle point and the reagent vibrational energies for adiabatically correlating vibrational states. Thus from a transition state point of view one can write

$$E_0 = E_c + E_0^{\ddagger} - E_0^{r},$$

where E_0^{r}, $E_0^{\ddagger}$ and E_0^{p} are the zero point energies for BC, ABC, and AB, respectively, as portrayed in Fig. 24.

Therefore if accurate potential energies are available, one can calculate E_0 and compare it with the experimental values. Conclusions on (1) the

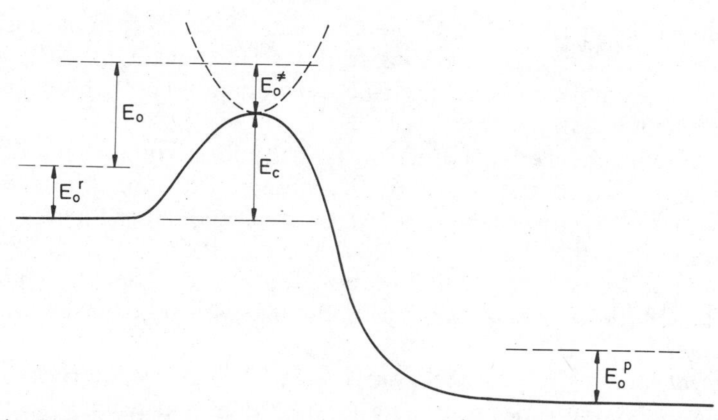

Figure 24. Typical energy profile for transition state configuration. E_0^r, $E_0^{\ddagger}$, and E_0^p are zero-point energies as indicated.

energy partition between motions parallel and perpendicular to the reaction coordinate during the reaction coordinate, and (2) the (vibrationally) adiabatic nature of the threshold can be obtained, providing insight into and a better understanding of the reaction dynamics. In this sense and with regard to point (1) it should be stated that what is most interesting concerning the motions perpendicular to the reaction coordinate is the adiabatic character of the bending motion, since there is no zero-point energy associated with this motion in the reagents or products. (These bending motions correlate with AB or BC rotations.)

Figure 25 shows the total reaction probability as a function of reagent

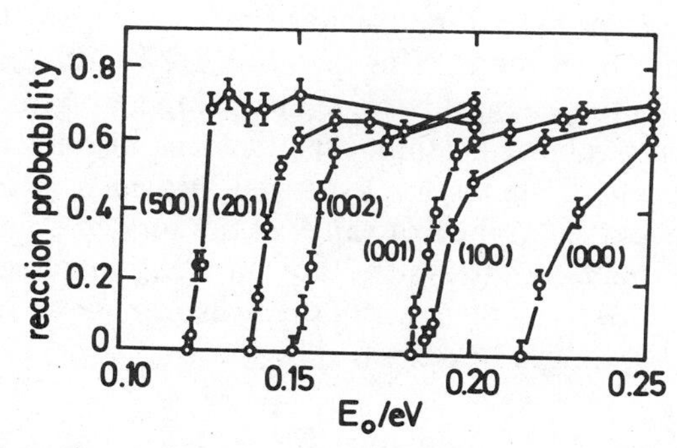

Figure 25. Total reaction probability as a function of reagent translational energy E_T for collinear reaction $O + CS_2$ with CS_2 initially in several vibrational states as indicated. The relative energies of these states are 0.0, 0.08, 0.20. 0.36, 0.40 and 0.40 eV, respectively. From Ref. 174.

translational energy for collinear $O + CS_2$ with CS_2 initially in the indicated vibrational states.[174] It is interesting to point out that the lowering of the threshold for excitation of the (100) state is comparable to that for the (001) state, despite the fact that $\nu_3 \gg \nu_1$.

Also, the threshold for (500) is substantially lower than that for (002), even though these states are nearly degenerate. Therefore the trajectory analysis reveals that the asymmetric stretch is vibrationally adiabatic during the reaction. It is not mixed with motion along the reaction coordinate, and thus it does not promote the reaction. On the other hand, asymmetric stretch is vibrationally adiabatic during the reaction. It correlates with the CS product.

A recent study[175] of the reaction $H + H_2 \rightarrow H_2 + H$ for $H_2(v = 0, 1)$ using quasi-classical trajectories and accurate quantum reactive scattering calculations has shown that the vibrationally adiabatic threshold is the correct one, that is, quantum threshold is obtained by the quasi-classical trajectory method only if the adiabatic bending motion is incorporated. This is so even when the motional time scale is not appropiate because the bending period is much larger than the duration of the reactive collision ($\tau \sim 10^{-14}$ s). The bending period τ_B is about 10^{-14} s (at the saddle point) and increases to infinity at infinite separation of the reagents or products. What is the origin of these three-dimensional quantum thresholds? Schatz[175] has indicated that the additional constraint is the uncertainty principle which forces the minimum energy in the bending motions to be the bending zero-point energy (in the harmonic approximation). This is in fact equivalent to the constraint imposed by the vibrationally adiabatic theory as it is incorporated in the quasi-classical trajectory calculations.

To summarize, Table VII displays the types of information that can be obtained from reaction threshold measurements.

It is well known that light atom exchange reactions may be affected by tunneling in the threshold range. This quantum effect admits reactions at energies smaller than the classical barrier. Tunneling may be of importance for hydrogen exchange and its isotopic reactions and also for the reactions $H + HL \rightarrow HH + L$ if L stands for hydrogen, provided that the barrier crest is likely to be located in the exit valley of the surface. In these circumstances the reduced mass of the system moving along the reaction path at the saddle point is essentially the mass of the hydrogen atom. In this region the reaction cross section can be described by

$$\sigma_R = \sigma_0 P_t(E),$$

where σ_0 is a scaling factor (normalization cross section factor) and $P_t(E)$ the penetration factor. Jakubetz has proposed the following expression for

TABLE VII
Information Available from Reaction Threshold Measurements

Process (experimental and/or calculation)	Information that can be obtained
Endothermic reactions, exoergic neutral–neutral reactions	Energetics information; electroaffinities of target molecule; comparison with activation energies
Complex stable formation	Binding energies of radicals or reaction intermediates
Reaction with oriented molecules	Angle dependence of energy barrier (pair anisotropy); different reaction mechanisms upon angle of attack
Accurate threshold determination plus near-threshold behavior	Comparison with accurate quantum values: 1. Vibrational adiabaticity during reaction; adiabatic barriers 2. Energy partitioning between parallel and perpendicular motion along reaction coordinate 3. Tunneling effect: test of accurate potential energy surface and quantum theories

this penetration factor[174]

$$P_t(E) = \left\{1 + \exp\left[\frac{2\pi(E_0 - E)}{\hbar|v^*|}\right]\right\}^{-1}$$

This equation has been obtained for the simple inverted parabola

$$V(s) = E_0 - 2\pi\mu|v^*|^2 s^2$$

as a model for the shape of the barrier along the reaction paths, s being measured from the saddle point.

Figure 26 shows experimental versus calculated cross sections using the above equation for the $K + FH \rightarrow KF + H$ system for the classically forbidden processes. It seems that the $v = 0$ cross section of the above reaction is dominated by tunneling.

Exact quantum reaction probabilities and reaction cross sections have been calculated as a function of collision energy for the reactions $H + H_2 \rightarrow H_2 + H$ and $D + H_2 \rightarrow DH + H$. For the H_3 system quantum and quasi-classical calculations over a wide collision energy used the potential energy surface of Porter and Karplus and, more recently, the Siegbahn–Liu–Truhlar–Horowitz (SLTH) potential energy surface, which

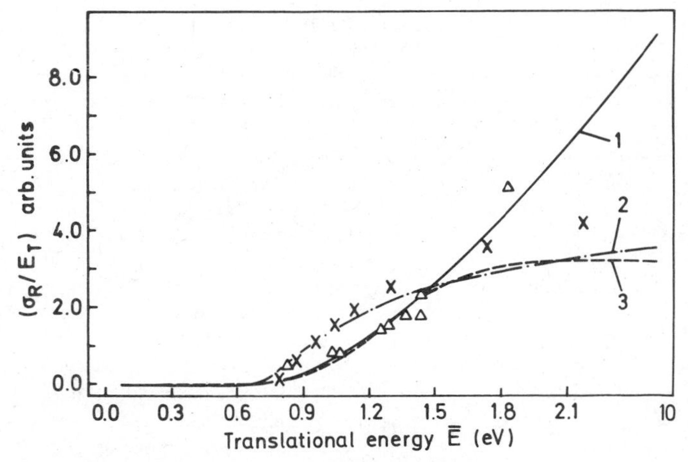

Figure 26. Excitation function for the reaction K + HF(v = 0) → KF + H (△). The crosses give the properly scaled result of phase space calculations. Line 1, threshold law calculation from $\sigma_R \sim E_T - E_0$; line 2, line-of-centers model calculation; line 3, tunneling model calculation from Ref. 54. Adapted from Ref. 54.

is believed to be the most accurately known one for any neutral reactive system. The accuracy of the quantum calculations is estimated to be 2% for the collinear probabilities, 5% for the three-dimesnional system $H + H_2(v = 0)$, and 15% for the three-dimesnional system $H + H_2(v = 1)$.

Figure 27 displays quasiclassical and quantum total integral reactive cross sections associated with the initial states v, j = (0, 0) and (1, 0). Note that for $v = 0$ there is good agreement of the quasi-classical and the quantum results down to a total energy of 0.59 eV. Below this tunneling effect the quantum cross section bends away, extending the reactive trajectories to 0.51 eV. An effective cross section threshold is defined as the value of E_T where the linear position of $\sigma_R(E_T)$ versus E_T extrapolates to zero. The classical and the quantum effective thresholds are $E_f(C) = 0.56$ eV and $E_f(Q) = 0.57$ eV. It is also shown that agreement between classical and quantum results is much worse for effective thresholds $v = 1$, concluding that the classical value is affected by a probable error of roughly 0.06 eV.

Recently a quasi-classical trajectory cross section calculation versus collision energy has been done for the same system, but using the SLTH potential, and the result was compared with that for the Porter and Karplus (PK) surface.[178]

The SLTH cross sections lie below the PK results at all collision energies studied (0.31 to 1.5 eV). However, the differential cross sections (product angular distributions in the center of mass) calculated at $E_{c.m.} = 0.65$ eV

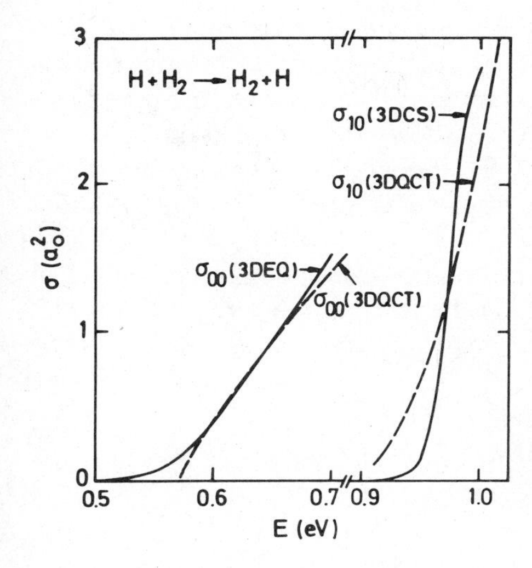

Figure 27. Total integral reactive cross sections σ_{00}, σ_{10} for the reaction $H + H_2(v = 0, 1) \rightarrow H_2 + H$ as a function of total energy. Left—comparison of three-dimensional exact quantum (3DEQ) and quasi-classical (3DQCT) results. Right—3DQCT results are now compared to three-dimensional coupled states approximation (3DCS) but for reactions $v = 1$. Adapted from Refs. 175 and 177. See text for comments.

have the same shape for both PK and SLTH surfaces. This prompted the suggestion that molecular beam experiments, which are able to prove relative angular distributions with more precision than an absolute cross section, might not be able to detect the difference between the two surfaces mentioned. Let us hope that highly structured differential cross sections (see dynamic resonances in Section IV.C) could do it.

At present only three molecular beam experiments have been carried out on the H_3 system, two for the reaction $D + H_2$ at collision energies of $E_T = 0.48$[179b] and 1.0 eV[179a] and a third for $H + T_2$ at an energy of 0.70 eV.[50] The observed differential cross sections have shown that whereas at 0.48 eV backward scattering of HD was observed, it scattered predominantly sideways as the collision energy was raised to 1.0 eV[179a] This trend was confirmed by classical trajectory calculations performed on the SLTH potential energy surface.

C. What Determines the Post-Threshold Behavior?

1. *Expected Behavior.* As presented in the section on post-threshold laws and from phase space theory considerations, the expectation for the collision energy dependence of a complex mechanism with (without) a reaction threshold is a positive (negative) energy dependence. Bernstein and co-workers[180] have measured both differential and integral reaction cross sections for the reactions

$$RbF(J) + K \rightarrow Rb + KF, \qquad \text{exoergic}$$

$$CsF(J) + K \rightarrow Cs + KF, \qquad \text{endoergic.}$$

These two processes are members of the so-called complex mode MX + M′ reactions, occurring via a long-lived ($\tau \sim 1$ ps) complex. The reactive branching ratio as well as the integral reaction cross section were measured as a function of collision energy. Figure 28 shows these excitation functions. Note the increase (decrease) in σ_R for the endoergic (exoergic) reaction with increasing E_T, as can be expected from statistical theory considerations in a complex formation.

Once experimentalists have measured a translational threshold, or even less, if they have just observed that the reaction cross section rises with the collision energy, they almost automatically try to fit the data to some empirical expression of the form

$$\sigma_R \sim \frac{(E_T - E_0)^n}{E_T^m}$$

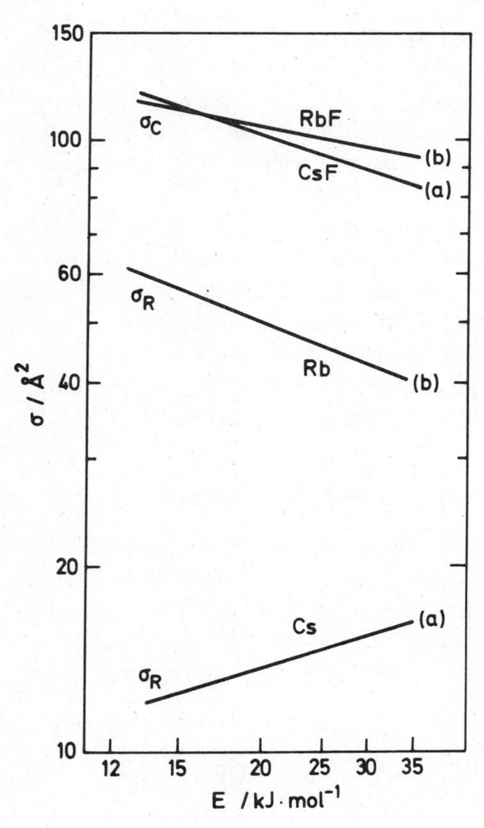

Figure 28. Log–log plot of cross sections for reactions of CsF, RbF + K versus collision energy. Least-squares lines have been drawn to represent experimental data. Upper lines—cross section σ_C for capture (complex formation) for (*a*) CsF + K and (*b*) RbF + K. Lower lines—cross section σ_R for reaction for (*a*) Cs formation and (*b*) Rb formation. Adapted from Ref. 180.

to see what kind of interpretation, if any, can be obtained from the fitted parameters n or m. In Table III several post-threshold laws were recommended, depending on the kind of model used. Care should be taken in deconvoluting the post-threshold data since the velocity spread of the reagents (more important for thermal neutral beams) is a serious problem which affects the quality of the deduced excitation function. Many of the post-threshold data now available, however, can be well represented by the line-of-centers model, like the example shown in Fig. 29 for the reaction $C_2H_5Br + K \rightarrow KBr + C_2H_5$.

When a complex mechanism is observed, the most common interpretation of the n parameter is the number of active degrees for the internal motion of the transition state (see Section III.A.2). Figure 30 shows the excitation function for the chemi-ionization channel for Ba (Sr) + $SF_6 \rightarrow BaF^+$ (SrF^+) + SF_5.[56] The full lines are the results of RRKM calculations for the number of active degrees indicated. It is interesting to note that the best fit is obtained for $s = 18$, that is, when all the vibrational degrees of freedom are involved.

A concave-up behavior has also been predicted by steric models developed by Smith[132] and extended by Levine and Bernstein.[133] They found a quadratic form $\sigma_R \sim (E_T - E_0)^2/E_T$ in qualitative agreement with the results found from early trajectory calculations for the H + H_2 reaction and, more recently, with quasi-classical calculations for the H + D_2 reaction.[172] The

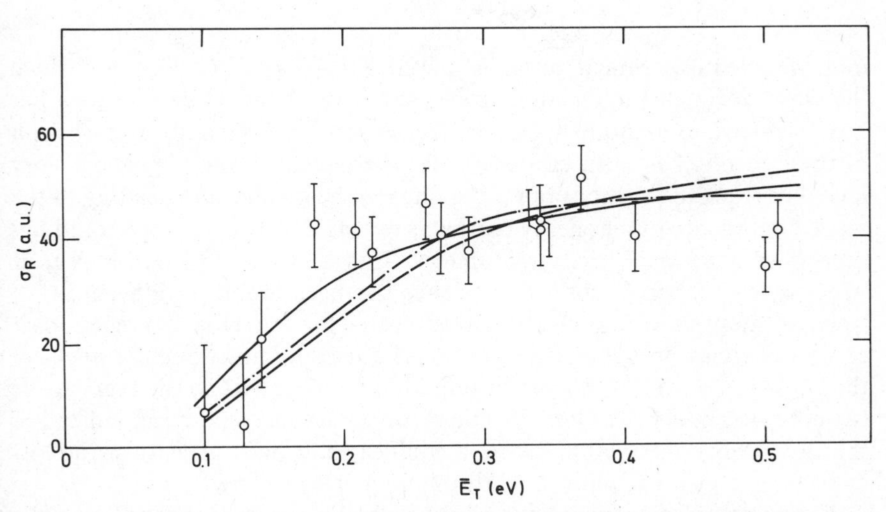

Figure 29. $\sigma_R(E_T)$ for the reaction $K + C_2H_5Br \rightarrow KBr + C_2H_5$. Solid line, is the best fit to experimental data from Ref. 49; broken and dotted lines, same fit to experimental data, using Eqs. (2) and (4) of Table III, respectively. In all cases the best E_0 value obtained was 0.12 ± 0.02 eV.

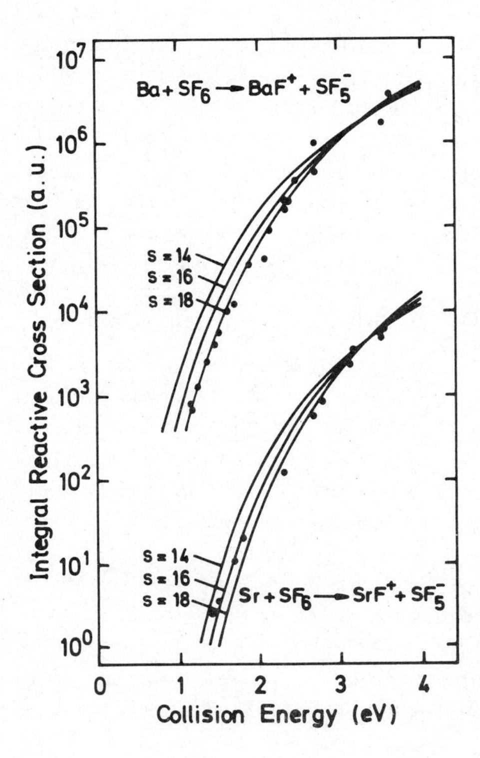

Figure 30. Excitation function for the chemi-ionization channel for Ba, Sr + $SF_6 \rightarrow BaF^+, SrF_5^-$. Solid lines, results from statistical model with the number of active degrees (vibrational degrees) as indicated. Adapted from Ref. 56.

most detailed experiment so far is for the Rb + CH_3I reaction, for which the dependence of the reaction cross section upon the angle of attack has been deduced, as mentioned earlier (see Section V.B). Recently an inversion of these results has been carried out by Bernstein[173] via a kinetic theory model of reactive collisions of rigid monospherical molecules, leading to the angle dependence of the energy barrier as well as an estimate of the colliding-pair anisotropy (asphericity) parameter. This is shown in Fig. 31. A near linear $\sigma_R(\cos\gamma)$ versus $\cos\gamma$ dependence has been found by Blais et al.[172] from an analysis of the classical trajectories for the H + D_2 reaction at several energies. In spite of the proposed quadratic functionality, many of the M + RX $\rightarrow$ MX + R post-threshold laws measured so far seem to be better represented by the line-of-centers model, which could be an indication that long-range interaction favors a collinear approach at least in the low collision energy range of the post-threshold now considered.

Experimental determinations of the collision energy dependence of the reaction cross section with (fixed) oriented reagents will clarify this and other interesting aspects of reaction dynamics.

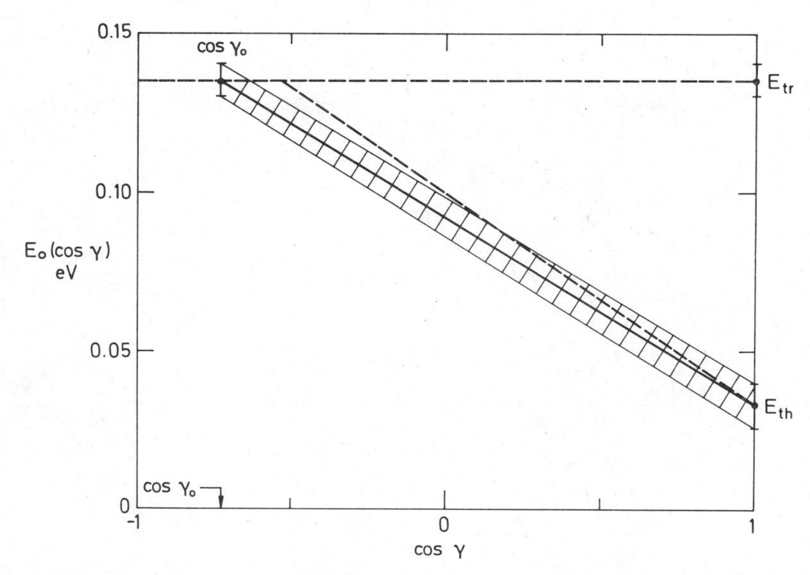

Figure 31. Orientation dependence of the activation energy barrier for the Rb + CH_3I reaction from the inversion procedure carried out by Bernstein.[173] Solid line, best present analysis; cross-hatched zone indicates uncertainty. From Ref. 173.

The excitation functions of K + HCl and K + FH for hot and cold HCl and HF have been measured by several laboratories,[39,46,54] and these are displayed in Figs. 32 and 33. A deconvolution analysis indicated an enhancement upon vibrational excitation. Phase space calculations gave good agreement for the ClH reaction when corrections for the steric factor were included.[54] From this good agreement was concluded that the results contained only little information on the potential energy surface. However, important deviations occurred for the same phase space treatment of the K + FH reaction. It was indicated that dynamic restrictions, in addition to steric effects, take place for this reaction. These differences between the ClH and FH reactions were associated with important differences in the potential energy surfaces of both systems. Zeiri and Shapiro[64b] have found a dramatic change in the barrier position from the exit (K + HF) to the entrance (K + HCl) valley. This could explain the stronger vibrational enhancement in the K + HF case. Moreover recrossing trajectories (similarly to those described in Section V.D for the reaction Sr + FH) can occur in association with a large curvature effect in the region near the transition state, which could in fact produce the maximum in the K + FH reaction cross section (Q_{max} = 0.78 eV).

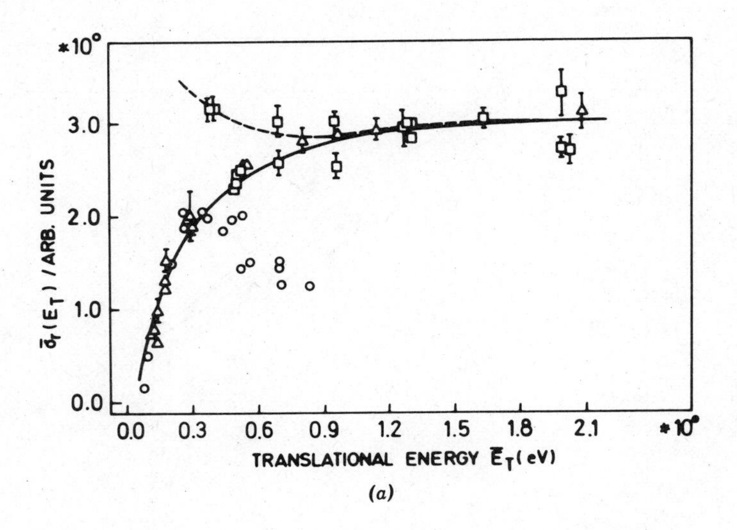

(a)

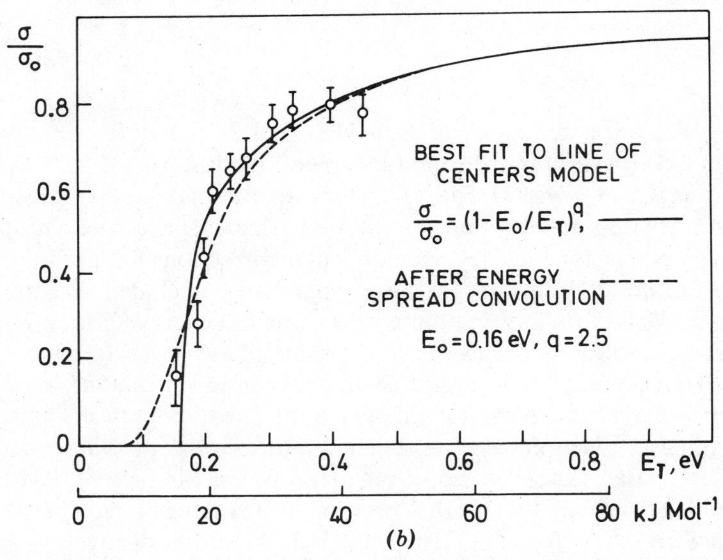

(b)

Figure 32. (*a*) Apparent integral reaction cross section as a function of mean translational energy E_T for K + HCl → KCl + H. △ and □ refer to cold and hot HCl ovens, respectively; ○ from Refs. 54, 39, and 46. Dashed line to assist visualization; solid line results of phase space calculations from Ref. 54. (*b*) Excitation function for the reaction $Xe(^3P_{0,2}) + HCl \rightarrow H + XeCl(B)$ is shown for comparison. Reproduced by courtesy of J. P. Simons.

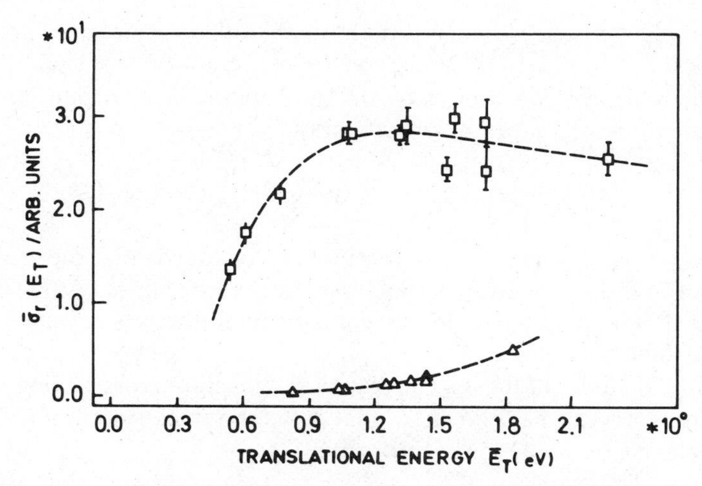

Figure 33. Same as Fig. 32*a*, but for the reaction K + HF → KF + H. Data points from Ref. 54

With regard to the K + ClH excitation function, Heismann and Loesch[54] have indicated that the decline in $\sigma_R(E_T)$ obtained by Brooks and co-workers[39] and shown in Fig. 32, in contradition with the continuous rise, can be ascribed to different rotational temperatures of the HCl in the two experiments. In this direction specific rates $k(v, j)$ measurements for these systems carried out by the infrared fluorescence depletion method[204] and the molecular beam laser experiment[46b] for the K + ClH($v = 1, j$) → ClK + H system have shown that the reaction cross section decreases as the rotational energy increases.[46b] Therefore since in the experiment of Brooks and co-workers the high collision energy is achieved by seeding and heating the HCl, it has been argued that a possible consequence of rising oven temperatures is a rising rotational temperature, which diminishes the reaction cross section.

2. *Some Kinematic and Angular Momentum Effect in the Excitation Functions.* Although no general study, in an isolated manner, of the systematic effect of kinematics on the reaction cross section and its collision energy dependence is known, some well-defined trends have been reported.

1. For a given set of initial and final conditions the reaction cross section seems to be enhanced in step with the reduced mass of the product. This effect can be justified from a wide range of considerations, of which follow (see also threshold laws).

(a) From phase space theory it follows that the requirement that the complex be able to dissociate over the final-state orbital angular momentum barrier, restricts the maximum final orbital angular momentum number L_f by the condition[124] ($V(r) = -c_f r^{-6}$)

$$L_f(L_f + 1)\hbar^2 \leqslant 6\mu_f c_f^{1/3}(1/2E_f)^{2/3},$$

from which an enhanced cross section can be expected as the attractive forces and the reduced mass increase, since more states of the product orbital momentum would be populated (at the same total energy available).

(b) A simple model of the dependence of the reaction cross section on the collision energy has shown[130] that a two-body collision model leads to an excitation function of the form

$$\sigma_R(E_T) = \pi R_c^2 \frac{\mu_f}{\mu_1}\left(\frac{E_T - V^{\ddagger}}{E_T}\right).$$

See Ref. 130 for details. A simple collision model[103] obtained a similar kinematic dependence on the μ_f/μ_i ratio.

A very interesting example of kinematic effects has been studied by Polanyi and co-workers[83] for the excitation function of all isotopic variants of the abstraction reaction H′ + H″Br → H′H″ + Br. The observed order of reactivity in the isotopic series designed H′H″ was (D, H) ⩾ (D, D) > (H, H) > (H, D). A simple model was devised to illustrate these effects. The main assumption was that for abstraction to occur the attacking A atom must approach BC by a nearly collinear path at the B end of BC such that one only needs to calculate the effective cone of acceptance of BC, taking into account the alignment of A–B–C, in order to get the reactive cross sections. As in other studies, they rationalized the favorable kinematics (D, H) as compared with (H, D) in terms of lengthened reaction time for D atom reaction (compared with H) and diminution in the time required for HBr (compared with D) to rotate into the preferred alignment for reaction.

2. High curvature in the minimum reaction path induces extensive product vibrational excitation for exothermic reactions and higher energy threshold for endothermic reactions.

Duff and Truhlar[181] have studied the effect of the curvature of the minimum reaction path on the dynamics in endothermic chemical reactions and the product energies in exothermic reactions. They clearly pointed out that the dynamics not only depend on a consideration of the saddle point location and magnitude, the percentage of attractive, repulsive, and mixed release, and the skew angle, but also on the location of the region with large reaction path curvature. They have confirmed and tested the Hofacker and Levine theory[182-184] concerning the conversion of reagent translational energy to product vibration. The main requirement was that the reaction path curvature be high where the local translational energy along the reaction path is high. They correlated the percentage of vibrational energy with the most negative value of the product of the curvature times the local translational energy, which was found to be an important rule in causing vibrational nonadiabaticity in exothermic reactions.

For the endothermic direction of the reaction the curvature was found to have important effects on the threshold behavior. Thus a high minimum energy path curvature in the region with potential energy below the barrier height seemed to be associated with higher threshold energies for these endothermic reactions. Thus these higher thresholds could in general be attributed to reflection of the surface before the saddle point, rather than to recrossing of the saddle point region. These effects were observed for endothermic reactions for all mass combinations except where a heavy atom is abstracted from a heteronuclear diatom that has one light and one heavy atom.

Polanyi and co-workers[185,186] have explored these kinematic effects on an endothermic energy surface for the following mass combinations: (1) LL + L; (2) LH + H → L + HH. At a total reagent energy of twice the barrier energy they found that the vibrational energy was 10^2 and < 10 times as effective as the translational energy for cases (1) and (2), respectively. The enhanced efficiency of translational energy in the second case with respect to the first was associated with the fact that it is easy for a fast moving light atom L to escape during the extended time spent close to one another. Furthermore they have studied the effect of different categories of endothermic potential energy surfaces of type II that is, both having a late barrier crest, one with a sudden rise to the barrier crest (type IIS) and one showing a gradual rise (type IIG)]. In the reverse view, surface IIS would correspond to a more attractive energy release than would type IIG because of its higher curvature of the minimum energy path in the region linking the entrance valley of the surface to the exit valley. They found that on the IIS surface the sudden change of direction of the minimum energy path gives rise to a bend that is difficult to negotiate at high speed. Consequently $\sigma_R(E_T)$ diminishes at enhanced collision energies. These authors have suggested this kinematic effect to explain the anticipated peak in $\sigma_R(E_T)$ for the reaction HF + Sr, as will be shown later (see Fig. 42). No quantitative trajectory calculation has been carried out on this system.

3. With respect to kinematic effects in differential excitation functions, a detailed study has been carried out by Siegel and Schultz[187] for reactions of the type H + HL → HH + L via trajectory calculations. Essentially three kinematic effects were found.

(a) The reaction probability $P(b)$ as a function of the impact parameter b is a step function of the form $P(b) = 1\,(0)$ for $b \leqslant (>)\, b_{\text{max}}$.
(b) The rotational angular momentum of the product molecule, J', is, to a very good approximation, equal to the orbital angular momentum of the reactions, $J' \simeq L$. Note that this close correlation was used to obtain the excitation function from the product rotational energy distributions.
(c) There exists a strong correlation between the product vibrational energy E'_{vib} and the impact parameter as well as between the product translational energy and the impact parameter.

These authors have shown that the main reason for the existence of these kinematic effects is the very small mass of the C atom in comparison with the masses m_A and m_B or, in other words, the very small moment of inertia I_{BC} of the molecule BC compared with that of the systems AB and A–BC, considering BC as a single particle.

Concerning the angular momentum constraint in reaction dynamics, there is no systematic and general treatment to account for its influence in the collision energy dependence of the reaction cross section, even though excellent treatments have considered its role in the product distribution. Several phase space calculations have shown little influence of such a constraint on the excitation function.[188] A study by Truhlar[188c] was, on the other hand, able to explain a maximum in $\sigma_R(E_T)$ by phase space calculations with angular momentum conservation requirements. Also a simple model calculation describing collision two-body cross sections was able to model the excitation function for the reaction M + RX → MX + R once allowance was made for angular momentum conservation (see Section V.D).

An interesting example[93] of angular momentum implications in the collision energy dependence of reaction dynamics occurs in the endothermic reaction $Cl + CH_3I \rightarrow ClI + CH_3$, which shows the complex mechanism in the post-threshold range of collision energies. Application of RRKM theory would indicate that the more energetically favorable dissociation of the collision complex should be the preferred pathway. If, however, dissociation back to reactants displaces the angular momentum primarily into orbital angular momentum while dissociation into products yields significant ICl product angular momentum, a compensation for the endoergicity, due to a minimum angular barrier, would occur, facilitating the ICl production. The reaction cross sections predicted by a modified RRKM calculation, where the total angular momentum of the complex is equally partitioned between product orbital and rotational angular momentum, is in good agreement with the experimental excitation function, as shown in Fig. 34.

3. *Effects of Other Energy Modes.* Electronic excitation of reagents is a very productive way to experiment and learn about molecular reaction dynamics since excited potential surfaces are generally involved in addition to the ground state surface. A high price has to be paid because of the difficulty introduced by the so-called "electronic" problem,[21] together with the "nuclear" problem in interpreting the molecular reaction dynamics. Excellent monographs and studies are available which treat this problem.[21,22] As far as we are concerned here, very few studies have appeared showing the translational energy dependence of the reaction cross section for different electronically excited reagents, even though very elegant experiments on electronically excited reagents have clearly shown an important

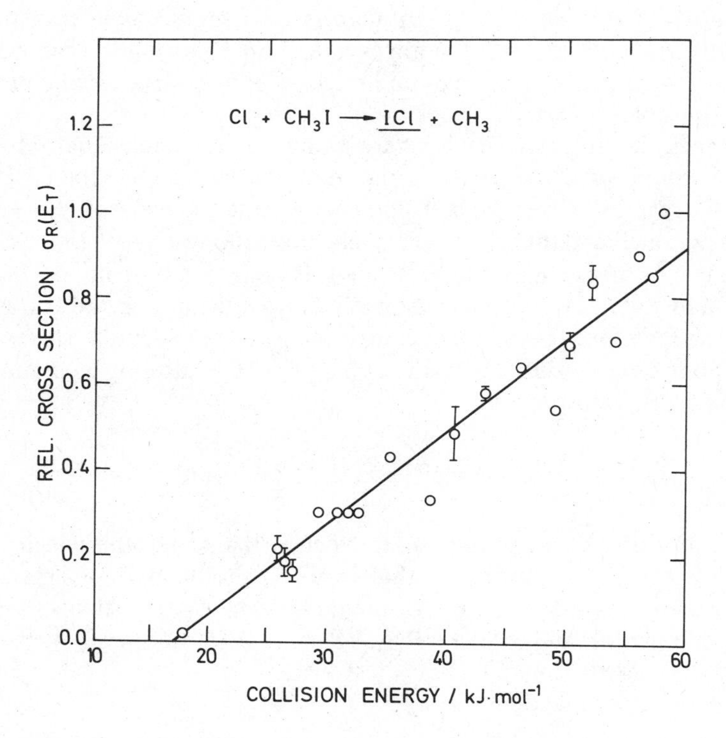

Figure 34. Excitation function for the reaction. $Cl + CH_3I \rightarrow ClI + CH_3$. Data and model calculations (solid line) adapted from Ref. 93.

enhancement of the reaction cross section, but without substantial change of collision energy.

One of the best chemiluminescence reactions studied so far is that of $NO + O_3 \rightarrow NO_2^* + O_2$ (see Table II for details). Different laboratories[57,64a,65a] have essentially shown a strong translational and rotational energy dependence in the reaction cross section. Stolte and co-workers[57] have studied this dependence with regard to the fine structure of the reaction $NO(\Pi_{3/2}, \Pi_{1/2})$ and have reported a branching ratio

$$\frac{\sigma(\Pi_{3/2})}{\sigma(\Pi_{1/2})} = 0.9 \pm 0.2,$$

concluding that there is no obvious difference in chemiluminescence reactivity for both fine structure states.

It appears that the reaction is direct but essentially shows two different modes, characterized by two different transition states, as displayed in Fig. 35, where a laboratory contour map is also shown. One transition state involves NO attack, with the nitrogen end approaching the central O atom of the O_3; the other involves an attack of NO to strip off one of the outer O atoms of the O_3.

Komornicki et al.[189] have also carried out three-dimensional Monte Carlo calculations on the three low-lying electronic surfaces for the $F(^2P_{3/2}, {}^2P_{1/2}) + H_2$ reaction. Over the collision energy range considered ($E_T = 0.1$ to 1 eV) they concluded that electronic quenching followed by reaction was a very significant process, since the $F(^2P_{1/2}) + H_2$ reaction does not correlate to the ground state of the product. Nonadiabatic transition in the entrance channel was therefore predicted as contributing to the reaction. Moreover close-coupling calculations of inelastic $F(^2P_{1/2}) + H_2$ scattering have shown that the near-resonant process[190]

$$F(^2P_{1/2}) + H_2(J = 0) \rightarrow F(^2P_{3/2}) + H_2(J = 2)$$

was about an order of magnitude more efficient than any other electronic quenching process. This suggested that $F(^2P_{1/2})$ should be more reactive with p-H_2 than with n-H_2 because of the larger $J = 0$ population in p-H_2. Recent experimental studies of $F(^2P_{1/2}, {}^2P_{3/2}) + n$-$H_2$ and p-H_2 have not confirmed these predictions.[151]

The nonadiabatic reactions

$$F(^2P_{3/2}, {}^2P_{1/2}) + HBr\,(DBr) \rightarrow HF\,(DF) + Br(^2P_{1/2}, {}^2P_{3/2})$$

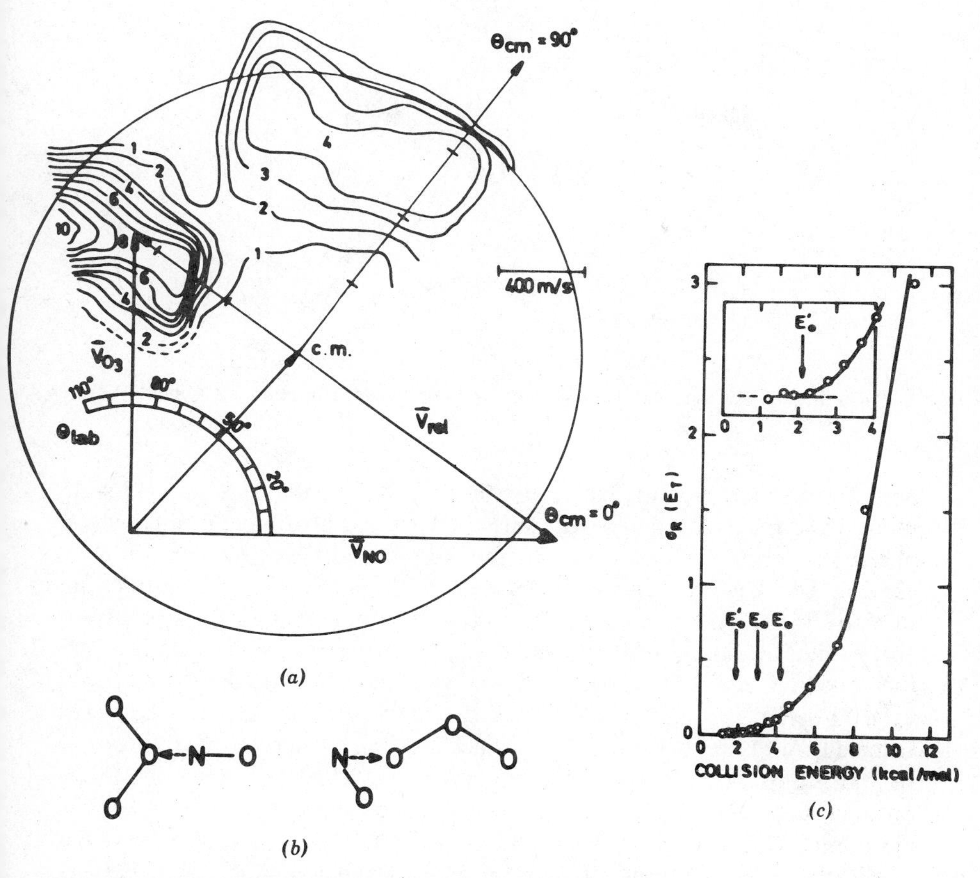

Figure 35. (*a*) Center-of-mass contour map of the NO_2 flux distribution obtained by a 1-newton diagram transformation of the laboratory data from Ref. 57. (*b*) Proposed mechanisms for $NO + O_3 \rightarrow NO_2^* + O_2$ from Refs. 57 and 62. (*c*) Phenomenological reaction cross section as a function of collision energy. Insert shows threshold region. E_0'—onset of $\sigma_R(E_T)$; E_0—true threshold energy, corrected for the energy distributions in beam and target E_a—Arrhenius activation energy. Adapted from Ref. 64.

have been studied in Polanyi's group by crossed molecular beam experiments.[81] A correlation diagram for FH–Br is shown in Fig. 36 for the HBr ground state (coplanar motion is assumed). Three doubly degenerate potential energy surfaces result when the several ^2P states of the F atom are combined with the $^1\Sigma$ state of HBr. As can be seen, excited atoms can be formed (1) directly from the adiabatic path through the $2^2A'$ surface or (2) via

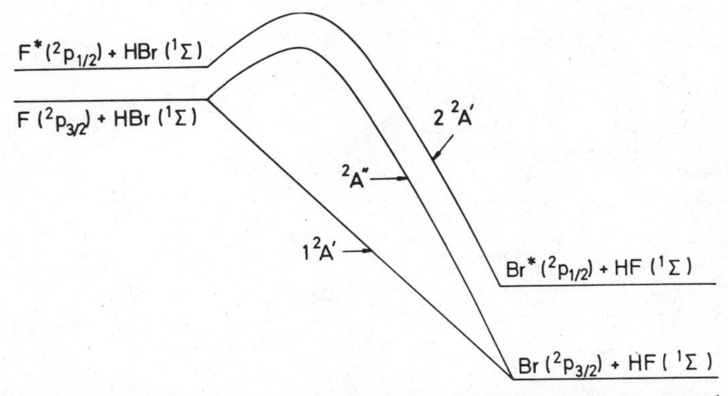

Figure 36. Correlation diagram for FH–Br. Coplanar motion is assumed.

nonadiabatic interactions between either the $^2A''$ or the $1^2A'$ ground state surface and the $2^2A'$ excited state surface. These authors have studied which of the three processes gave rise to the electronically excited Br* atom by varying the collision energy of the reagents without changing the F* fraction, or conversely, by varing the F* fraction without changing the collision energy significantly. They conclude that a nonadiabatic process leads to the Br* product since they observed no variation in the branching ratio $\sigma_R(\text{Br}^*)/\sigma_R(\text{Br})$ as the [F*]/[F] fraction was varied. A hop from the lower surface $1^2A'$ to the adjacent upper surface $2^2A'$ was then required to produce Br*. This hop can only occur in configurations at which the upper and lower surfaces approach one another in energy, which they do asymptotically as the reagents approach or the products separate. They concluded that the nonadiabatic transition takes place in the exit valley configuration FH–Br since a hop in the entrance channel would involve a barrier on the upper surface, precluding the reaction.[81]

In the presentation described so far we have not considered the slope of the reaction cross section at the threshold limit. Let us now make a few comments on this point. From a simple empirical point of view, using the general form

$$\sigma_R \sim \frac{(E_T - E_0)^n}{E_T},$$

two different limits of the slope of the reaction cross section can be obtained in the limit of $E_T \to E_0$, depending on the n value. For example, for $n > 1$,

$$\sigma_R' \to 0, \qquad E_T \to E_0$$

and for $n = 1$,

$$\sigma'_R \to 1/E_0, \qquad E_T \to E_0,$$

where σ' stands for $d\sigma/dE_T$. Therefore, provided that the microcanonical transition state theory gives a good description of the observed post-threshold data, and no less important, that accurate enough data are available on the near-post-threshold region, the observed slope can help in the diagnosis of the number of degrees of freedom (one only or several) that are active in the transition state.

Energy selectivity in the promotion of chemical reactions is a very important question and can also affect the post-threshold behavior. Figure 37 shows different post-threshold behaviors calculated for the $O + H_2(v) \to OH + H$ system as a function of the vibrational state of the reagents.[150] Note how as the vibrational energy is increased, the threshold goes down, but also the threshold slope increases. It is interesting to point out that the limiting slope described above will now be

$$\sigma'_R \to \frac{1}{E_0 - \alpha E_v} \quad \text{instead of} \quad \frac{1}{E_0},$$

where we just have accounted for a vibrational energy fraction α used to surmount the barrier. (This sort of parametrization has been widely used in the literature; see, for example, Ref. 191.) Within this simple approach we

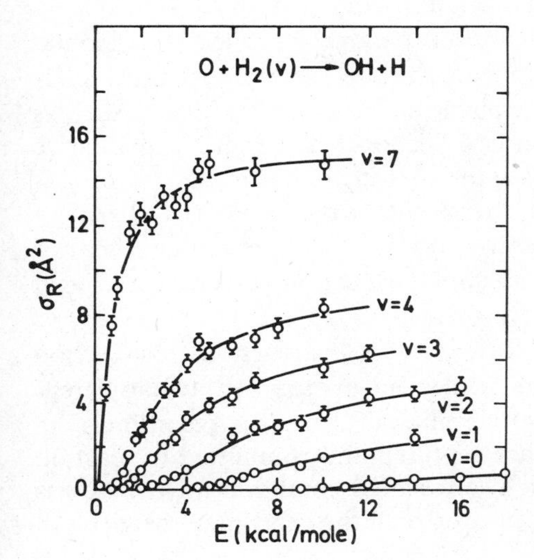

Figure 37. Reaction cross section σ_R as a function of collision energy E_T for the reaction $O + H_2(v) \to OH + H$, v as indicated. From Ref. 150.

can account for the fact that the threshold slope rises as the threshold energy diminishes as a consequence of the vibrational enhancement. In fact Fig. 22 in Section 5.B.1 would be a limiting example of this trend as a result of the threshold disappearance (within the measured range) as the N_2O molecule was excited in its bending motion via heating up the nozzle.

4. Absence of Reaction Threshold: A Continuous Decline in $\sigma^R(E_T)$. A Vibrational or Translational Capture? This behavior has been observed for many alkali plus halogen reactions in atom exchange with an alkaline earth in either chemiluminescence or chemi-ionization processes or in neutral exchange reactions.[22] Figure 38 shows typical examples of this behavior. As shown in Table III, the easiest model to account for this decline in $\sigma_R(E_T)$ is the centrifugal barrier model, which gives an excitation function $\sigma_R \sim (E_T)^{-2/n}$, with n being the long-range potential panameter. For those systems where harpooning is likely to occur at large distances, the excitation function exhibits the common feature that the cross sections fall with increasing translational energy, and rather than tending to zero, they level out at a given value of collision energy. It has been shown[22,136] that if one considers the harpooning potential simply covalent and flat for $R > R_c$, where R_c is the (covalent-ionic) crossing distance, and ionic for $R < R_c$, one would expect a collision cross section given by πR_c^2, that is, independent of energy, which would correspond to the high energy region of the cross section. However, due to the interaction of covalent and ionic configurations,[22] the avoided interaction is rounded off, rendering the potential quite attractive even for $R > R_c$. Consequently the lower the collision energy, the more effectively will this part of the potential pull the reactants in large impact parameter collisions ($b > R_c$) within the harpooning distance R_c.

For reactions involving small impact parameters, such as rebound reactions, and others with close-range collisions, several dynamic interpretations have been proposed to explain the cross section failure, such as the so-called recrossing mechanism (see Section V.D). From a qualitative point of view it is equivalent to start at the crossing radius R_c and to account for the cross section enhancement due to the attractive part of the long-range potential, or to begin at an early location of the centrifugal barrier and account for recrossing trajectories until the location of such a centrifugal barrier reaches the crossing radius R_c.

Gislason and co-workers[137a] in their study of the reaction cross section for alkali atoms with bromine molecules have proposed two different mechanisms: translational and/or vibrational capture. Accordingly in translational capture the particles are captured and committed to reaction before the electron transfer occurs. The essential condition for reaction is that kinetic energy is high enough to overcome the centrifugal barrier. An

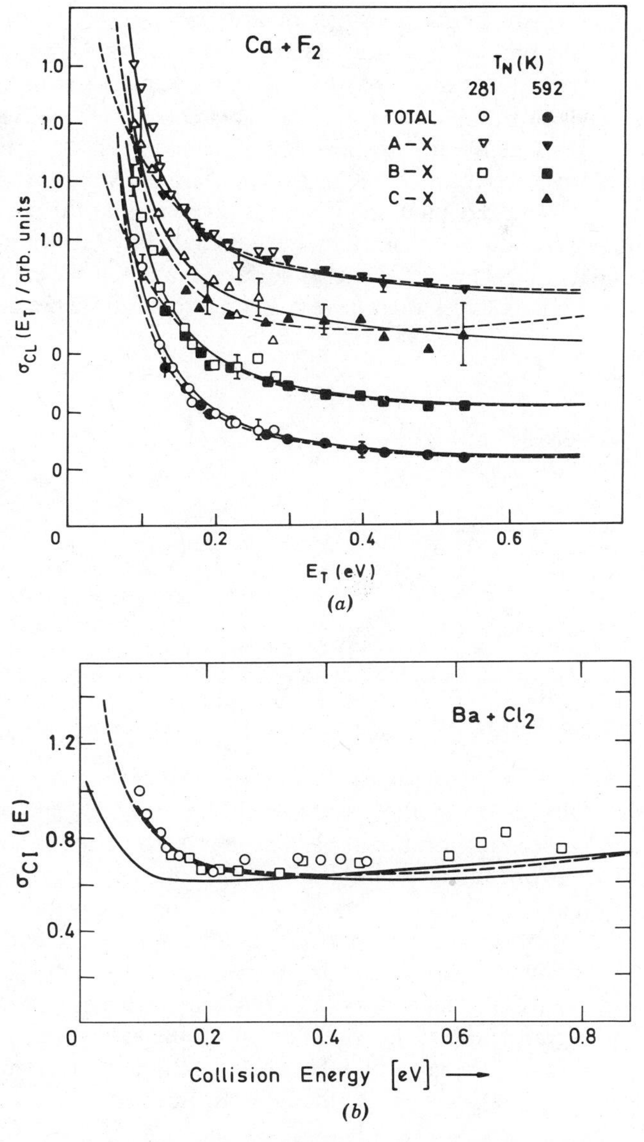

Figure 38. (*a*) Chemiluminescence and (*b*) chemi-ionization excitation functions for the systems (*a*) Ca + F_2 and (*b*) Ba + Cl_2. (*a*) Experiments in the Ca + F_2 cross section do not depend strongly on reactant vibration. Dashed curves, fits of modified Eu function to the cross section of the individual states; solid lines are drawn through points. All excitation functions are arbitrarily normalized to $\sigma_{CL} = 1$ at $E_T = 0.09$ eV and are presented vertically displayed for clarity. (*b*) ○, nozzle temperature $T = 281$ K; □, $T = 587$ K; solid curves, fits by model described in Ref. 136 (see also Table III, electron jump models); dashed cruves, data fits to Eu's Born approximation. Adapted from Refs. 22 and 136.

alternative mechanism (more pausible for low collision energies) was also proposed for this type of reaction, thermal vibrational capture. In this mechanism the electron transfer occurs before the centrifugal barrier along the translational coordinate is overcome, and therefore it is vibrational motion that is responsible for the reaction. The energy dependence of the total reactive cross reaction for K + Br_2 measured by van den Meulen et al.[40] is shown in Fig. 39, including other results as well as the prediction of the vibrational translational capture model calculated by Gislason et al.[137a] by using a semiempirical potential energy surface. Note that the experimental cross sections of van den Meulen have a stronger energy dependence than

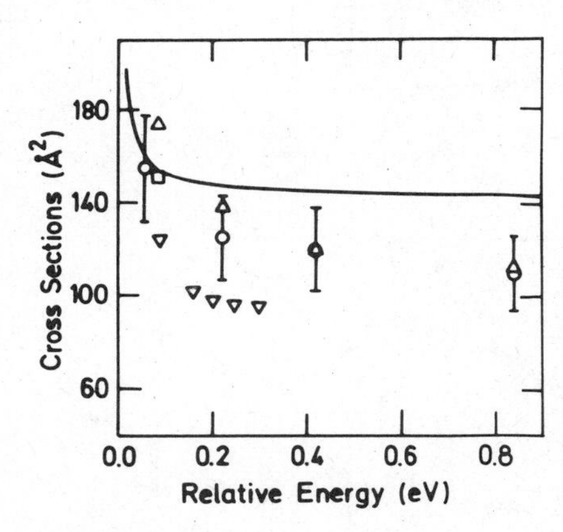

Figure 39. Total reactive cross sections for K + Br_2 plotted against relative collision energy. Solid line, theoretical calculations from Ref. 137b; ○, experimental results from Ref. 40; □, thermal measurement; ▽, results of the Anderson–Herschbach analysis; △ taken from the surface-hopping trajectory calculations of Evers et al.[167a] Adapted from Ref. 137b.

the predicted behavior. Several explanations were given to explain this disagreement. One disputes the assumption that the Br_2 molecule will instantly vibrate over the vibrational barrier, that is, while the time required for Br_2 to complete one full vibration is 0.1 ps, the time required for K + Br_2 to move 1 Å ranges from 0.1 to 0.004 ps as the collision energy increases from 0.1 to 1.0 eV. Therefore the assumption should only be valid at energies below 0.1 eV. It would be very interesting to measure the excitation function for several state-selected reagents besides performing trajectory studies to determine the overall effect of this breakdown on the reactive cross section and to assess the role of the vibrational capture mechanism in these neutral–neutral interactions as well as in the aforementioned chemi-ionization processes.

D. Maximum in the Collision Energy Dependence of the Reaction Cross Section: The Recrossing Mechanism

Since the experimental determination of a maximum in the K + $CH_3I \rightarrow$ KI + CH_3 excitation function around $E_T = 0.18$ eV, considerable theoretical attention was attracted trying to explain it. The main approach was dedicated to the incorporation of a proper recrossing mechanism for the cross section decline as the collision energy increases. To this end different recrossing mechanism have been reported. (1) A classical model[118] based on a three-body collinear collision ascribed the falloff in σ_R to reflection from the repulsive potential to the reactant valley as E_T increases. (2) A simple orbiting model[135] reproduces the observed behavior by a reduction in reactive trajectories from centrifugal barrier repulsion as E_T increases. (3) A hard-sphere model[103] was also used, and it attributes the falloff to a reduction in the number of allowed states through angular momentum conservation. Figure 40b shows the experimental data plus some of these model calculations.

In a more quantitative way LaBudde et al.[192] used the diatomics-in-molecules (DIM) approximation to build a set of diabatic potential surfaces for the $KICH_3$ system (CH_3 as a single particle). Where diabatic surfaces intersected, the lower surface was chosen as a way of selecting the lower "pseudoadiabatic surface" on which quasi-classical trajectories were run. Then the excitation functions were computed for the following processes:

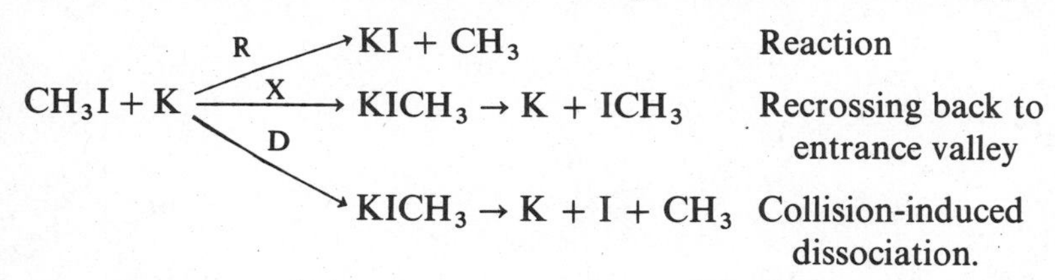

However, only qualitative agreement with experiment was obtained.

A different approach from a microcanonical transition state point of view stated that the above different explanations may reduce to the same fact, namely, that the transition state location moves from the entrance channel barrier (at low energy) to the product valley as E_T increases. In other words, the preceding statements are different ways of describing the unique fact of the reactive flux reduction allowed by a potential energy surface topology. The main idea was therefore that *the shape evolution of* σ_R *with* E_T *is a measure of the collision energy dependence of the transition state location.*[130] Figure 40a shows a type of data representation where the excitation function of the $K + CH_3I$ reaction is modeled by microcanonical transition state analysis. (See Ref. 130 for details.) According to this model, only two transition state locations are required, depending on the collision range studied, to describe the entire excitation function. Have these transition states been located by using appropriate methods? For the $K + CH_3I$ system, not yet. A search for trapped trajectories carried out by Pollak and Pechukas for the H_3 system has found different transition states.[163] At lower energies these periodic trajectories traverse the saddle point region on the potential energy surface.

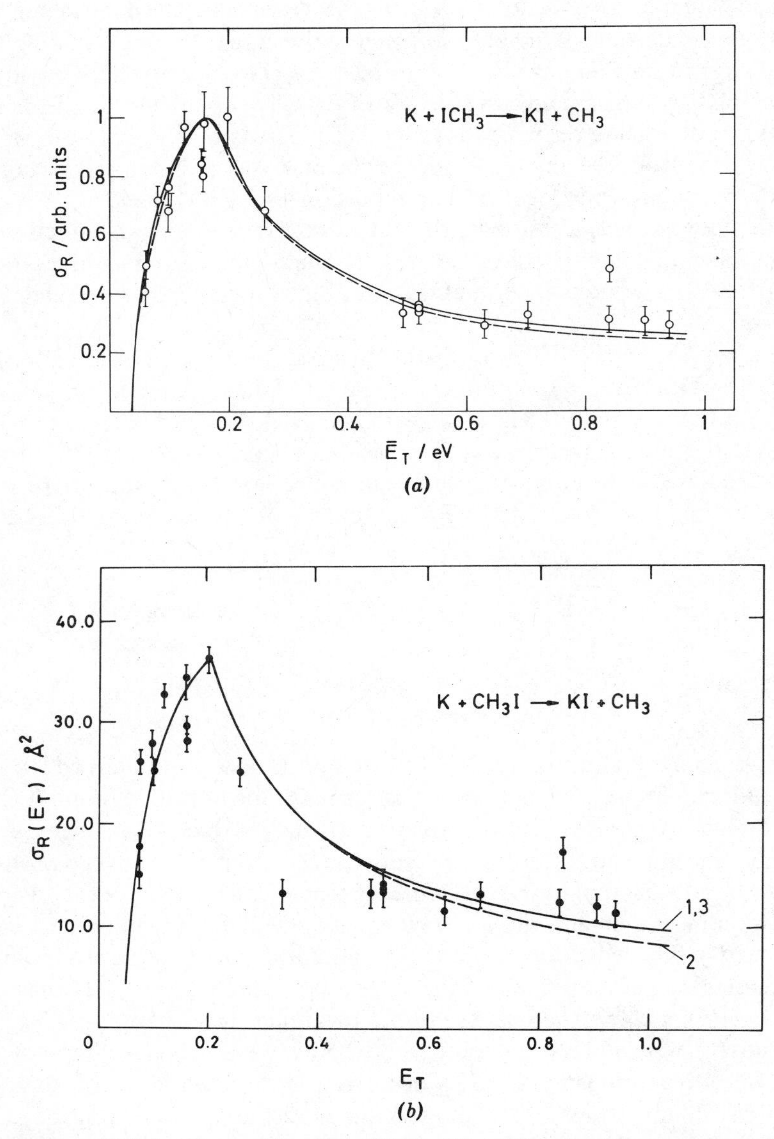

Figure 40. $\sigma_R(E_T)$ for the reaction $K + CH_3I \rightarrow KI + CH_3$. Symbols, experimental data from Ref. 25. (*a*) Dashed line, microcanonical transition state model calculation[130]; solid line, modified Eu model.[37] Adapted from Ref. 53. (*b*) Data same as (*a*). Curve 1—Harris and Herschbach model; curve 2—Shin's model predictions; curve 3—modified electron jump.[202] See Table III for model expressions and text for comments.

As E_T increases, this no longer occurs, but the transition state location moves toward two similar transition states in the entrance and exit valleys. Going back to the $KICH_3$ system, potential energy surfaces of the extended surfaces were built and a search for the minimum flux of reactive trajectories was carried out. In the collinear surface the transition state stays close to the saddle point as the collision energy increases.[131] Therefore care should be taken in interpreting the $\sigma_R(E_T)$ shape on the basis of these "transitions" in the transition state location.

Other statistical models have been proposed to account for the maximum appearence in $\sigma_R(E_T)$. One is that of Brooks and co-workers[39]

The basic idea of the model is to assume that the rate of transition from given initial reactant states to final product states W is only a function of the total energy available to the products. Then the total cross section is proportional to the density of the product states $\rho(E)$ and is given by

$$\sigma_R(E = \rho(E)W(E)/v, \qquad E > 0,$$

where v is the relative velocity of the reactants. Typically $W(E)$ is obtained by using phase space theory, and then the state-to-state rate is determined empirically,

$$W = W_0 \exp[-\beta(E^{1/2})].$$

The fits used adopt the form

$$\sigma_R(E_T) \sim (E)^S \exp[-\beta(E)^{1/2}]/(E_T)^{1/2},$$

where $(E)^S$ represents the density of the product states, the exponential term is the average state-to-state transition probability, and $E_T^{1/2}$ is proportional to the relative velocity. The data fit obtained for the K + ClH excitation function[39] is $S = 2.2$ and $\beta = 1.36$ (kcal/mol)$^{1/2}$. For Sr + FH the best least-squared fit gave[79] $S = 2.1$ and $\beta = 1.1$ (kcal/mol)$^{1/2}$. The above functionality is quite flexible, and one can pay more attention to the dynamically interesting variation in the state-to-state rate constant, but unfortunately it does not incorporate explicit interaction potentials. Therefore any dynamic explanation in terms of characteristic topological features of the potential energy surface is missing.

To provide some qualitative understanding of the shape of the excitation function with a little more dynamics, let us consider a simple microcanonical variational transition state approach. We shall take only one active vibration at the transition state location corresponding to the motion perpendicular to the reaction coordinate. If we consider internally cold reagents (i.e. the BC reagent molecule in $v = j = 0$), the reactant density of states will be $\rho(E_T) = A_T E_T^{1/2}$, and the reaction cross section is[130]

$$\sigma_R = \frac{\hbar^2}{8\pi\mu E_T} N^{\ddagger}(E_T, s),$$

where $N^{\ddagger}(E_T, s)$, as mentioned before, stands for the number of internal states available at the transition state located at s along the reaction coordinate.

Let us now locate our transition state at the reactant valley in the collinear potential energy surface, as shown in Fig. 41. This location is typical of an early barrier case. Since the transition state is so close to the asymptotic reactant zone, one may expect that vibrational adiabaticity holds well[140e] and the transition state internal distribution will be that of the reactants (in this case $v^{\ddagger} = J^{\ddagger} = 0$), and therefore $N^{\ddagger}(E, s) \to 1$. This collinear (vibrationally adiabatic) reaction cross section (actually it would be just a reaction probability) will then be

$$\sigma_R(0) \equiv P_R = \begin{cases} \sim \dfrac{1}{E_T}, & E_T > E_0 \\ 0 & E_T \leqslant E_0 \end{cases}$$

or just the former expression if no energy threshold is present.

To average this simple expression over any range of angles of attack γ, one can use a linear dependence of the type[133]

$$E_{th}(\gamma) = E_0 + E'(1 - \cos\gamma).$$

Then the reaction cross section can be obtained,[133]

$$\sigma_R(E_T) = \frac{\int_{-1}^{+1} d(\cos\gamma)\, \sigma_R(\gamma)}{\int_{-1}^{+1} d(\cos\gamma)},$$

assuming that at any collision energy the maximum angle of attack is given by

$$E_T = E_0 + E'(1 - \cos\gamma_{max}).$$

So if one assumes, for simplicity, that the above relation holds from 0 to π, a maximum in $\sigma_R(E_T)$ would appear for $E_{max} = E_0 + 2E'$, and therefore

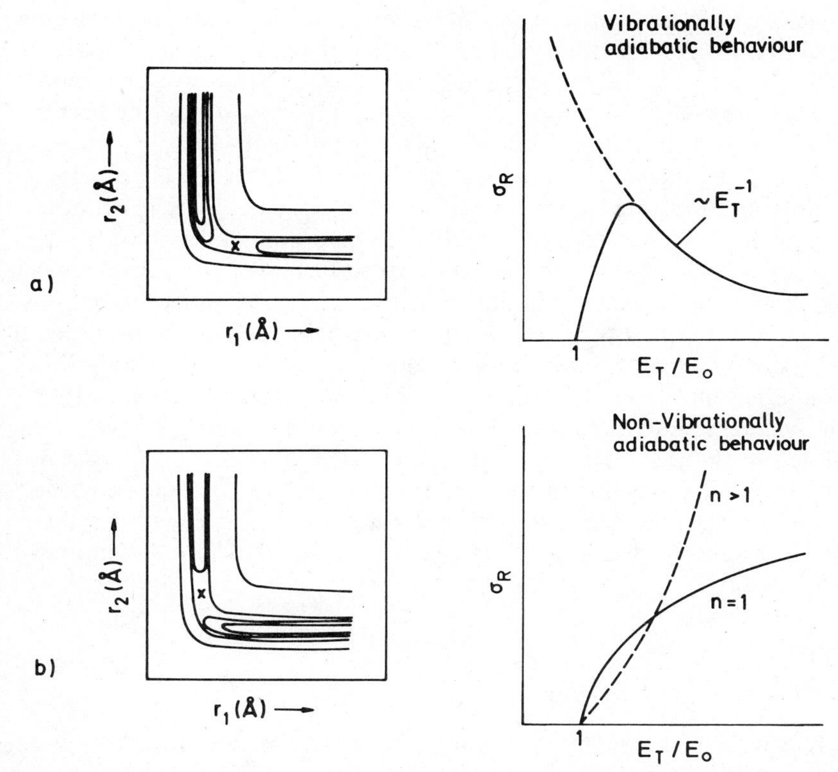

Figure 41. (*a*) Typical representation of collinear potential energy surface of an early barrier type. In the right side the shape evolution of σ_R with E_T is shown assuming a vibrationally adiabatic behavior. Provided an energy threshold is present, the maximum in $\sigma_R(E_T)$ would be a consequence of the reaction barrier dependence upon the angle of attack. See text for details. (*b*) Same as (*a*). Now the potential energy surface is of a late type and vibrational adiabaticity is not likely to occur. Complex dynamics and/or extensive energy mixing may take place, causing a post-threshold rise with $n = 1$ or $n > 1$, depending upon the actual number of degrees in the (dynamics) "transition state." See text for comments.

$$\sigma_R \sim \begin{cases} \dfrac{E_T - E_0}{E_T}, & E_T < E_{max} \\[2ex] \dfrac{E_{max} - E_0}{E_T}, & E_T > E_{max}. \end{cases}$$

Figure 41*a* also shows these vibrationally adiabatic reaction probabilities and reaction cross sections. From a qualitative point of view deviations from this functionality may arise, either because of the appearance of recrossing trajectories (negative deviations) or because of the failure of the vibrational adiabaticity which will enhance the cross section (positive deviations).

Let us move the transition state out of the reactant toward the product valley, as shown in Fig. 41. Now the vibrational adiabaticity is a stronger assumption; a large curvature along the reaction coordinate, the possibility of a potential well, and so on, can destroy the (original) local reactant distribution of the internal energy modes. Thus even though the nonrecrossing assumption is now quite accurate (since the transition state is located near the product valley), energy exchange between internal modes and translational motion along the reaction coordinate is to be expected. Additional assumptions will be required to calculate the flux of reactive trajectories (the internal microcanonical partition function of the activated complex), such as energy randomization among the internal modes (as could be expected in the case of a deep well in the potential energy surface), if one wants to avoid a dynamic calculation, which sooner or later will be needed. One thing is clear, the expected behavior is no longer vibrationally adiabatic as in the previous case. This (vibrationally) nonadiabatic cross section can now be given by

$$\sigma_{\mathrm{R}} \sim \frac{(E - E_0)^n}{E_{\mathrm{T}}},$$

where extensive energy mixing among the internal modes of the transition state was accounted for.

Figure 41*b* shows a prototype of this typical excitation function for this limiting case. One can go from the simple line-of-centers model ($n = 1$) with E_0 as reaction threshold to a more general post-threshold law (see Section V.C).

In conclusion, suppose that experimentalists measuring an excitation function find a maximum in their reaction cross section. What can be said about it? What kind of interpretation can be made? What sort of experimental suggestions can be given to provide further evidence for the proposed (mechanistic) explanation?

In general no statistical justification has been given to explain such maxima in the excitation function. In the closest approach the function was decomposed into a coupling of a dynamic and a statistical factor, but this parametrization, in terms of state-to-state reaction cross sections, transfers the main question to this last quantity without any clear dynamic answer, since it has not been connected with the potential energy surface topology. Only empirical equations are being used to account for its collision energy dependence.

From a dynamic point of view a clear division can be made depending on whether the reaction has a late barrier (typical of an endothermic case) or an early one (an exothermic reaction). In the former case recrossing may arise before the crossing of the barrier crest. Clear examples of this behavior seem to be important, and have been shown in potential energy surfaces having a large curvature path before the saddle point, as was shown for the Sr + FH and K + FH reactions. Figure 42 shows some examples of this family of reactions. A more general kinematic correlation between the energy value where $\sigma_R(E_T)$ adopts a maximum and parameters such as the skew angle and the scaling factor has not been found.

Table VIII summarizes the skew angles and scaling factors for several reactions where a maximum in $\sigma_R(E_T)$ was found and no correlation appears to exist between these factors and the ratio of maximum to threshold energy.

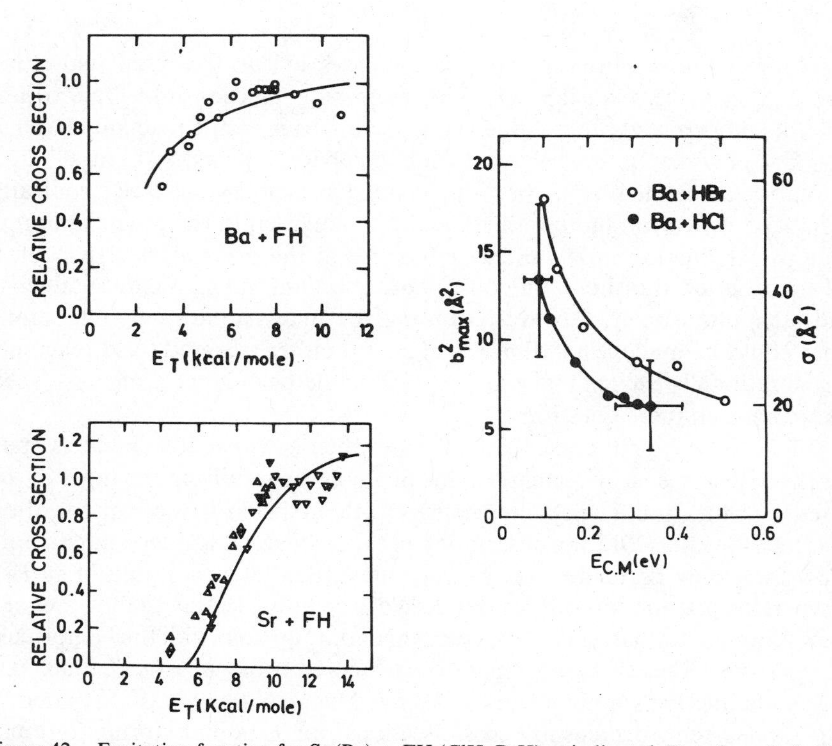

Figure 42. Excitation function for Sr (Ba) + FH (ClH, BrH) as indicated. Data from Refs. 79, 80, and 84. Solid lines—line-of-centers model (Ba + FH), statistical fitting of Ref. 39 (Sr + FH), best fits (right panel). Note the different shapes of the Ba reactions (that is, HCl versus FH case) even though both reactions are exoergic.

TABLE VIII.
Kinematic Factors and Maxima in Excitation Functions for Several Reactions

Reaction	E_{max}/E_0	F	β (deg)
$Ba + FH \rightarrow BaF + H$	3.0	0.24	77.8
$K + CH_3I \rightarrow KI + CH_3$	4.7	0.67	80.9
$K + C_2H_5Br \rightarrow KBr + C_2H_5$	2.90	0.90	72.8
$K + ClH \rightarrow ClK + H$	6.5	0.23	83.0
$Na + CH_3I \rightarrow NaI + CH_3$	2.8	0.83	82.7
$Sm + N_2O$ (cold) $\rightarrow SmO^*$ (blue) $+ N_2$	5	0.84	40.7
$Sm + N_2O$ (cold) $\rightarrow SmO^*$ (red) $+ N_2$	5	0.84	40.7
$Ba + N_2O$ (cold) $\rightarrow BaO^* + N_2$	>1.2	0.84	41.0
$Xe^* + CH_3Br \rightarrow XeBr(B, C) + CH_3$	8.6	0.50	71.8
$Ga + F_2 \rightarrow GaF(^3\Pi) + F$	<1.3	0.80	51.2
$Ba + Br_2 \rightarrow BaBr^+ + Br^-$	<1.6	0.89	55.8

For exoergic reactions (or early barrier cases) at least two main qualitative explanations are plausible. (1) The recrossing mechanism. Trajectories, after having crossed the saddle region, recross back to the reactant channel. (2) The post-maximum behavior. This behavior follows a "natural" E_T^{-1} evolution associated with vibrationally adiabatic motion (also electronically adiabatic if electron jump occurs), and the reduction in the premaximum is due to a reduction in steric factor because of the orientation dependence of the reaction threshold. Of course both mechanisms may play significant roles simultaneously. Obviously among the many interesting suggestions that could be made, excitation functions for either oriented fixed reactants or vibrationally excited reagents, or both, would be of great promise to shed more light on these questions.

So far we have not considered the case when new reaction channels open as the collision energy increases. This might cause significant changes in the excitation function. The obvious example is the collision-induced dissociation process. (See Ref. 201 for a description of this subject.) It is interesting to note how these new channel reactions may affect the shape evolution of $\sigma_R(E_T)$, even in the presence of collision-induced dissociation. In Fig. 14 the onset of a new rebound mechanism for the collision of the Hg atom with the I_2 molecule was shown. Whereas below 1.6 eV the observed behavior is Arrhenius like, a well-defined maximum occurs at 3.0 eV. This maximum was attributed to the concurrent decreasing cross sections for complex formation and increasing cross sections for direct (rebound) abstraction, since the complex mechanism alone could not account for this maximum, but rather the combination of the complex and the direct mechanism. This mechanistic conclusion was a posteriori because no indication of the change in mechanism was apparent from the shape of the excitation function alone. Most of the dynamic and mechanistic information given by the differential center-of-mass reaction cross section is lost in the angular integration leading to the total reaction cross section. This interesting study was proposed as evidence that differential reaction cross sections will continue to be essential for the dynamic studies of elementary chemical reactions, in particular to avoid ambiguity in interpreting the collision energy dependence of the total reaction cross section.

The reaction of Eu with O_2 also seems to proceed via two different reaction mechanisms.[91] At low energies a long-lived complex mechanism seems to prevail, whereas at higher energies an additional rebound mechanism seems to take place. A qualitative explanation for these two different channels is the dependence of the energy barrier on the orientation of the colliding partners. While the Eu insertion shows a barrier of 0.15 eV, the collinear collision needs $E > 0.6$ eV for reaction to occur.[91]

E. More about the $\sigma_R(E_T)$ Shape: Bimodalities

We shall refer to bimodality in $\sigma_R(E_T)$ as the onset of a maximum or a minimum. In Section V.D we have discussed different examples and mechanisms to explain the maximum appearance in the excitation function. Here we shall comment on the minimum in the reaction cross section. It often appears as a secondary rise after the post-maximum decline in $\sigma_R(E_T)$. Figures 21 and 43 show several examples of experimentally reported minima in the reaction cross sections. As far as the appearance of the maximum is concerned, this translational feature has also attracted considerable theoretical attention.[122]

Several statistical models have accounted for the appearance of the minimum in $\sigma_R(E_T)$ as a result of the product of a dynamic and a statistical factor, which decreases and increases, respectively, with increasing energy, resulting in the minimum in question.[119] We should cite the modified cross section due to Eu (see Table III) and the atom–diatom collision model.[122] In these models the energy where the minimum is reached is given by

$$E_{\text{min}} = (n - 1)Q_{\text{max}}, \qquad \text{Eu's model}$$

$$E_{\text{min}} = \tfrac{1}{2}n(n - 1)Q_{\text{max}}, \qquad \text{atom–diatom collision model.}$$

Figure 43 shows a detail of the Rb + CH_3I cross section minimum with both statistical models fitting. The result is quite good, but the reader should keep in mind that "acceptable fits do not constitute proof of the reality."[202] In addition it is difficult to predict such a minimum for other systems.

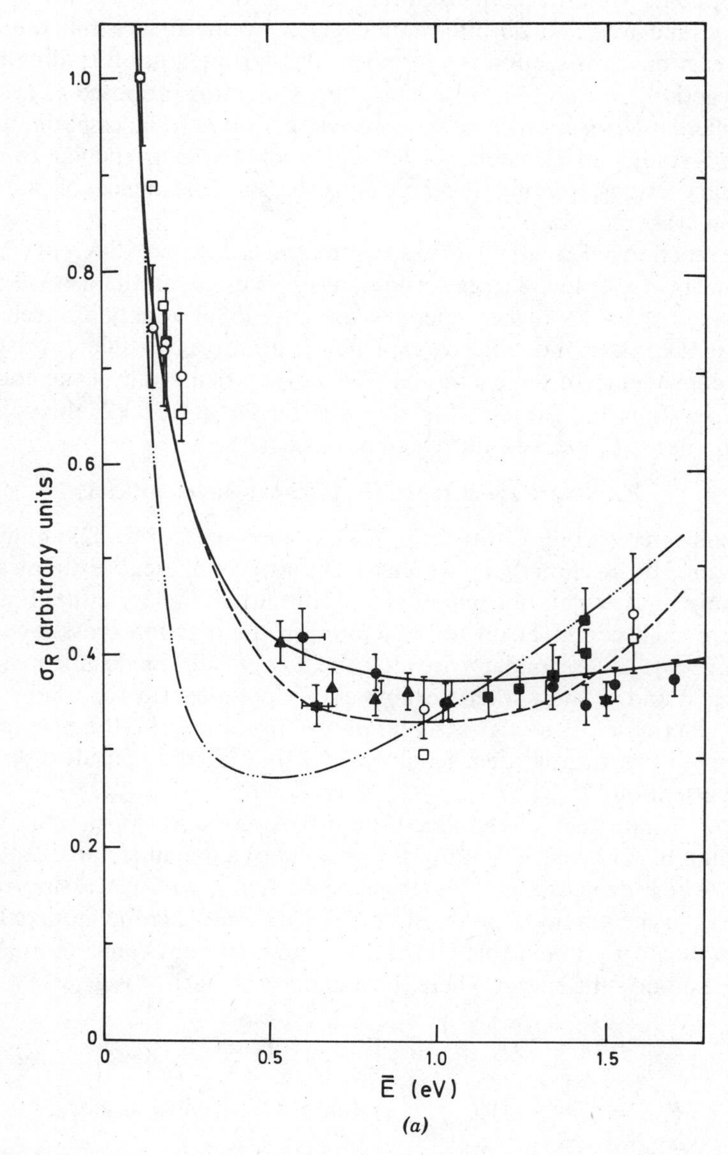

Figure 43. Excitation function for the Rb + CH_3I reaction. Data from Refs. 37 and 42. (*a*) Solid line, modified equation of Eu (Ref. 37) from Ref. 119; broken line, from Ref. 122; triple dot, from Refs. 102 and 109. Adapted from Ref. 122. (*b*) Solid line, nonadiabatic model calculation from Ref. 202; dashed line, same but without the nonadiabatic contribution. Adapted from Ref. 202.

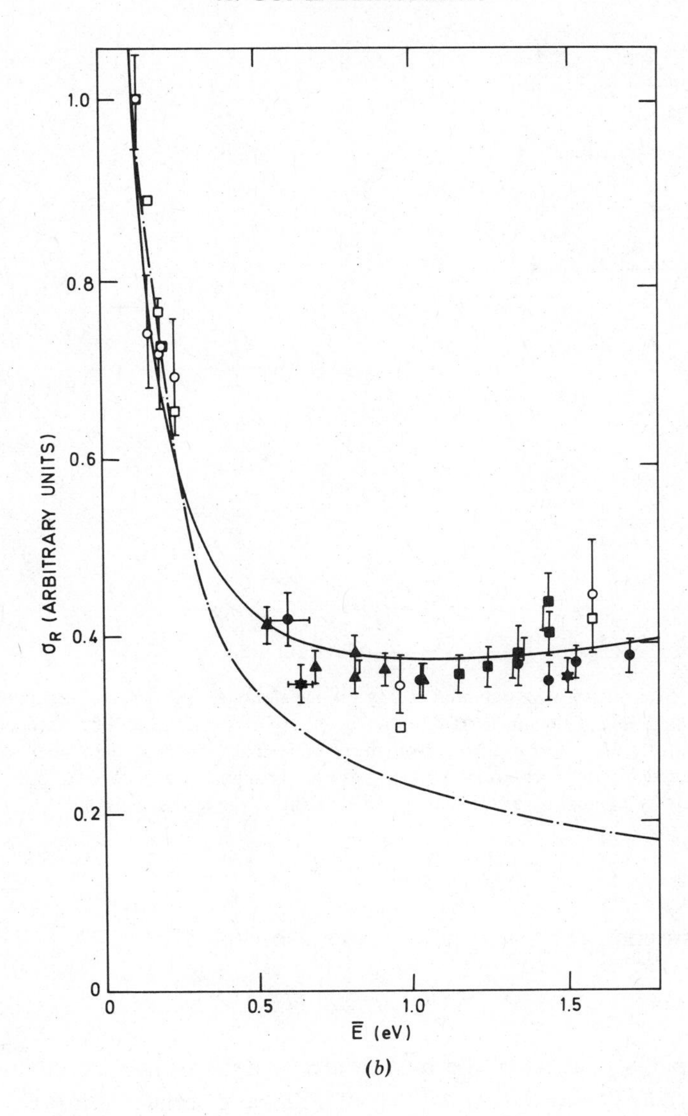

(*b*)

Dynamic explanations for these bimodalities have also been reported.[136] A clear dynamic model was successful in fitting the measured bimodality in the $Ba + Cl_2 \rightarrow BaCl^+ + Cl^-$ chemi-ionization excitation function. The essential features of this model can be visualized with the aid of Fig. 44. Essentially it consists of a harpooning interaction in the entrance channel followed by a diabatic transition in the exit channel potential energy surface.

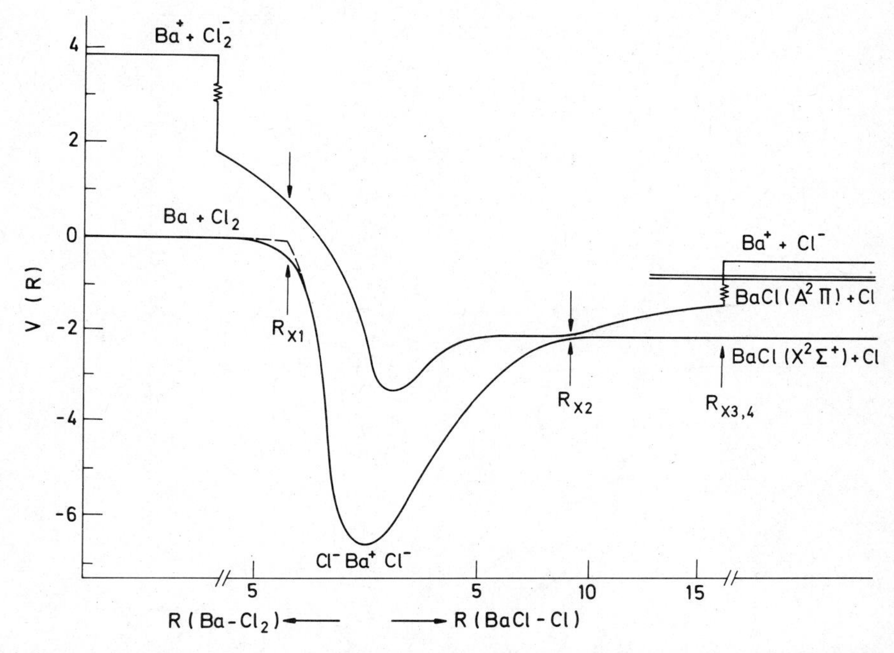

Figure 44. Semiquantitative potential energy surface along the reaction coordinate (extended correlation diagram in C_s symmetry) for Ba + Cl_2. Adapted from Ref. 136. The avoided crossing distances R_{X1}, R_{X2}, R_{X3} were evaluated from the primitive harpooning model. Chemi-ionization is characterized first by harpooning at R_{X1}, passage through the ground state $BaCl_2$ energy well, secondly by a diabatic transition at R_{X2}, and finally by polar dissociation.

The cross section can be written as[136] (see also Table III)

$$\sigma_R(E_T) = \sigma_{harp} P(E_T),$$

for example, a product of the monotonically decreasing harpooning cross section $\sigma_{harp}(E_T)$ and the probability of a diabatic surface hop $P(E_T)$ which rises with E_T. Figure 38 shows this chemi-ionization excitation function, including the present dynamic model and Eu's Born approximation/statistical theory. As was pointed out, the fit of the statistical model is much better than that of the dynamic model, but one is left wondering whether this is accidental or not. With respect to the more modest success of the dynamic model, there are also some advantages. (1) It provides a clear physical picture correlating the observation with different topological features of $\sigma_R(E_T)$. (2) It can be improved since the main assumptions are very well established. (3) It provides quantitative information about the harpooning potential.

Such diabatic jumps as the collision energy increases have also been considered in neutral–neutral interactions for reactions occuring via the electron jump mechanism.[202] As shown in Fig. 45, the onset of the diabatic channel for $E_T > E_0$ would increase the reaction cross section due to a depletion in the recrossing mechanism as E_T increases. Figure 43 shows the $Rb + CH_3I$ bimodality fitted by this modified electron jump model calculation. The dashed lines represent simple model predictions, neglecting the diabatic contribution.

Important kinematic factors have also been invoked in interpreting this high energy rise in the reaction cross section. Levine and co-workers[108–111] have applied information theory[112] coupled with a minimal transfer-of-momentum constraint to calculate reaction cross sections. The ingredients of their theory are the normalized translational energy distribution in the reaction products as the starting point. Then they introduced the constraint to obtain the joint distribution in translational energy via microscopic reversibility, that is, the probability matrix (for i → f transition) given by[109]

$$P_{\text{if}}(E'_T, E_T) = P_{\text{if}}(E'_T/E_T)P_{\text{if}}(E_T),$$

where $P_{\text{if}}(E'_T/E_T)$ is the normalized distribution of the product translational

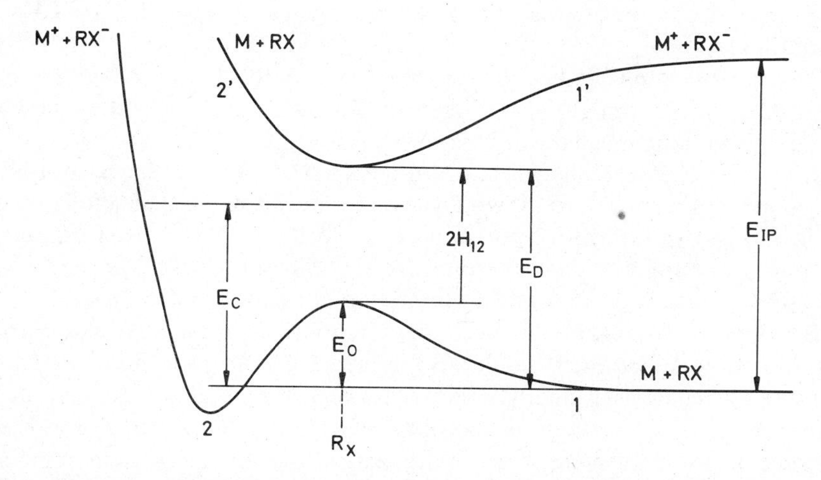

Figure 45. Schematic of the one-dimensional model. Reaction proceeds from the ground state well 2 or the excited state well 1′2′. Ion-pair formation occurs for $E_T > E_{IP}$ along 1′. The different thresholds are indicated.

energy E'_T at a given initial translational energy E_T. This distribution takes the form given in Section III.A.1).

Once $P(E'_T, E_T)$ is known, one obtains $P(E_T)$ by integrating over E'_T. Then the reaction cross section is obtained,[109]

$$\sigma_R = \left(\frac{\mu}{2E_T}\right)^{1/2} \frac{\rho(E)\rho(E_T)}{\rho(E_T; E)},$$

where $\rho(E)$ is the total density of states at the energy E and $\rho(E_T; E)$ is the density of states for the reactants when their translational energy is E_T, E being the total energy. Both densities can be evaluated from prior distributions.[166a]

These authors found that the high translational energies enhance the reaction cross section if and only if $\mu_i/\mu_f > 2$, and consequently their minimum prediction, is a mass-dependent function. The above criterion for the high E_T increase of $\sigma_R(E_T)$ is only valid for repulsive release. They claimed that for stripping-like reactions, where $dE'_T/dE_T = \cos^2 \beta < 1$, one would thus expect to see such a minimum in $\sigma_R(E_T)$ as a fairly general feature if $\beta > 45$.

In the Fig. 43 such a minimum prediction from these information-theoretic arguments is also shown for the reaction Rb + CH_3I, and it lies at a lower energy (0.5 eV less than the experimental value). However, the M + CH_3I systematic trend seems to be very attractive since no minima have appeared so far in the K or Na excitation functions, which is consistent with the proposition that it would only appear for the Rb and Cs reactions. It would be very interesting to see whether such a minimum takes place for the Cs + CH_3I excitation function.

The trend of a lower value of E_{min} as one goes in the direction Na to K to Rb to Cs, if it exits, might also be interpreted by using the modified (close-range) harpooning mechanism illustrated in Fig. 45.

It is obvious that as one goes toward the Cs reaction case, the asymptotic energy separation (for fixed M) would also be reduced, and so is E_D, following the lowering of the ionization potential as one goes from Na toward Cs atoms. Therefore one might expect the onset of the diabatic channel precisely in the (above) same sequence, and so the minimum appearance.

Besides (electronic) nonadiabatic contributions, like these just mentioned, vibrational nonadiabaticity could also influence this cross section enhancement at higher translational energies. Although no detailed treatment has appeared, it is interesting to speculate that translational to vibrational energy transfer in the close range of interactions would increase the number of internal states of the transition state. In the context of our previous model (see Section V.D) this is the same as to establish that $N(E - V^{\ddagger})$ would no longer be 1, but some higher value as a result of significant energy transfer.

Even considering a very simple hard-sphere collision model, one could use the relationship $E_v/E_T \simeq \sin^2(2\beta)$ to account for this effect. Our excitation function due to this (vibrationally) nonadiabatic contribution will change from an E_T^{-1} behavior to a constant or even increasing function of E_T, depending on the extent of the proposed nonadiabaticity. Note that from this qualitative modeling this transition is also mass dependent, like the above theoretical information description.

In conclusion, what can be summarized about the appearance of the minimum in $\sigma(E_T)$? It seems clear that for chemi-ionization processes a sole explanation associates the minimum with an energy-increasing probability of a diabatic jump in the outer harpooning of the potential surface.

For neutral–neutral interactions both statistical and dynamic explanations (including kinematic effects) have been proposed to account for the observed features. It seems that nonadiabatic contributions (either electronic or vibrational or both) may play an important role in the minimum appearance in $\sigma_R(E_T)$, which is also often observed in the excitation functions of many ion–molecule reactions.[193] However, more experimental and theoretical work is required to clarify this point. Vibrational state-selected excitation functions will, for example, give a clearer answer to this matter. Differential excitation functions before and after the energy minimum value would also provide a precise clarification of the dynamic origin of such a phenomenon.

VI. COLLISION ENERGY DEPENDENCE OF TRANSITION STATE SPECTROSCOPY

One of the main goals in the investigation of reaction dynamics is to understand the processes during the close encounters of the particles. Tremendous excitement has been generated in the past few years concerning the probe of the transition state.[194] Among the few experiments that have presented evidence that the transition state of a reaction can be manipulated or detected, we should cite an elegant series of experiments[194,195] that have reported optical emission in the far wings around the Na D line from the $NaNaF^{\ddagger}$ electronically excited transition state in the reaction $F + Na_2$ or from photo excitation of the NaKCl reaction complex. Polanyi and Wolf have generated the emission spectra of the transition state by using three-dimensional Monte Carlo trajectories on different LEPS potential energy surfaces.[195]

The experimental and theoretical calculations are shown in Fig. 46. Figure 46*b* refers to the experimental wing spectrum (corrected for black-body radiation and Na_2^* emission), and Fig. 46*a* shows computed wing emission intensities, relative to D-line emission at the two collision energies $E_T = 5$ and 35 kcal/mol. Note, for example, that a change in relative translational

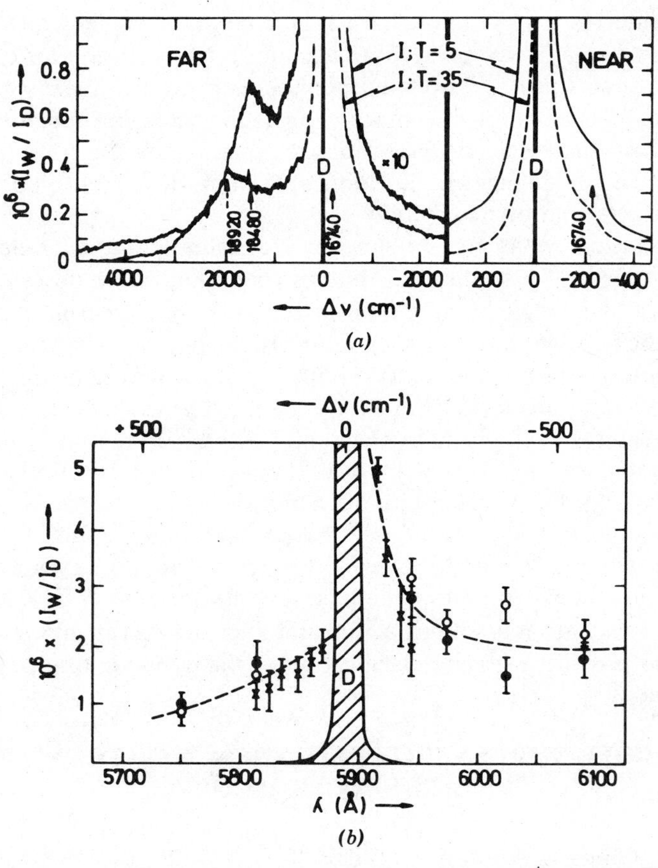

Figure 46. Experimental and theoretical calculations for the NaNaF‡ emission. (*a*) Computed wing emission intensities, relative to D-line emission (labeled D). Two collision energies were used, $E_T = 5$ and 35 kcal/mol. Adapted from Ref. 195. (*b*) Experimental wing spectrum from Ref. 194.

energy lowers the intensity in the near wing but results in an increase for the far wing. It was also observed (not shown in the figure) that changes in the LEPS parameters dramatically shift or create peaks in the wing spectrum.

The important point to be stated is that changes in experimental conditions (such as the collision energy) should mirror the hypersurface, and the collision energy spectra of the transition state appear to be of great value in probing the "three particle" range of the potential.

Finally, Fig. 47 shows the absorption spectra for H_3 calculated by Sathyamurthy and co-workers[197] for different collision energies as indicated.

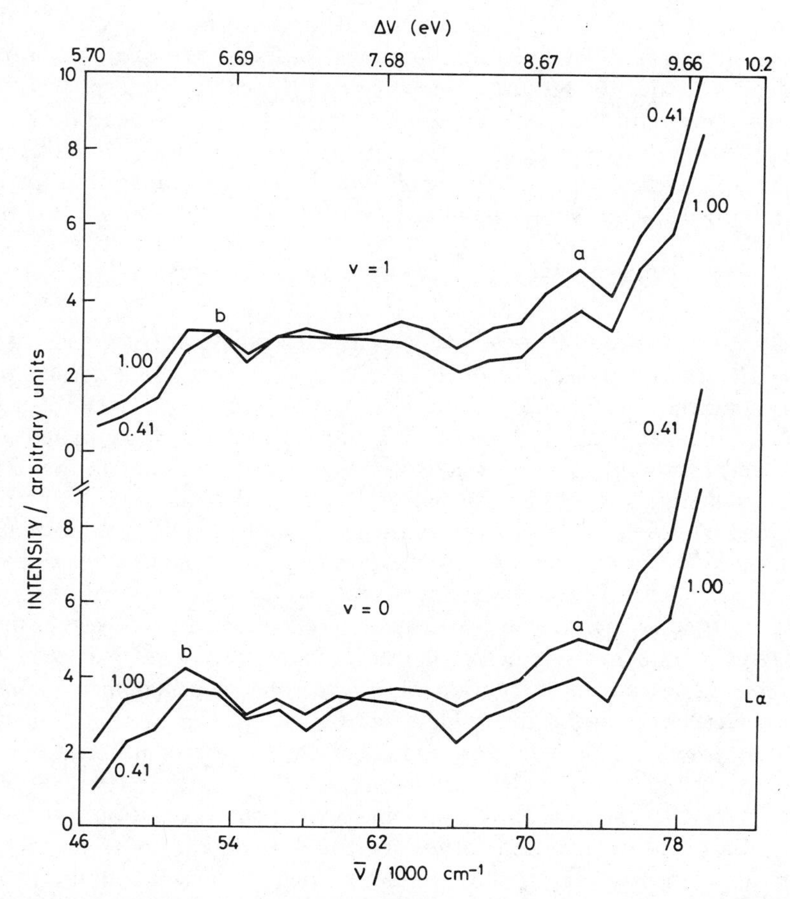

Figure 47. Absorption spectra for $H_3^{\pm}$ under different initial conditions. Average collision energies are indicated against each spectrum. L_α—Lynam-α line. Adapted from Ref. 197.

L_α represents the Lyman α line. These authors used a time-dependent wave mechanical calculation of the absorption spectrum for the collinear reaction. Note that as the collision energy increases, the intensity of the peak close to the L_α line decreases, and that of the peak in the lower frequency region increases. This trend is in good agreement with other calculations by Polanyi and co-workers on the same system.[198]

Two simple remarks are in order to conclude this section.

1. Improvements in the detection systems need to be made so that the structure in the wing spectra can be seen.

2. We shall repeat that whereas emission or absorption spectra of the transition state involve two unknown surfaces, reactive resonances deal only with the ground electronic potential energy surface. Thus we may expect that resonances should provide a very sensitive probe of the strong coupling region of the potential energy surface. It is like the familiar comparison of electronic versus vibrational spectra of an unknown system. Many imaginative and exciting results are yet to come.

VII. CONCLUDING REMARKS AND SUGGESTIONS

Nearly 80 excitation functions have already been measured, as displayed in Table II. More than 60% of them belong to the chemiluminescence and chemi-ionization groups, where the total photon flux or ion yield can be extracted independently of the scattering angle. Therefore spectrum analysis together with total photon flux collection turn out to be some of the most powerful methods for collecting data for total and differential excitation functions. This field shows great promise in the understanding of molecular reaction dynamics, and many different experimental extensions are desirable: (1) Very well defined collision energy. Rotatable beam sources, for instance, could be used to perform experiments at very low and well established energy. Accurate measurements of translational thresholds will then be available. (2) Electronic excitation of at least one reagent either by low-voltage discharge (metastable alkaline earth) or by laser excitation at the collision volume. This extension combined with vibrational excitation of the reagents will provide the best information on energy selectivity as a function of translational, electronic, and vibrational excitation. (3) High resolution of the chemiluminescence spectra as a function of translational excitation of the reagents is still very much needed.[68,69,73] Undoubtedly with all these data many reaction mechanisms and theories could be checked and developed to assist in the interpretation of the "nuclear and electronic" problems of reaction dynamics.

Many of these extensions are also needed and desirable for dark product channels in neutral–neutral collisions. However, the study of the influence of translational energy upon reactive scattering cross sections still poses great difficulties. If the product reactions have low ionization potentials, which is not the general case, high efficiency can be achieved in product detection, but the (tedious) procedure of integration over laboratory scattering angles cannot be bypassed. In favorable kinematic conditions the product laboratory velocity distribution can be obtained from known reagent conditions. In these cases density detection techniques can be used, and almost 20 excitation functions have already been measured by these methods.

It is, however, important to stress that despite (1) the elegant and complete work done in the reaction $Hg + I_2 \rightarrow HgI + I$ (see Section II) and (2) the capability of the Fourier transform Doppler spectroscopy (which is not used very much), *the crucial problem to convert density to flux product distribution still remains as one of the major difficulties in measuring excitation functions for (dark) neutral–neutral interactions.* Any effort in this direction to overcome this difficulty will be of great importance to facilitate the first step of data collection for a large number of excitation functions.

Who can resist the revolutionary impact of laser techniques and their applications. To some extent there is an (erroneous) impression that experiments that do not use one or two lasers are probably obsolete. In particular one should note the beauty and capability of laser–molecular beam interactions[3] in producing state-selected reactant or probing product state distributions and even more recently in probing the unknown transition state.[3] As far we are here concerned some novel directions should be mentioned. (1) The laser vaporization technique produces fast beams of materials having a low vapor pressure, such as Zn and Ho. New data on their reaction dynamics are now available. (2) The laser molecular photofragmentation method produces fast light atom beams such as H and D, in well-resolved kinetic energy and are therefore suitable to study the influence of translational energy upon reactive scattering as well as to determine high translational thresholds accurately. Also many basic and practical applications will be available with these laser techniques, such as the study of carbon atom reactions (combusion processes) by using laser vaporization.

A. Differential Excitation Functions

To summarize, many differential excitation functions have been measured extensively for direct and complex reactions. More emphasis has been placed on the product angular velocity distribution than on internal (vibrorotational) states. Common features observed for direct reactions are increased forward scattering as the collision energy increases and energy adiabaticity for either translational or vibrational modes, in good agreement, with the Polanyi rules. Well tested methods, such as DIPR-DIP, seem to recover the main observed features in reaction dynamics. Exceptions of the above rules are often explained via secondary (complex) collision dynamics, typically at low collision energy, such as the proposed migratory dynamics in the $Ba + CH_3Br$ reactive collisions.

One thing that is needed is to obtain detailed product internal distributions as a function of the collision energy over a wide energy range. This will complete the total energy evolution of $\sigma_R(E_T)$, which is necessary to obtain a clear and unambigous picture of the reaction dynamics. Unfortunately this has only been accomplished for a few systems, but a great effort needs to be made, essentially from the experimental point of view, to extend the capability of internal product detection to systems that do not fluoresce. Synchrotron radiation and related techniques will obviously play an important role in this direction.

The collision energy dependence of the complex reaction dynamics has been studied in a few cases, showing the appearance of direct reaction mechanisms as translational energy increases. These kinds of experiments provide an excellent opportunity to understand the energy randomization assumption, in addition to testing and developing unified theories (that is, a global approach to direct and complex reactions), such as the one proposed by Miller,[129] which does not seem to have been applied to realistic systems in this context. We should keep in mind that the best way to understand complex and direct reaction dynamics is to follow its evolution by changing experimentally the translation energy in the same reacting system. More (experimental) differential excitation functions are necessary in this particular direction.

One of the greatest excitements has been the "measurement" of the reactive resonances provided by the vibrationally state-resolved differential cross section for the reaction $F + H_2$ and its isotopic hydrogen mixtures. This interesting way to perform vibrational spectroscopy of the transition state represents a direct way to learn about the unknown part of the potential energy surface. Within the encouraging interplay between theory and experiment, the observations of oscillatory behavior, reactive resonances, and so on, provide an exceptional challenge to check potential energy calculations and quantum theories on molecular reactive scattering.

B. On Thresholds and the Shape of Excitation Functions

The translational energy threshold provides a very rich and important piece of information from both the energetic and the dynamic points of view. First of all, accurate determinations are required. This can only be achieved by using a beam–beam arrangement having both reagents well defined and a very narrow velocity distribution. The use of supersonic beams using the seeding technique or a rotatable source capability seems to be the more adequate procedure for this purpose. For an endothermic reaction or high energy threshold one could use a beam–gas arrangement provided that the thermal energy spread of the gas is negligible with regard to the actual value of the threshold to be measured. Under these circumstances the laser photofragmentation technique can be of great help since it may provide light atom "beams" in a well-defined translational energy range.

Once a reaction threshold has been accurately determined, many excellent opportunities open: (1) energetic information, (2) electroaffinites of the target molecule, (3) comparison with quantum predictions and checking of tunneling as well as vibrational adiabaticity, and so on (see Table VII).

In the coming years it will be necessary to measure accurate thresholds, providing a challenge for quantum calculations. In fact this would also be more critical with the experimental determination of post-threshold behavior.

First, technical care is needed to measure the post-threshold behavior accurately. Typically for thermal neutral beams the velocity spread of the reagents is still a serious problem in most studies carried out so far. Either a merging (fast) supersonic beam (with narrow velocity distribution) or a possibility of velocity selection for both beams will be required to guarantee the quality of the post-threshold data. Once these post-threshold data become available, a new world opens. There is a real need to check the huge number of dynamic and statistical models described in Table III and Section III over a wide energy range, that is, to test tunneling models, aligment collision models, steric factors, and many others which claim that they essentially govern the post-threshold behavior.

Post-threshold data can now be compared with accurate quantum calculations. In spite of the lengendary statement made by Dirac in 1929 that the "underlying physical laws necessary for the mathematical theory of a large part of physics and the whole of chemistry are thus completely known . . . ," it has only been possible since 1975 to perform accurate three-dimensional calculations for a real system, such as $H + H_2 \rightarrow H_2 + H$ due to Schatz and Kuppermann.[177] (See the very interesting comments in the introduction of their paper.) Because of the large expenditure of computer time, which obviously increases with the collision energy, such accurate quantum calculations will be limited to a small number of systems and to the low energy range. Therefore for the coming years the prime emphasis will be on the development of accurate but efficient approximate quantum methods and techniques. Well-resolved post-threshold data, in addition to full, accurate three-dimensional calculations, are very much needed to check all these quantum approximations.

With regard to the already measured post-threshold data, which are the factors that determine the post-threshold behavior? The easiest picture arises when no or little dynamics is involved in the reaction. On statistical grounds the expectation is a positive (negative) collision energy dependence for an endoergic (exoergic) reaction. In a classical approach the positive or negative slope of such dependence is controlled either by the number of degrees of freedom involved in the transition state configuration (positive slope) or by the exact nature of the long-range potential for the exoergic reactions.

Indeed, no energy selectivity can be expected, and the enhancement of the reaction cross section should depend on the total energy, regardless of the mode that was actually excited.

As dynamic contributions become more important, things start to become complicated, as expected, since a detailed knowledge of the potential surface is not typically available. For reactions showing an energy barrier (endoergic, exoergic), not only the barrier locations (early or late barrier), but also the extent of the curvature along the minimum energy path before the actual saddle point is important. In fact, kinematic effects have been shown to produce significant recrossing trajectories because high curvature effects diminishing the translational enhancement of the reaction cross section may take place.

Steric factors seem to be important in governing the post-threshold law. Available models predict a quadratic dependence on the excess energy $E_T - E_0$, which is in good agreement with several trajectory calculations, but not in direct reaction mechanisms, such as those followed by electron jump processes. It seems that in normal (not oriented) reactive collisions, aligment of the reagents, at least at low collision energy, through the bending potential can determine the reactive interaction and therefore the post-threshold behavior. Several hydrogen (isotopically substituted) excitation functions have been explained by these aligment models. Of course without excitation function measurements with oriented molecules it seems difficult to elucidate these dynamic effects, and such measurements would be of great promise in their interpretation.

Angular momentum partition between orbital and rotational momenta of the product as a function of collision energy can be an important constraint in the collision energy dependence of the reaction dynamics, but more experiments and theoretical work are still required to gain insight. Undoubtely excitation functions for product rotational distribution (carried out by laser-induced fluorescence), even in favorable kinematic conditions, in addition to product polarization, would be of great value to clarify this aspect of reaction dynamics.

The (translational threshold and) post-threshold excitation function dependence upon vibrational and/or electronic excitation provides highly valuable information about the "chemistry behind the studied reactions." Besides the energy selectivity available from the threshold evolution upon other modes of excitation, the changes on the limiting slope of σ_R as E_v is varied can be a good diagnosis for direct versus complex mechanisms, as discussed. Moreover vibrational state selected excitation functions would be of great importance to see whether vibrational or translational capture is the crucial pathway leading to products for those reactions showing little or no threshold at all.

Unfortunately few examples are available for excitation functions with reagents in selected electronic states. There is no doubt that it would provide the best way to understand the nuclear and electronic problems already present in (excited) reaction dynamics. Electronic excitation via laser techniques in the actual cross beam volume and reactions with metastable electronically excited species would be two clear examples of desirable work to be pursued over the coming years.

Several times we have referred to the dynamic interpretation of reaction dynamics when there was a simultaneous presence of two mechanisms with different energy barriers such as, for example, the abstraction or the insertion mechanism. The measurement of excitation functions at fixed reagent orientations will provide valuable information to clarify this aspect of reaction dynamics, which is so important even in typical rate constant experiments. With regard to this new direction, the inversion obtained by Bernstein on the orientation dependence of the energy barrier for the reaction $Rb + CH_3I$ is encouraging.

Is the shape of the excitation function a route to dynamics? Yes, but in many cases its interpretation needs additional information from differential reaction cross sections. For example, the experimentalist needs to know whether or not a new reaction channel is open as the collision energy is increased. A central idea proposed that the shape evolution of $\sigma_R(E_T)$ is a consequence of the collision energy dependence of the transition state location, but more work needs to be done in that direction to find such dividing surfaces in the often not very well known potential energy hypersurfaces.

From the body of data now available it seems that dynamic factors rather than statistical ones are reponsible for the maximum appearance in the excitation function.

Two attractive explanations of the maximum in $\sigma_R(E_T)$ were given: (1) the recrossing mechanism and (2) the orientation dependence of the energy barrier, both for exoergic reactions. In the case of exothermic reactions, kinematic factors governing the recrossing trajectories, beyond the saddle point, were also reported as a possible explanation of the appearance of the maximum in the excitation function.

With regard to the minimum in $\sigma_R(E_T)$, electronic or vibrational nonadiabaticity was given as a major source of such post-minimum rising of $\sigma_R(E_T)$. Kinematic factors were also involved, via an information-theoretical analysis, in order to explain such a minimum. An interesting possibility is the observation of oscillatory behavior in the collision energy dependence of the reaction cross section. *Clearly, it is still premature to exclude broad and easily observable structures at higher energies due to interference effects.*

There is no doubt that highly resolved differential reaction cross sections with the possible observation of reactive resonances together with the collision energy dependence of the transition state spectroscopy will provide very rich information. In addition simultaneous total and differential (reactive and nonreactive) excitation function measurements carried out at collision energies before and after the particular value at which the actual maximum or minimum occurs will give us the precise information still missing in our interpretations. Also experiments with reagents in well-selected vibrational and electronic states or with oriented molecules will be extremely useful. They will clarify our knowledge of the close range of the potential energy surface, and subsequently of the molecular reaction dynamics and the collision energy dependence. They will provide us with great excitement for the coming years.

Acknowledgment

I wish to thank Prof. R. B. Bernstein for his comments on the manuscript and his continuous scientific stimulus, and Prof. M. Menzinger for many stimulating discussions for supplying several interesting references and comments on the manuscript. I should also like to express my thanks to all my colleagues, F. J. Aoiz, V. Saez Rábanos, V. J. Herrero, E. Verdasco, F. L. Tabares, L. Bañares, and J. Alonso, for many helpful discussions and their important contributions to our joint efforts.

The author's research in this area has been supported in part by the Comisión Asesora and the Dirección de Política Científica of Spain.

The author appreciates the kind permission from the Editor of the J. Chem. Phys., Chem. Phys., Chem. Phys. Lett., Molec. Phys. and Plenum Publishing Corp. to reproduce several figures (as referred to) of the manuscript.

REFERENCES

1. M. A. D. Fluendy and K. P. Lawley, *Chemical Applications of Molecular Beam Scattering*, Chapman and Hall, London, 1973.
2. R. D. Levine and R. B. Bernstein, *Molecular Reaction Dynamics*, Clarendon, Oxford, 1974.
3. R. B. Bernstein, *Chemical Dynamics via Molecular Beam and Laser Techniques*, Clarendon, Oxford, 1982.
4. I. W. M. Smith, *Kinetics and Dynamics of Elementary Gas Reactions*, Butterworth, London, 1980.
5. R. B. Bernstein, Ed., *Atom–Molecule Collision Theory, A Guide for the Experimentalist*, Plenum, New York, 1979.
6. (a) W. H. Miller, Ed. *Dynamics of Molecular Collisions*, Plenum, New York, 1976; (b) M. S. Child, *Molecular Collision Theory*, Academic, New York, 1974.
7. D. R. Herschbach, *Adv. Chem. Phys.* **10**, 319 (1966).
8. K. J. Laidler and J. C. Polanyi, *Progr. Reaction Kinet.* **3**, 1 (1965).
9. (a) J. C. Polanyi, *Disc. Faraday Soc.* **44**, 293 (1967); (b) **55**, 293 (1973).

10. J. L. Kinsey, *MTP International Review of Science, Phys. Chem. Ser.* 1, vol. **9**, J. C. Polanyi, Ed., Butterworth, Oxford, 1972, p. 123.
11. D. R. Herschbach, *Disc. Faraday Soc.* **55**, 233 (1973).
12. J. P. Toennies, *Physical Chemistry, An Advanced Treatise*, vol. VIA, H. Eyring, D. Henderson, and W. Jost, Eds., Academic, New York, 1974, p. 237.
13. J. M. Farrar and Y. T. Lee, *Ann. Rev. Phys. Chem.* **25**, 357 (1974).
14. J. C. Polanyi and J. L. Schreiber, *Physical Chemistry, An Advanced Treatise*, vol. VIA, H. Eyring, D. Henderson, and W. Jost Eds., Academic, New York, 1974, p. 383.
15. R. B. Bernstein, *Adv. At. Mol. Phys.* **15**, 167 (1979),
16. M. R. Levy, *Dynamics of Reactive Collisions, Progr. Reaction Kinet.*, vol **10**, Pergamon, New York, 1979.
17. R. Grice, *Disc. Faraday Soc.* **67**, 16 (1979).
18. R. Grice, *Molecular Scattering, Physical and Chemical Applications, Adv. Chem. Phys.*, vol. 30, K. P. Lawley, Ed., Wiley, 1975, p. 247.
19. S. H. Bauer, *Chem. Rev.* **78**, 147 (1978).
20. J. Dubrin, *MTP Int. Rev. of Science, Phys. Chem. Ser.* 1, vol **9**, J. C. Polanyi, Ed., Butterworth, Oxford, 1972.
21. M. Menzinger, "Electronic Chemiluminescence in Gases," in *Potential Energy Surfaces*, K. P. Lawley, Ed., Wiley, New York, 1980.
22. M. Menzinger, in *Gas Phase Chemiluminescence and Chemionization*, A. Fontijn, Ed., North-Holland, Amsterdam, 1985; M. Menzinger, private communication.
23. S. R. Leone, *Ann. Rev. Phys. Chem.* **35**, 109 (1984).
24. D. G. Truhlar and B. C. Garrett, *Ann. Rev. Phys. Chem.* **35**, 159 (1984).
25. M. E. Gersh and R. B. Bernstein, *J. Chem. Phys.* **55**, 4461 (1971); **56**, 6131 (1972).
26. G. Hall, K. Liu, M. J. Aulitte, C. F. Giese, and W. R. Gentry, *J. Chem. Phys.* **81**, 5577 (1984)
27. (a) A. M. Moutinho, A. W. Kleyn, and J. Loss, *Chem. Phys. Lett.* **61**, 249 (1975); (b) M. N. Hubers, A. W. Kleyn, and J. Loss, *Chem. Phys.* **17**, 303 (1976).
28. (a) B. G. Wicke, *J. Chem. Phys.* **78**, 6036 (1983); (b) B. G. Wicke, S. P. Tang, and J. F. Frichtenicht, *Chem. Phys. Lett.* **53**, 304 (1978); (c) S. P. Tang, N. G. Utterback, and J. F. Frichtenicht, *J. Chem. Phys.* **64**, 3833 (1976).
29. (a) J. Dubrin, *MTP Int. Rev. of Science, Phys. Chem. Ser.* 1, vol. 9, J. C. Polanyi, Ed., Butterworth, Oxford, 1972; (b) E. E. Marinero, C. T. Rettner, and R. N. Zare, *J. Chem. Phys.* **80**, 4142 (1984); (c) D. P. Gerrity and J. J. Valentini, *J. Chem. Phys.* **79**, 5202 (1983); (d) K. Kleinermanns and J. Wolfrum, *J. Chem. Phys.* **80**, 1446 (1984).
30. R. D. Levine and R. B. Bernstein, *Chem. Phys. Lett.* **105**, 467 (1984).
31. (a) R. M. Harris and J. F. Wilson, *J. Chem. Phys.* **54**, 2088 (1971); D. Beck, E. F. Greene, and J. Ross, *J. Chem. Phys.* **37**, 2895 (1962); (b) E. F. Greene, A. L. Moursund, and J. Ross, *Adv. Chem. Phys.* **10**, 135 (1966); (c) J. L. Kinsey, G. H. Kwei, and D. R. Herschbach, *J. Chem. Phys.* **64**, 1914 (1976).
32. (a) J. P. Toennies, *Physical Chemistry, An Advanced Treatise*, vol. VIA, H. Eyring, D. Henderson, and W. Jost, Eds., Academic, New York, 1974, p. 237; S. A. Pace, H. F. Pang, and R. B. Bernstein, *J. Chem. Phys.* **66**, 3635 (1977); V. Sáez Rábanos, E. Verdasco, V. J. Herrero, and A. González Ureña, *J. Chem. Phys.* 81, 5725 (1984). (b) J. B. Anderson and J. B. Fenn, *Phys. Fluids* **8**, 780 (1965); (c) J. B. Anderson, R. P. Andres, and J. B. Fenn, *Adv. Chem. Phys.* **10**, 275 (1966).

33. M. E. Gersh and R. B. Bernstein, *J. Chem. Phys.* **55**, 4461 (1971), **56**, 6131 (1972).

34. (a) E. H. Taylor and S. Datz, *J. Chem. Phys.* **23**, 1711 (1955); (b) S. Datz and E. H. Taylor, *J. Chem. Phys.* **25**, 389 (1956), **25**, 395 (1956).

35. (a) K. T. Gillen, C. Riley, and R. B. Bernstein, *J. Chem. Phys.* **50**, 4019 (1969); (b) K. T. Gillen, A. M. Rulis, and R. B. Bernstein, *J. Chem. Phys.* **54**, 2831 (1971); T. T. Warnock and R. B. Bernstein, *J. Chem. Phys.* **49**, 1878 (1968); (c) A. M. Rulis, Ph.D. Thesis, University of Wisconsin, Madison (1972); (d) A. M. Rulis and R. B. Bernstein, *J. Chem. Phys.* **57**, 5497 (1972); (e) R. B. Bernstein and A. M. Rulis, *Disc. Faraday Soc.* **55**, 293 (1973).

36. G. Rotzoll, R. Viard, and K. Schügerl, *Chem. Phys. Lett.* **35**, 353 (1975); M. Pauluth and G. Rotzoll, *J. Chem. Phys.* **81**, 1515 (1984).

37. (a) H. E. Litvak, A. González Ureña, and R. B. Bernstein, *J. Chem. Phys.* **61**, 738 (1974); (b) **61**, 4091 (1974).

38. F. J. Aoiz, V. J. Herrero, and A. González Ureña, *Chem. Phys.* **59**, 61 (1981).

39. J. G. Pruett, F. R. Grabiner, and P. R. Brooks, *J. Chem. Phys.* **63**, 1173 (1975).

40. A. van den Meulen, A. M. Rulis, and A. E. de Vries, *Chem. Phys.* **7**, 1 (1975).

41. F.J. Aoiz, V.J. Herrero and A. González Ureña, *Chem. Phys. Lett.* **74**, 398 (1980).

42. S. A. Pace, H. F. Pang, and R. B. Bernstein, *J. Chem. Phys.* **66**, 3635 (1977).

43. V. J. Herrero, F. L. Tabarés, V. Sáez Rábanos, F. J. Aoiz, and A. González Ureña, *Mol. Phys.* **44**, 1239 (1981).

44. (a) H. F. Pang, K. T. Wu, and R. B. Bernstein, *J. Chem. Phys.* **54**, 2088 (1977); (b) K. T. Wu, H. F. Pang, and R. B. Bernstein, *J. Chem. Phys.* **61**, 1064 (1978); (*c*) K. T. Wu, *J. Phys. Chem.* **83**, 1043 (1979).

45. J. H. Birely, E. A. Entemann, R. R. Herm, and K. R. Wilson, *J. Chem. Phys.* **51**, 5461 (1969).

46. (a) M. W. Geiss, H. Dispert, T. L. Budzynski, and P. R. Brooks, in P. R. Brooks and E. Hayes, Eds., *ACS Symp. Ser.* **56**, 103 (1977); (b) H. H. Dispert, M. W. Geiss, and P. R. Brooks, *J. Chem. Phys.* **70**, 5317 (1979).

47. V. J. Herrero, F. L. Tabarés, V. Sáez Rábanos, F. J. Aoiz, and A. González Ureña, *Mol. Phys.* **44**, 1239 (1981).

48. V. J. Herrero, V. Sáez Rábanos, and A. Gonzáles Ureña, *J. Phys. Chem.* **88**, 2399 (1984).

49. V. J. Herrero, V. Sáez Rábanos, and A. González Ureña, *Mol. Phys.* **47**, 725 (1982).

50. G. H. Kwei and V. W. S. Lo, *J. Chem. Phys.* **72**, 6265 (1980).

51. (a) A. Persky, *J. Chem. Phys.* **50**, 3835 (1969); (b) L. R. Martin and J. L. Kinsey, *J. Chem. Phys.* **46**, 4834 (1967).

52. V. Sáez Rábanos, E. Verdasco, A. Segura, and A. González Ureña, *Mol. Phys.* **50**, 825 (1983).

53. V. Sáez Rábanos, E. Verdasco, V. J. Herrero, and A. González Ureña, *J. Chem. Phys.* **81**, 5725 (1984).

54. F. Heismann and H. J. Loesch, *Chem. Phys.* **64**, 43 (1982).

55. P. J. Dagdigian, *Chem. Phys. Lett.* **55**, 239 (1978).

56. U. Ross, H. J. Meyer, and Th. Schulze, *Chem. Phys. Lett.* **84**, 359 (1984).

57. D. Van den Ende, S. Stolte, J. B. Cross, G. H. Kwei, and J. J. Valentini, *J. Chem. Phys.* **77**, 2206 (1982); D. Van den Ende and S. Stolte, *Chem. Phys. Lett.* **76**, 13 (1980); S. Stolte, *Ber. Bunsenges. Phys. Chem.* **86**, 413 (1983).

58. C. T. Rettner and J. P. Simons, *Disc. Faraday Soc.* **67**, 109 (1979); J. P. Simons, Private communication.

59. R. J. Hennessy, Y. Ono, and J. P. Simons, *Mol. Phys.* **43**, 181 (1981).
60. R. J. Hennessy and J. P. Simons, *Mol. Phys.* **44**, 1027 (1981).
61. C. K. Kahler and Y. T. Lee, *J. Chem. Phys.* **73**, 5122 (1980).
62. C. C. Kahler, E. Husell, C. Merten Upshur, and W. H. Green, Jr., *J. Chem. Phys.* **80**, 3644 (1984).
63. T. J. Odiorne and P. R. Brooks, *J. Chem. Phys.* **51**, 4676 (1969).
64. (a) A. E. Redpath and M. Menzinger, *Can. J. Chem.* **49**, 3063 (1971); (b) Y. Zeiri and M. Shapiro, *J. Chem. Phys.* **75**, 1170 (1981); (c) A. E. Redpath and M. Menzinger, *J. Chem. Phys.* **62**, 1987 (1975).
65. (a) A. E. Redpath, M. Menzinger, and T. Carrington, *Chem. Phys.* **27**, 409 (1978); (b) M. Menzinger and A. Yokozeki, *Chem. Phys.* **22**, 273 (1977).
66. (a) A. Yokozeki and M. Menzinger, *Chem. Phys.* **20**, 9 (1977); (b) D. J. Wren and M. Menzinger, *Disc. Faraday* **67**, 97 (1979); (c) *J. Chem. Phys.* **63**, 4557 (1975); (d) K. T. Alben, A. Auerbach, W. M. Ollison, J. Weiner, and R. J. Cross, Jr., *J. Am. Chem. Soc.* **100**, 3274 (1978).
67. M. Menzinger and D. J. Wren, *Disc. Faraday Soc.* **67**, 142 (1979).
68. D. M. Manos and J. M. Parson, *J. Chem. Phys.* **69**, 231 (1978).
69. D. M. Manos and J. M. Parson, *J. Chem. Phys.* **63**, 3575 (1975).
70. H. J. Mayer, Th. Schulze, and U. Ross, *Chem. Phys.* **90**, 185 (1984).
71. E. K. Parks, L. G. Pobo, and S. Wexler, *J. Chem. Phys.* **80**, 5003 (1984).
72. E. K. Parks, S. H. Sheen, and S. Wexler, *J. Chem. Phys.* **69**, 1190 (1978).
73. R. H. Schwenz and J. M. Parson, *J. Chem. Phys.* **76**, 4439 (1982).
74. Th. Schulze, H. J. Meyer, and U. Ross, *Chem. Phys. Lett.* **112**, 563 (1984).
75. S. H. Sheen, G. Dimoplon, E. K. Parks, and S. Wexler, *J. Chem. Phys.* **68**, 4950 (1978).
76. H. W. Cruse, P. J. Dagdigian, and R. N. Zare, *Disc. Faraday Soc.* **55**, 277 (1973).
77. (a) G. P. Smith, J. C. Whitehead, and R. N. Zare, *J. Chem. Phys.* **64**, 2632 (1976); (b) G. P. Smith and R. N. Zare, *J. Chem. Phys.* **67**, 4912 (1977).
78. J. G. Pruett and R. N. Zare, *J. Chem. Phys.* **64**, 1174 (1976).
79. A. Gupta, D. S. Perry, and R. N. Zare, *J. Chem. Phys.* **72**, 6237 (1980).
80. A. Gupta, D. S. Perry, and R. N. Zare, *J. Chem. Phys.* **72**, 6250 (1980).
81. J. W. Hepburn, K. Lim, R. G. Macdonald, F. J. Northrup, and J. C. Polanyi, *J. Chem. Phys.* **75**, 3353 (1981).
82. J. W. Hepburn, D. Klimek, K. Lim, J. C. Polanyi, and J. C. Wallace, *J. Chem. Phys.* **69**, 4311 (1978).
83. J. W. Hepburn, D. Klimek, K. Lim, R. G. MacDonald, F. J. Northrup, and J. C. Polanyi, *J. Chem. Phys.* **74**, 6226 (1981).
84. A. Siegel and A. Schultz, *J. Chem. Phys.* **72**, 6227 (1980).
85. T. Munakata, Y. Matsuni, and T. Kasuya, *J. Chem. Phys.* **79**, 1698 (1983).
86. B. E. Wilcomb, T. M. Mayer, R. B. Bernstein, and R. W. Bickes, Jr., *J. Am. Chem. Soc.* **98**, 4676 (1976).
87. (a) M. M. Oprysko, F. J. Aoiz, M. A. McMahan, and R. B. Bernstein, *J. Chem. Phys.* **78**, pt. II, 3816 (1983); T. M. Mayer, B. E. Wilcomb, and R. B. Bernstein, *J. Chem. Phys.* **67**, 3507 (1977); (b) A. Freedman, T. P. Parr, R. Behrens, Jr., and R. R. Herm, *J. Chem. Phys.* **70**, 5251 (1979).

88. (a) A. R. Clemo, F. E. Davidson, G. L. Duncan, and R. Grice, *Chem. Phys. Lett.* **84**, 509 (1981); (b) A. R. Clemo, G. L. Duncan, and R. Grice, *J. Chem. Soc. Faraday Trans. 2*, **78**, 1231 (1982); (c) D. P. Fernie, D. J. Smith, A. Durkin, and R. Grice, *Mol. Phys.* **46**, 41 (1982). (*d*) A. Durkin, D. J. Smith and R. Grice, *Mol. Phys.* **46**, 1251 (1982); ibid, **46**, 55 (1982). (*e*) A. Durkin, D. J. Smith, S.M.A. Hoffmann and R. Grice, *Mol. Phys.* **46**, 1261 (1982).

89. M. J. Coggiolo, J. Anes, J. J. Valentini, and Y. T. Lee, *Int. J. Chem. Kinet.* **8**, 605 (1976).

90. F. E. Davidson, A. R. Clemo, G. L. Duncan, R. J. Browett, J. H. Hobson, and R. Grice, *Mod. Phys.* **46**, 33 (1982).

91. R. Dirscherl and K. W. Michel, *Chem. Phys. Lett.* **43**, 547 (1976).

92. J. M. Farrar and Y. T. Lee, *J. Chem. Phys.* **63**, 3639 (1975).

93. S. M. A. Hoffmann, D. J. Smith, A. González Ureña, T. A. Steele, and R. Grice, *Mol. Phys.* **53**, 1067 (1984).

94. J. E. Mosch, S. A. Safron, and J. P. Toennies, *Chem. Phys.* **8**, 304 (1975).

95. (a) J. J. Valentini, M. J. Coggiolo, and Y. T. Lee, *Disc. Faraday Soc.* 232 (1977); *J. Am. Chem. Soc.* **98**, 853 (1976); (b) C. H. Becker, P. Casavecchia, P. W. Tiedermann, J. J. Valentini, and Y. T. Lee, *J. Chem. Phys.* **73**, 2833 (1980).

96. J. L. Kinsey, *J. Chem. Phys.* **66**, 2560 (1977).

97. E. J. Murphy, J. H. Brophy, G. S. Arnold, W. L. Dimpft, and J. L. Kinsey, *J. Chem. Phys.* **70**, 5910 (1979).

98. M. A. Eliason and J. O. Hirschfelder, *J. Chem. Phys.* **30**, 1426 (1959); A. F. Jones and D. L. Missell, *J. Phys. A* **3**, 462 (1970).

99. (a) I. Rusinek and R. E. Roberts, *Chem. Phys.* **1**, 392 (1973); (b) R. E. Roberts and C. I. Nelson, *Chem. Phys. Lett.* **25**, 278 (1974).

100. D. G. Truhlar and D. A. Dixon, in *Atom–Molecule Collision Theory, A Guide for the Experimentalist*, R. B. Bernstein, Ed., Plenum, New York, 1979, chap. 18.

101. (a) A. Henglein, *Adv. Chem. ser.* vol. 58, P. J. Ausloos, Ed., Am. Chem. Soc., Washington, D. C., 1966, p. 63; (b) R. E. Minturn, S. Datz, and R. C. Becker, *J. Chem. Phys.* **44**, 1149 (1966).

102. H. Kaplan and R. D. Levine, *Chem. Phys. Lett.* **39**, 1 (1976).

103. A. González Ureña and F. J. Aoiz, *Chem. Phys. Lett.* **51**, 281 (1977).

104. J. L. Magee, *J. Chem. Phys.* **8**, 687 (1940).

105. (a) P. J. Kuntz, *Mol. Phys.* **23**, 1035 (1972); (b) P. J. Kuntz, *Trans. Faraday Soc.* **66**, 2980 (1970).

106. M. T. Marron, *J. Chem. Phys.* **58**, 153 (1973).

107. (a) D. R. Herschbach, *Disc. Faraday Soc.* 55, 233 (1973); D. G. Truhlar and D. A. Dixon, in *Atom–Molecule Collision Theory, A Guide for the Experimentalist*, R. B. Bernstein, Ed., Plenum, New York, 1979, sec. 4.5; (c) P. J. Kuntz, M. H. Mok, and J. C. Polanyi, *J. Chem. Phys.* **50**, 4623 (1969); (d) M. G. Prisant, C. T. Rettner, and R. N. Zare, *J. Chem. Phys.* **81**, 2699 (1984).

108. H. Kaplan and R. D. Levine, *Chem. Phys.* **63**, 5064 (1975).

109. H. Kaplan and R. D. Levine, *Chem. Phys.* **18**, 103 (1976).

110. A. Kafri, Y. Shimoni, R. D. Levine, and S. Alexander, *Chem. Phys.* **13**, 323 (1976).

111. A. Kafri, E. Pollak, R. Kosloff, and R. D. Levine, *Chem. Phys. Lett.* **33**, 201 (1975).

112. (a) R. B. Bernstein and R. D. Levine, *Adv. At. Mol. Phys.* **11**, 215 (1975); (b) R. D. Levine and R. B. Bernstein, *Accts. Chem. Res.* **7**, 393 (1974).
113. (a) E. P. Wigner, *Phys. Rev.* **73**, 1002 (1948); (b) T. F. O'Malley, *Phys. Rev.* **137**, A 1668 (1965).
114. P. Pechukas and J. C. Light, *J. Chem. Phys.* **42**, 3281 (1965).
115. (a) D. G. Truhlar, *J. Chem. Phys.* **51**, 4617 (1969); (b) W. R. Gentry, *J. Chem. Phys.* **81**, 5737 (1984); (c) D. R. Herschbach, in *Chemiluminescence and Bioluminescence*, M. J. Cotmier, D. M. Hercules, and J. Lee, Eds., Plenum, New York, 1973, p. 29.
116. R. D. Levine and R. B. Bernstein, *J. Chem. Phys.* **56**, 2281 (1972).
117. B. C. Eu and W. S. Liu, *J. Chem. Phys.* **63**, 592 (1975).
118. (a) H. K. Shin, *Chem. Phys. Lett.* **34**, 546 (1975); (b) **38**, 253 (1976); (c) **45**, 533 (1977).
119. (a) B. C. Eu, *J. Chem. Phys.* **60**, 1178 (1974); (b) *Chem. Phys.* **5**, 95 (1974).
120. R. B. Bernstein and B. E. Wilcomb, *J. Chem. Phys.* **67**, 5809 (1977).
121. M. Menzinger and A. Yokozeki, *Chem. Phys.* **22**, 273 (1977).
122. A. González Ureña, V. J. Herrero, and F. J. Aoiz, *Chem. Phys.* **44**, 81 (1979).
123. J. C. Light, *Disc. Faraday Soc.* **44**, 14 (1967).
124. P. Pechukas, J. C. Light, and C. Rankin, *J. Chem. Phys.* **44**, 794 (1966).
125. E. E. Nikitin, *Theor. Expt. Chem.* (*USSR*) **1**, 5 (1965).
126. S. A. Safron, N. D. Weinstein, D. R. Herschbach, and J. C. Tully, *Chem. Phys. Lett.* **12**, 564 (1972).
127. R. A. Marcus. *J. Chem. Phys.* **45**, 2138 (1966).
128. (a) G. Worry and R. A. Marcus, *J. Chem. Phys.* **67**, 1636 (1977); (b) J. M. Parson, K. Shobatake, Y. T. Lee, and S. A. Rice, *Disc. Faraday Soc.* **55**, 4658 (1972).
129. W. H. Miller, *J. Chem. Phys.* **65**, 2216 (1976).
130. A González Ureña, *Mol. Phys.* **52**, 1145 (1984).
131. J. Alonso and A. González Ureña, unpublished.
132. I. W. M. Smith, *J. Chem. Ed.* **59**, 9 (1982).
133. R. D. Levine and R. B. Bernstein, *Chem. Phys. Lett.* **105**, 467 (1984).
134. G. T. Evans, R. S. C. She, and R. B. Bernstein, *J. Chem. Phys.* **82**, 2258 (1985).
135. R. M. Harris and R. D. Herschbach, *Disc. Faraday Soc.* **55**, 121 (1973).
136. M. Menzinger and D. Wren, *Chem. Phys. Lett.* **81**, 599 (1981).
137. (a) E. M. Goldfield, A. M. Kosmas, and E. A. Gislason, *J. Chem. Phys.* **82**, 3191 (1985); (b) E. A. Gislason and J. G. Sachs, *J. Chem. Phys.* **62**, 2678 (1975).
138. (a) H. Eyring, S. H. Lin, and S. M. Lin, *Basic Chemical Kinetics*, Wiley, New York, 1980; (b) K. H. Lau, S. H. Lin, and J. Eyring, *J. Chem. Phys.* **58**, 1261 (1973).
140. (a) D. G. Truhlar and B. C. Garrett, *Accts. Chem. Res.* **13**, 440 (1980); (b) D. G. Truhlar, W. L. Hase, and J. T. Hynes, *J. Phys. Chem.* **87**, 2664 (1983); (c) B. C. Garrett and D. G. Truhlar, *J. Phys. Chem.* **83**, 1079 (1979); (d) D. G. Truhlar and B. C. Garrett, *Ann. Rev. Phys. Chem.* **35**, 159 (1984); A. González Ureña, *Mol. Phys.* 52, 1145 (1984); (e) I. W. M. Smith, *J. Chem. Soc. Faraday Trans. 2* **77**, 747 (1981).
141. D. G. Truhlar and J. T. Muckerman, in *Atom–Molecule Collision Theory, A Guide for the Experimentalist*, R. B. Bernstein, Ed., Plenum, New York, 1979, chap. 16.
142. R. E. Wyatt, in *Atom–Molecule Collision Theory, A Guide for the Experimentalist*, R. B. Bernstein, Ed., Plenum, New York, 1979, chaps. 15 and 17.

143. C. J. Ashton, J. T. Muckerman, and F. E. Schubert, *J. Chem. Phys.* **81**, 5786 (1984).
144. J. P. Toennies, W. Helz, and G. Woet, *Chem. Phys. Lett.* **44**, 5 (1976).
145. D. K. Bondi, J. N. L. Connor, J. Manz, and J. Rommelt, *Mol. Phys.* **50**, 467 (1983).
146. R. E. Wyatt, J. F. McNutt, and M. J. Redmon, *Ber. Bunsenges. Phys. Chem.* **86**, 437 (1982).
147. R. D. Evans, *The Atomic Nucleus*, McGraw-Hill, New York, 1955.
148. N. F. Mott and H. S. W. Massey, *The Theory of Atomic Collisions*, Oxford University, Oxford, 1971.
149. J. A. Kaye and A. Kuppermann, *Chem. Phys. Lett.* **77**, 573 (1981).
150. M. Broida and A. Persky, *J. Chem. Phys.* **80**, 3687 (1984).
151. D. M. Neumark, A. M. Wodtke, G. N. Robinson, C. C. Hayden, and Y. T. Lee, *J. Chem. Phys.* **82**, 3045 (1985).
152. D. M. Nuemark, A. M. Wodtke, G. N. Robinson, C. C. Hayden, K. Shobatake, R. K. Sparks, T. P. Schafer, and Y. T. Lee, *J. Chem. Phys.* **82**, 3067 (1985).
153. W. B. Miller, S. A. Safron, and D. R. Herschbach, *Disc. Faraday Soc.* **44**, 108 (1967).
154. D. D. Parrish and D. R. Herschbach, *J. Am. Chem. Soc.* **95**, 6133 (1973).
155. D. E. Fitz and P. Brumer, *J. Chem. Phys.* **70**, 5527 (1979).
156. (a) C. Hiller, J. Manz, W. H. Miller, and J. Rommelt, *J. Chem. Phys.* **78**, 3850 (1983); (b) J. Manz and J. Rommelt, *Chem. Phys. Lett.* **81**, 179 (1981).
157. (a) V. K. Babamov, V. Lopez, and R. A. Marcus, *J. Chem. Phys.* **78**, 5621 (1983); (b) J. M. Alvariño, O. Gervasi, and A. Lagana, *Chem. Phys. Lett.* **87**, 254 (1982); (c) V. Aquilanti, S. Cavalli, and A. Lagana, *Chem. Phys. Lett.* **93**, 179 (1982).
158. D. G. Truhlar and A. Kuppermann, *J. Chem. Phys.* **52**, 384 (1970); **56**, 2232 (1972); S. F. Wu and R. D. Levine, *Mol. Phys.* **22**, 991 (1971).
159. G. C. Schatz, J. M. Bowman, and A. Kuppermann, *J. Chem. Phys.* **63**, 674 (1975).
160. A. Kuppermann, in *Potential Energy Surfaces and Dynamics Calculations*, D. G. Truhlar, Ed., Plenum, New York, 1981.
161. J. M. Laumay and M. LeDourneut, *J. Phys. B*, **15**, L455 (1982).
162. N. C. Blais and D. G. Truhlar, *J. Chem. Phys.* **58**, 1090 (1973).
163. E. Pollak and P. Pechukas, *J. Chem. Phys.* **69**, 1218 (1978).
164. E. Pollak, *J. Chem. Phys.* **74**, 5586 (1981).
165. E. Pollak and M. S. Child, *Chem. Phys.* **60**, 23 (1981).
166. (a) E. Pollak, *J. Chem. Phys.* **76**, 5843 (1982); (b) S. Ron, E. Pollak, and M. Baer, *J. Chem. Phys.* **79**, 5204 (1983).
167. (a) C. W. A. Evers, A. E. de Vries, and J. Loss, *Chem. Phys.* **29**, 339 (1978); (b) R. Sayos, A. Aguilar, J. M. Lucas, A. Solè, and J. Virgili, *Chem. Phys.* **93**, 265 (1985).
168. J. A. Stockadale, F. J. Davies, R. N. Compton, and C. E. Klots, *J. Chem. Phys.* **60**, 4279 (1974).
169. L. G. Christophorou, J. G. Carter, P. M. Collins, and A. A. Christodoulides, *J. Chem. Phys.* **54**, 4691 (1971).
170. M. Menzinger and R. Wolfgang, *Angew. Chem. Int. Ed.* **8**, 438 (1969).
171. R. L. Le Roy, *J. Phys. Chem.* **73**, 4338 (1969).
172. N. C. Blais, R. B. Bernstein, and R. D. Levine, *J. Phys. Chem.* **89**, 10 (1985).
173. R. B. Bernstein, *J. Chem. Phys.* **82**, 3656 (1985).
174. G. C. Schatz, *Faraday Disc. Soc.* **67**, 140 (1979).

175. G. C. Schatz, *J. Chem. Phys.* **79**, 5386 (1983).
176. W. Jakubetz, *J. Am. Chem. Soc.* **101**, 298 (1973).
177. G. C. Schatz and A. Kuppermann, *J. Chem. Phys.* **65**, 4668 (1976).
178. H. R. Mayne and J. P. Toennies, *J. Chem. Phys.* **70**, 5314 (1979).
179. (*a*) R. Gotting, H. R. Mayne, and J. P. Toennies, *J. Chem. Phys.* **80**, 2230 (1984). (*b*) J. Geddes, H. F. Krause and W. C. Fife, *J. Chem. Phys.* **56**, 3298 (1972).
180. S. Stolte, A. E. Proctor, and R. B. Bernstein, *J. Chem. Phys.* **65**, 4990 (1976).
181. J. W. Duff and D. G. Truhlar, *J. Chem. Phys.* **62**, 2477 (1975).
182. G. L. Hofacker and R. D. Levine, *Chem. Phys. Lett.* **9**, 617 (1971).
183. R. D. Levine, *Chem. Phys. Lett.* **10**, 510 (1971).
184. G. L. Hofacker and R. D. Levine, *Chem. Phys. Lett.* **15**, 165 (1973).
185. D. S. Perry, J. C. Polanyi, and C. Woodrow Wilson, Jr., *Chem. Phys.* **3**, 317 (1974).
186. J. C. Polanyi and N. Sathyamurthy, *Chem. Phys.* **33**, 287 (1978).
187. A. Siegel and A. Schultz, *J. Chem. Phys.* **76**, 4513 (1982).
188. (a) J. G. Pruett, F. R. Grabiner, and P. R. Brooks, *J. Chem. Phys.* **63**, 1173 (1975); (b) J. R. Krenos and J. C. Tully, *J. Chem. Phys.* **62**, 420 (1975); (c) D. G. Truhlar, *J. Am. Chem. Soc.* **97**, 6310 (1975).
189. A. Komornicki, T. F. George, and K. Morokuma, *J. Chem. Phys.* **65**, 4312 (1976).
190. F. Rebentrost and W. A. Lester, Jr., *J. Chem. Phys.* **67**, 3367 (1971).
191. R. B. Bernstein, *ACS Sympm. Ser.* **56**, 3 (1977).
192. R. A. La Budde, P. J. Kuntz, R. B. Bernstein, and R. D. Levine, *J. Chem. Phys.* **59**, 6286 (1973).
193. M. McFarland, D. L. Albritton, F. C. Fehsenfeld, E. E. Ferguson, and A. L. Schmeltekopf, *J. Chem. Phys.* **59**, 6620 (1973).
194. H. J. Foth, J. C. Polanyi, and H. H. Telle, *J. Phys. Chem.* **86**, 5027 (1982).
195. J. C. Polanyi and R. J. Wolf, *J. Chem. Phys.* **75**, 5951 (1981). T. C. Maguire, P. R. Brooks and R. F. Curl, Jr. *Physical Review Letters*, **50**, 1918 (1983).
196. H. Telle, *Acta Phys. Polonica* **A63**, 223 (1983).
197. P. M. Agrawal, V. Mohan, and N. Sathyamurthy, *Chem. Phys. Lett.* **114**, 343 (1985).
198. H. R. Mayne, J. C. Polanyi, N. Sathyamurthy, and S. Raymor, *J. Phys. Chem.* **88**, 4064 (1984).
199. D. J. Wren and M. Menzinger, *Chem. Phys.* **66**, 85 (1982).
200. I. N. Batcha and N. Sathyamurthy, *J. Chem. Phys.* **76**, 6147 (1982).
201. P. J. Kuntz, in *Atom–Molecule Collision Theory, A Guide for the Experimentalist*, R. B. Bernstein, Ed., Plenum, New York, 1979, chap. 21; D. J. Diestler, *ibid.*, chap. 20.
202. A. Gonzalez Ureña and M. Menzinger, *Chem. Phys.* **99**, 437 (1985).
203. D. L. Bunker and E. A. Goring-Simpson, *Disc. Faraday Chem. Soc.* **55**, 93 (1973).
204. B. A. Blackwell, J. C. Polanyi, and J. J. Sloan, *Chem. Phys.* **30**, 299 (1978).

AUTHOR INDEX

SUBJECT INDEX